PHILOPONUS

PHILOPONUS

and the Rejection of Aristotelian Science

edited by
Richard Sorabji

Cornell University Press
Ithaca, New York

First published 1987 by Cornell University Press

Library of Congress Cataloging-in-Publication Data

Philoponus and the rejection of Aristotelian science.

Bibliography: p.
Includes index.
1. Philoponus, John, 6th cent.   I. Sorabji, Richard.
B673.J64P44   1987   186'.4   86-29116
ISBN 0-8014-2049-0

Printed in Great Britain

Contents

Contributors

Dr Wolfgang Bernard, Seminar für Klassische Philologie, Johannes Gutenberg Universität, Mainz
Professor Henry Chadwick, Magdalene College, Cambridge
Professor David Furley, Department of Classics, Princeton
Professor Philippe Hoffmann, C.N.R.S., Paris
Mr Lindsay Judson, Christ Church, Oxford
Dr Charles Schmitt, late of the Warburg Institute, London
Dr David Sedley, Christ's College, Cambridge
Professor Richard Sorabji, Department of Philosophy, King's College, London
Dr Christian Wildberg, Gonville and Caius College, Cambridge
Professor Michael Wolff, Universität Bielefeld, Abteilung Philosophie
Dr Fritz Zimmermann, Oriental Institute, Oxford

Abbreviations

CAG = *Commentaria in Aristotelem Graeca*, ed H. Diels, Berlin 1882-1909.
LSJ = H.G. Liddell and R. Scott, *A Greek-English Lexicon*, rev. H.S. Jones, Oxford, 1968.
OSAP = *Oxford Studies in Ancient Philosophy*.
RE = Pauly-Wissowa, *Realencyclopädie der klassischen Altertumswissenschaft*, Stuttgart, 1893-.

aet = *de Aeternitate Mundi contra Proclum*
in An Post = *in Analytica Posteriora*
in Cat = *in Categorias*
in Cael = *in de Caelo*
in DA = *in de Anima*
in Meteor = *in Meteorologica 1*
in Phys = *in Physica*
Opif = *de Opificio Mundi*

Preface

There is no general treatment of John Philoponus at book length,[1] despite the influence he exerted on philosophy and more particularly on science. Galileo mentioned him in his early writing more often than Plato, and inherited from him, without mention, the impetus theory whose introduction Thomas Kuhn has called a scientific revolution.

Philoponus' chief claim to fame is his massive attack on the Aristotelian science of his day, referred to in the title of this book, and his provision of alternative theories which helped to fuel the Renaissance break away from Aristotle. But there are many other facets to his work, as this volume will show. Only recently, with the studies of van Roey, has the vigorous and startling character of his contributions to Christian doctrine become more apparent. Philoponus is also our earliest source for, even if he is not the originator of, various philosophical ideas that were offered as a means of interpreting Aristotle, not of refuting him.

The study of Philoponus has been impeded by the shortage of translations. This deficiency is due to be remedied by the translation into English of most of his commentaries on Aristotle and most of his works on the eternity of the world, as part of a larger series covering the ancient commentators on Aristotle, edited by Richard Sorabji.

All the chapters in this book are new, except for the inaugural lecture (Chapter 9), which I apologise for reprinting virtually unrevised and with the original lecture context still apparent. It seemed desirable, however, that so crucial a part of the controversy should be represented.

The collection originated in a conference on Philoponus held at the Institute of Classical Studies in London in June 1983, which provided an opportunity for interested parties to pool knowledge from the many different disciplines that are relevant to his work. Chapters 2, 3, 4 and 6 are drawn from the conference, while two other conference papers, those of Henry Blumenthal and Richard Sorabji, are being incorporated into books in preparation (see Bibliography). Sorabji's main suggestions, however, are

[1] However, after this book had gone to press, there appeared a Ph.D. dissertation on Philoponus in Dutch submitted to the Catholic University of Louvain by Koenraad Verrycken (1985). Its subject is God and the world in the philosophy of Philoponus, and it detects in him a transition from Ammonius' doctrine to a Christian doctrine of creation. Its findings are to be represented in English in a publication by the Belgian Royal Academy, and, I hope, in a projected anthology of articles on the Ancient Commentators on Aristotle. See also Chapter 1, n. 260a.

included in Chapter 1 in the discussion of matter and extension (pp 18 and 23). The remaining chapters, apart from the inaugural lecture, were solicited or written for the volume, two of them (5 and 12) having been delivered first at a seminar on Ancient Science at the Institute of Classical Studies.

Chapter 1 offers a general account of Philoponus, which should not be taken as committing other contributors. It is followed by two chapters on religion. Henry Chadwick's depiction of Philoponus' contributions to Christian doctrine is almost the only general account, and certainly the first to make use of the new findings of van Roey. Philippe Hoffmann provides a salutary reminder of how Christianity could look to a devout pagan: an irreverent and ungodly position, which elevates the corpse of Christ above the divine heavens. Simplicius also believed that Philoponus did not understand how to write commentaries on Aristotle, something which he himself did with a view to displaying the agreement between Aristotle and Plato, and to directing the reader through a course of studies that would lead him to God. Enemies of Philoponus will find the quotations from Simplicius a splendid source of invective, but the invective needs to be understood in the context which Hoffman provides.

Chapters 4 and 5 are concerned with impetus theory. Michael Wolff traces the origins of the modern study of the theory, and suggests an original analysis of what is going on in Philoponus. Fritz Zimmermann's note on work in progress throws light on the route of transmission of Philoponus' impetus theory, which has baffled previous commentators despite the important findings of Pines. Pines detected impetus theory in Avicenna and in many other Islamic sources. Zimmermann argues that it could have been transmitted to the Latin West, when Ghazali's summary of Avicenna was translated into Latin in the second half of the twelfth century. For further comment see pp 11-12.

Chapters 6 and 7 are concerned with space. In the absence of any complete translation, it is useful to have David Furley's summary of Philoponus' influential Corollaries on Place and on Void. Furley will himself be providing translations of these in due course. David Sedley illuminates the text by asking what is meant by 'the force of vacuum' and by the claim that space might be vacuous, 'so far as depended on it'.

In Chapter 8, Wolfgang Bernard makes an addition to the literature on later Greek treatments of self-consciousness. Excessively slim, it has hitherto ignored the passage of Philoponus which Bernard discusses. Whether one takes Philoponus to be elaborating Aristotle, or once again rejecting him, depends on whether one thinks that Aristotle has left himself free to agree with Philoponus' view that our consciousness of our own vision is due to our faculty of reason. Philoponus himself says that Aristotle both agrees and disagrees.

Chapters 9 to 11 take up Philoponus' views on the creation of the universe and its future destruction. Chapter 9 discusses the most spectacular of his arguments for creation, in which he maintains that the Aristotelian concept of infinity accepted by his pagan opponents rules out the beginningless past in which they believe. Philoponus also claims Plato's support for the idea that the universe began, and so has to consider whether Plato can consistently hold that the universe begins, but does not end. Lindsay Judson has elsewhere considered Aristotle's treatment of this subject, and now in Chapter 10 he reveals the subtlety of Philoponus' discussion of the relevant modalities. Christian Wildberg introduces the *contra Aristotelem*, a work whose considerable influence (see p 26) has been little researched because of the relative inaccessibility of the surviving fragments. This lack will soon be made good by Wildberg's collection and translation of them. In Chapter 11 he speaks of the Syriac fragment which he has investigated for the first time. The fragment shows that the *contra Aristotelem* originally contained two more books than was previously thought, and that the extra books were concerned with the Christian expectation of a new heaven and a new earth. This connects with Judson's theme, because it shows Philoponus occupying a position part way in the direction of Plato's: some world, even if not this one, will begin and then last without end. (For another example of something beginning without ending, see Chapter 1, n. 223 on the rational soul.)

In the final chapter, Charles Schmitt offers the fullest documentation to date of Philoponus' impact on Renaissance science with special reference to his views on space and vacuum, which are summarised by Furley in Chapter 6. He shows how influential was the sixteenth-century translation into Latin of the Aristotelian commentators, of Simplicius even more than of Philoponus. Their record of alternatives to Aristotelian science added momentum to the Renaissance break away from Aristotle.

It gives me pleasure to acknowledge several kinds of help. The conference on Philoponus was generously supported by the British Academy, the Centro Internazionale A. Beltrame di Storia dello Spazio e del Tempo and the Henry Brown Fund, and some of the editorial expenses were met by the National Endowment for the Humanities. The chapter by Hoffmann was translated from the French by Jennifer Barnes. A.P. Segonds, Christian Wildberg and Larry Schrenk gave me extensive assistance with the Bibliography, and Koenraad Verrycken allowed me to add items in proof from the bibliography of his dissertation (1985). The typing was meticulously performed by Mrs Dee Woods. I should also like to thank all those who contributed their interest and expertise to the conference and to subsequent discussions. Finally, Larry Schrenk played a special role, preparing the entire volume for press, carrying out the proof-reading and supplying the indexes.

*

While this book was in proof, we learnt of the early death of Charles Schmitt. His unique contribution in Chapter 12 is but one small reminder of what we have all lost.

R.R.K.S.

To the memory of
Charles Schmitt

CHAPTER ONE

John Philoponus

Richard Sorabji

John Philoponus, a Christian schooled in Neoplatonism in the sixth century
A.D., mounted a massive attack on the Aristotelian science of his day. The
attack was tailored to fit his Christian belief, a central contention being that
the matter of the universe had a beginning, as the orthodox conception of
creation required. This ramifying view was connected with not a few of his
other innovations in science, some far removed from his Christian concerns.
The interconnection of his ideas and the scale of his innovation are
impressive, but his eventual influence was delayed by theological controversy.

In the later part of his life, Philoponus turned to contentious matters of
Christian doctrine. Daring and logical again, he none the less fell foul of the
Christian authorities. In 680, a hundred years or so after his death, he was
anathematised for his views on the Trinity.[1] This had the ironical result that
his ideas were first taken up in the Islamic world, not in Christendom. In the
Latin West some became known by direct translation in the thirteenth
century, but some only as filtered through Arabic sources, so that they were
not attributed to him, and modern scholars have believed them to be
thirteenth-century discoveries. He came fully into his own in the West only
with the extensive Latin translations of the Renaissance. Then, with his
name made respectable by his defence of the Creation, he came to be
acknowledged by such thinkers as Galileo, and his ideas contributed to the
break away from Aristotelian science.

Philoponus' life extended probably from around 490 to the 570s. He lived
in Alexandria and studied philosophy under Ammonius, son of Hermeias, the
head of the Alexandrian Neoplatonist school.

Neoplatonism and Christianity

Neoplatonism did not have to be opposed to Christianity, as it was in Athens.
In Alexandria, by contrast, the Neoplatonists reached an understanding with
the Christians. Indeed, the head of the Athenian Neoplatonist school,
Damascius, accused Ammonius of making a sordid deal with the Christian

[1] For a Syriac rendering of the anathema, with Latin translation, see G. Furlani (1919-20a)
190-1.

rulers for financial gain.[2] Ammonius and three of his pupils, Philoponus, Elias and even the Athenian Simplicius, gave an unusual interpretation of Aristotle which made him seem close to Christianity. They thought that Aristotle recognised God as creator of the physical world, albeit in the special sense of being causally responsible for its beginningless existence, not in the sense of giving it a beginning.[3] We are told that Ammonius wrote a whole book to establish the point.[4]

Thanks to the accommodation between Neoplatonists and Christians, some of Ammonius' successors in the chair at Alexandria, namely Elias and David, were Christians. On the other hand, Philoponus did not obtain the chair, nor was Ammonius himself a Christian. In this last regard, too much has been made of a rather fanciful dialogue written by another Christian participant in Ammonius' seminars, Zacharias. In this dialogue, called *Ammonius* or *de Mundi Opificio*, Zacharias represents Ammonius as convinced by the Christian view that the physical world is of finite duration,[5] and as trapped into treating God as a Trinity.[6] However, even Zacharias does not claim that Ammonius was converted: instead, he blushed at the last concession, fell silent and changed the subject.[7]

It was not only in Alexandria that Neoplatonism and Christianity were able to co-exist. At the very same time, Boethius in Rome was producing philosophical works in the Neoplatonist mould and Christian works tinged with Neoplatonist logic. Because his greatest work *The Consolation of Philosophy* includes nothing specifically Christian, it has been thought that he finally turned away from Christianity, but this view has been convincingly refuted.[8] Over a hundred years earlier, another Latin author, Augustine, had been led to Christianity through reading Platonist works and had at first found it difficult to see any conflict between them.[9]

Things were quite different in Athens, where there was a history of antagonism. In the preceding century, the head of the school, Proclus (*c*.411-485), had attacked the Christians with eighteen arguments against their belief that the universe had a beginning. Then, in 529, the Christian Emperor Justinian stopped the Athenian Neoplatonists from teaching. Archaeology claims to have discovered their teaching premises and to have confirmed that philosophy did not continue there.[10] In Chapter 3, Philippe

---

[2] Damascius *Life of Isidore* at Photius 242 §292 (= Fr. 316 Zintzen).

[3] Simplicius *in Phys* (*Commentaria in Aristotelem Graeca*) 256,16-25; 1363,8-24; *in Cael* 271,13-21; Philoponus *in Phys* 189,10-26; *in GC* 136,33-137,3; cf 286,7; Elias *in Cat* 120,16-17.

[4] Simplicius *in Phys* 1363,8-12; *in Cael* 271,18-21; Farabi *Harmony of Plato and Aristotle* 24-5, ed Dieterici in *Alfarabi's philosophische Abhandlungen*, Leiden 1890, quoted in English by M. Mahdi (1967) 236-7. But Farabi drops out Ammonius' qualification that the creation was beginningless for Aristotle (p 39 in Dieterici, ch 11 in *Harmony*, ed Nader, Beirut 1968), basing his interpretation on the spurious work *The Theology of Aristotle*.

[5] Zacharias *Ammonius* or *de Mundi Opificio* PG 85, 1113B; 1116B.

[6] ibid. 1117B.

[7] See E. Evrard (1965) 592-8.

[8] See Henry Chadwick, *Boethius*, Oxford 1981, 247-53.

[9] See Sorabji (1983) 163-72; 302-5.

[10] Alison Frantz, 'Pagan philosophers in Christian Athens', *Proceedings of the American Philosophical Society* 119, 1975, 29-38. We can talk about a 'school' for the late Athenian

Hoffmann gives a vivid picture of how Christianity looked to Simplicius, one of the victims of this closure, for closure it was.[11] Simplicius went on to write, it is not known where, and to write voluminously, but teaching was not resumed in Athens, nor is any successor to Simplicius known. One of Hoffmann's main points is that the conflict between pagan Neoplatonism and Christianity was not a conflict between the irreligious and the religious. Simplicius sees himself as defending religion. Like all Neoplatonists, he believes in a God who created the world, but in the special sense, already mentioned, of being causally responsible for its beginningless existence. Like the Christian Boethius, he intersperses his philosophical work with prayer. He sees the intellectual discipline of commentary on Aristotle as leading people upward towards union with God. To him it seems irreverent and irreligious that Christians can make God like themselves, and can deny the divinity of the heavens, while in unseemly fashion they venerate the corpse of Christ and the relics of martyrs. But he sees Christianity as only a temporary phenomenon.

Philoponus is no more neutral than Simplicius in the controversy between Christianity and pagan Neoplatonism. But he does not show the same devoutness or the same animosity as Simplicius. His tone is rather logical and argumentative. Moreover, in the commentaries on Aristotle, his attacks on paganism are comparatively few and far between, so that much of the time we could be reading a pagan Neoplatonist commentator on Aristotle. When he does attack, Philoponus' speciality is turning the Neoplatonists' own views, in which he was steeped, against them on behalf of Christianity.

Philoponus and Ammonius

Philoponus wrote a large variety of works. One major way of doing philosophy at this time was by writing commentaries on Plato or Aristotle. Seven of Philoponus' commentaries on Aristotle are extant and bear his name, but of these, four are described as being 'from the seminars of Ammonius son of Hermeias' (*ek tôn sunousiôn Ammôniou tou Hermeiou*), although that description is in turn qualified in three cases by the phrase, 'with some personal reflections' (*meta tinôn idiôn epistaseôn*).[12] This raises the question of how far the commentaries represent Ammonius' ideas. However, comparison with the

Neoplatonists, who had a building of their own, despite the timely warnings of John Lynch and John Glucker against assuming there was anything as continuous as a school in certain other periods (John P. Lynch, *Aristotle's School*, Berkeley and Los Angeles 1972; John Glucker *Antiochus and the Late Academy*, *Hypomnemata* 56, Göttingen 1978).

[11] See Sorabji (1983) 199-200, commenting on an important paper by Alan Cameron 'The last days of the Academy at Athens', *Proceedings of the Cambridge Philological Society* 195 (n.s. 15) 1969, 7-29. The latest suggestion, based on inscriptional evidence, is that Simplicius resumed his teaching at Harrān (or Carrhae), just on the Turkish side of the modern border between Turkey and Iraq, and that a school continued there. See M. Tardieu, 'Ṣābiens Coraniques et "Ṣābiens" de Harrān' *Journal Asiatique* 274, 1986, 1-44, and in 'Les calendriers en usage à Harrān d'après les sources arabes et le commentaire de Simplicius à la *Physique* d'Aristote' in I. Hadot, ed, *Simplicius, sa vie, son oeuvre, sa survie*, Peripatoi vol. 15, Berlin, forthcoming.

[12] The four commentaries from Ammonius' seminars are *in An Pr*; *in An Post*; *in DA*; *in GC*. The last three warn of Philoponus' reflections. The remaining commentaries extant are *in Cat*; *in Phys*; *in Meteor*. Details in Évrard (1965). The role and organisation of commentaries will be more fully described in the anthology on Aristotelian commentators now in preparation.

commentaries on Aristotle ascribed to Ammonius himself and to his other pupils suggest Philoponus' independence,[13] and at times he points out that he is dissenting from Ammonius.[14] The phrase 'from the seminars' is presumably like the standard phrase 'from the voice' (*apo phônês*). It allows considerable latitude when the name of the compiler is given, as it is here, alongside the name of the original lecturer.[15] We shall in any case find that a number of Philoponus' innovative views of the physical world have their first expression in his commentaries on Aristotle, and no sources suggest that these views are also to be found in Ammonius, if we exclude the fanciful Zacharias. Moreover, we can see Philoponus' views shifting between commentaries. This has been made especially clear for the late commentary on Aristotle's *Meteorology*, where Philoponus moves away from his own earlier views and from views of Ammonius' other pupils.[16]

Philoponus' Christian faith

Philoponus seems to have been a Christian from the beginning. This is suggested by his name 'John'.[17] We also find his arguments for the distinctively Christian view that the universe had a beginning as early as 517, if that is the correct date for his commentary on Aristotle's *Physics*.[18] Indeed, in that commentary he says that he has already rehearsed the arguments elsewhere.[19] This view is distinctively Christian, because it is the view that matter itself had a beginning, not the view[20] ascribed to certain Middle Platonists, and probably held by Plato himself, that the present orderly arrangement of matter had a beginning. Philoponus' concern with creation is striking. It has been pointed out that in the commentary on *de Generatione et Corruptione* he goes out of his way to draw attention to Ammonius' ascription to Aristotle of belief in a creator God,[21] while conversely there are signs in

[13] See A. Gudeman-W. Kroll (1916) cols 1764-1795, esp 1775; L.G. Westerink (1964) 526-35, with Evrard's reply (1965); Tae-Soo Lee (1984) 43. An exception is urged by Westerink for the case of Philoponus' commentary on the mathematician Nicomachus, where he says that Philoponus copies another commentator, Asclepius. But even here Philoponus' additions are extensive.

[14] e.g. *in Phys* 583,13-584,4, where Ammonius is called 'the philosopher'.

[15] See M. Richard (1950) 191-222, who points out that two of the three Aristotle commentaries which we ascribe to Ammonius (*in Cat*; *in An Pr*) are described as being 'from the voice of Ammonius', without the name of any compiler.

[16] So one of the most important articles to have appeared on Philoponus, Étienne Evrard (1953). Evrard shows that Philoponus' view here shifts on the existence of a fifth element, on the explanation of the sun's heat and on whether the rotation of the elemental fire above us is supernatural, as Ammonius' other pupils said. See now further Verrycken (1985).

[17] This point and several of the others in this paragraph are made by Evrard (1953). See also Henry J. Blumenthal (1982). But on dating see also n 260a below.

[18] See pp 37-8 below for chronology, including Verrycken's revision (n 260a) of the standard dating of *in Phys*.

[19] Philoponus *in Phys* 55,26. From his calling the arguments *theorêmata*, Evrard (1953) conjectures that the earlier work may have been the lost *Summikta Theôrêmata*. But see n 260a below.

[20] I must here dissent from Henry J. Blumenthal (1982) 59; I must also thank him for allowing me to draw on his immense knowledge of the period.

[21] Evrard (1953) 354.

the commentary on Nicomachus that he suppresses Ammonius' further belief that the universe which God creates lacks beginning or end.[22] These points would hardly need making but for the thesis that appeared in an early encyclopaedia article according to which Philoponus was not yet a Christian when he wrote his commentaries on Aristotle.[23] This thesis is already refuted by the point that the commentary on Aristotle's *Meteorology* is later than at least one major Christian work, the *de Aeternitate Mundi contra Proclum*.[24] There is no need to indulge in discussion of whether Philoponus' references to angels, heresy, punishment after death, or pre-existence of the soul establish or exclude Christian belief. In fact they do neither, but this hardly matters beside the commitment to a beginning of the universe. One further intriguing fact is that two explicit references to the Trinity are assigned by a Syriac source to Philoponus' 'commentaries'.[25] But these, admittedly, are not to be found in any extant commentaries on *Aristotle*.

What was reserved for the later part of Philoponus' career was not his faith but the major part of his writing about his faith. By then he had become the leading intellectual first among the monophysites, who held that Christ had one nature, not two (human and divine), and later among the tritheists who viewed the Trinity as three substances or godheads. The earliest of these works (which I shall discuss below) was the *Diaetêtês* (or *Arbiter*), written around 552.

Philoponus' names

The name of 'Philoponus' is a nickname. It had been given to various philosophers because of its literal meaning: 'lover of work'. It had also been given to groups of Christian lay workers. Such workers are known to have lived together in something like a guild called a *Philoponeion*. But it remains uncertain which of these two reasons (if either) accounts for John Philoponus' name.[26] He was also known as the Grammarian (*grammatikos*), and we are told that he himself used this name.[27] He studied grammar under Romanus, and two of his books on grammar are extant. However, given their unimportance compared with his philosophical and theological work, it remains a matter of conjecture why the name 'Grammarian' was preferred. He may have held a post in grammar, and if so, it may have been a chair, or

[22] L. Taran has shown that in a discussion parallel to that of Asclepius, Philoponus suppresses reference to Ammonius' belief: (1969) 11.

[23] A. Gudeman-W. Kroll (1916).

[24] Evrard (1953).

[25] Michael the Syrian quoting in his Chronicle from Damian of Alexandria, who in turn quotes from Philoponus = fragments 29 and 30 of Philoponus in Syriac and in Latin translation, in A. van Roey (1980) 157-8; 162-3. There is a French translation by J.-B. Chabot, *Chronique de Michel le Syrien*, vol 4, Paris 1910.

[26] The connection with hard work is supported by A.P. Segonds (1981) 40 n 4, and by Henry J. Blumenthal in his book in preparation on the interpretation of Aristotle in late antiquity. The connection with lay workers is supported by J. Maspéro, *Histoire des Patriarches d'Alexandrie*, Paris 1923, 197 n 4; H.D. Saffrey (1954) 396-410; Michael Wolff (1978) 109-11.

[27] Simplicius *in Cael* 119,7.

he may have taught Greek grammar to the Coptic community in Alexandria.[28]

Attack on the Aristotelian world view

I shall now turn to Philoponus' attack on the Aristotelian world view, a view which had been inherited, with adaptations, by the Neoplatonist milieu in which he lived. In his attack, Philoponus both draws on earlier opposition to Aristotle and constructs highly original positions of his own. A remark of his opponent Simplicius has been misused.[29] Simplicius complains that Philoponus lengthens his text, to impress the uninitiated, by importing the interpretations of Alexander and Themistius. It has sometimes been inferred that Philoponus was not an original thinker.[30] Nothing could be further from the truth. In fact, Philoponus often cites Alexander and Themistius in order to disagree with them, and Simplicius is not complaining, at least not here,[31] of unoriginality. He would be more likely to disapprove of what we consider originality, as being an example of Philoponus' deviation from the proper role of a commentator, namely displaying the fundamental agreement of Plato and Aristotle.[32]

Philoponus reached some of his anti-Aristotelian positions only gradually, but by the time he had finished he was opposing traditional assumptions across a very wide front. I shall start with his views on the creation of the universe.

The creation of the universe

The idea of a creation of the universe is ambiguous. What differentiated Christians from pagan Greeks was their belief that matter itself had a beginning. Many Greeks endorsed the weaker thesis that the present orderly arrangement of matter had a beginning, but the view that matter itself did seemed to them absurd. Philoponus' great achievement[33] was to find a contradiction at the heart of pagan Greek philosophy. For the majority of Greek philosophers, and certainly the Neoplatonists, had accepted Aristotle's view that there cannot be a more than finite number of anything, nor can anything pass through a more than finite number. What Philoponus pointed out was that the universe would have had to pass through a more than finite number of years if the pagans were right that it had no beginning. What is more, if the number of years traversed was infinite by now, what would it be

---

[28] For the possibilities see e.g. Michael Wolff (1978) 108; Henry J. Blumenthal, book in preparation on the interpretation of Aristotle in late antiquity.

[29] Simplicius *in Phys* 1130,3-6.

[30] A. Gudeman-W. Kroll (1916) col 1773; O. Schissel von Fleschenberg (1932) esp 108.

[31] We shall see that Simplicius does accuse Philoponus of plagiarising Xenarchus, *in Cael* 25,23; 42,20.

[32] On this see Philippe Hoffmann in Chapter 3 below.

[33] Described in Chapter 9 below. The arguments appear at *in Phys* 428,14-430,10; 467,5-468,4; *aet* pp 9-11 and 619; *in Meteor* 16,36ff; *contra Aristotelem*, ap. Simplicium *in Phys* 1179,12-26. I have adapted the examples given.

by next year, and how many days would have been traversed? Unless they accepted the Christian belief in a beginning, they would be committed to something apparently absurd, the multiplication of infinity by 365, or the addition to it of successive numbers.

These difficulties had remained unnoticed through the 850 years since Aristotle, and it would be another 800 years before enough was understood about infinity to see how such multiplication and addition could in a sense be treated as innocuous. Given their own view of infinity, the pagan Greeks, despite Simplicius' replies, had no satisfactory answer. The only other uses I know of an infinity argument in connection with creation had concerned the beginning not of matter, but of its present orderly arrangement, and had been of a quite different character.[34] Philoponus' arguments were highly influential and I shall return to their influence below. They are nowadays most commonly encountered only in pale replicas of their original selves in Kant's *Critique of Pure Reason*.

Philoponus defended the Christian belief in a beginning of matter in a whole series of writings, some especially devoted to the subject.[35] But the most massive defence, at least of those extant, is provided by the *de Aeternitate Mundi contra Proclum* of A.D. 529, directed against the former Athenian Neoplatonist Proclus. 529 was the very year in which Justinian stopped the teaching activities of the pagan Neoplatonist school at Athens, where Simplicius was working. Simplicius' bitterness is understandable, although he found Philoponus' arguments not in the *de Aeternitate Mundi contra Proclum*, which he tells us he did not read,[36] but in the later, now fragmentary, *contra Aristotelem*, which presents at least one of the arguments[37] in a more arresting form, and in a later treatise[38] whose arguments turn on the idea of infinite force.

Dynamics unified by impetus theory

Philoponus was also to overturn Aristotle's dynamics, starting with the motion of projectiles. Aristotle had been puzzled as to what makes a javelin continue to move after it has left the hand. For such 'forced' or 'unnatural' motion he sought a cause external to the projectile, and he decided that

---

[34] Plotinus and Origen had argued in the third century A.D. that there cannot be a more than finite number of creatures created in the history of the universe. But rather then taking Philoponus' course and making the beginning of our present world the very beginning of everything, they postulated that the world's history was repeated in cycles, with the same or many of the same individual creatures recurring (Plotinus 5.7.1 (23-5); 5.7.3 (14-19); Origen *On First Principles* 2.9.1; cf 3.5.2-3). In the second century A.D., Galen, like Origen after him, used a premise about knowledge, that knowledge of infinity is impossible even for God. He too drew a conclusion only about our present world, that it cannot lack a beginning, if God's providential knowledge is to cover it; yet he thinks there are equal difficulties if it did have a beginning (Galen *On Medical Experience* ch 19, sec 3, translated into English from the surviving Arabic version by R. Walzer, see pp 122-3). I am grateful to Larry Schrenk for drawing my attention to Galen.

[35] See Bibliography for a list of those known.

[36] See Simplicius *in Cael* 135,27.

[37] The infinity argument, recorded by Simplicius *in Phys* 1179, 12-26.

[38] Recorded by Simplicius *in Phys* 1326-1336.

successive pockets of air behind the javelin received the power to push it onwards, not only when the thrower's hand was pushing the air, but even after his hand had come to rest.[39] In effect, the pockets of air were unmoved movers, although Aristotle does not put it that way, and he might prefer to say that they were no-longer-moved movers. Philoponus' innovation was to suggest instead that a force (*dunamis, ischus, hormê, energeia, archê*) could be implanted by the thrower directly into the javelin, and need not remain external to it in the air. This force came to be called an impetus and was still a commonplace in the time of Galileo. Thomas Kuhn, without mentioning Philoponus, has called the switch to impetus theory a shift of paradigm, in other words, a scientific revolution.[40]

Aristotle's theory of projectiles was ripe for replacement. He does not sufficiently explain why air should sometimes help motion, as with projectiles, while at other times[41] it creates resistance to motion and reduces speed. As to why it helps motion, he can only plead that air is light in relation to some bodies, even though it is heavy in relation to others.[42] Philoponus' contemporary, Simplicius, gets as far as raising the crucial question: why say the motion is impressed on the air, rather than on the missile, so forcing ourselves to make the air a mover?[43] But he is satisfied with Aristotle's explanation of the special propensity of air. The position of air as a no-longer-moved mover seems all the stranger because Aristotle is otherwise so sparing in explaining motion by reference to unmoved movers. Objects of desire can act as unmoved movers, but otherwise the only acknowledged case of an unmoved mover is the soul,[44] and this is still moved in the sense of being stirred by the object of desire.[45] Philoponus has enormous fun ridiculing Aristotle's appeal to pockets of air. If this were the mechanism, an army would not need to touch its projectiles. It could perch them on a thin parapet and set the air behind in motion with 10,000 pairs of bellows. The projectiles should go hurtling towards the enemy, but in fact they would drop idly down, and not move the distance of a cubit.

These ideas appear in Philoponus' *Physics* commentary,[46] and he applies his impetus theory to another question which had puzzled Aristotle: why the belt of elemental fire which Aristotle places below the stars rotates in a circle. Aristotle, writing before the Copernican revolution, thinks of the earth as stationary at the centre of the universe, surrounded by belts of water, air and fire and then by the transparent spheres which carry the stars. These

---

[39] Aristotle *Phys* 8.10, 267a2-12; cf *Cael* 3.2, 301b23-30.

[40] Thomas Kuhn, *The Structure of Scientific Revolutions*, Chicago 1962, 2nd ed 1970, 120.

[41] *Phys* 4.8, 215a24-216a11. I thank Daniel Law for this point.

[42] *Cael* 3.2, 301b23-30; used by Themistius *in Phys* 235,5-6; Simplicius *in Phys* 1349,30-2.

[43] Simplicius *in Phys* 1349,26-9, cited by H. Carteron, *La Notion de force dans le système d'Aristote*, Paris 1923, 11-32, translated into English in Jonathan Barnes, Malcolm Schofield, Richard Sorabji, eds, *Articles on Aristotle* I, London 1965, see p 170 n 39; and by S. Sambursky (1962) 72.

[44] Aristotle *DA* 1.3, 406a3; *Phys* 8.5, 258a7; a19.

[45] Aristotle *DA* 3.10, 433b13-19.

[46] Philoponus *in Phys* 641,13-642,20, translated in part in Morris R. Cohen and I.E. Drabkin (1958) 221-3, and in part in Sambursky (1962) 75. See p 37 and n 260a below for the date of the *Physics* commentary.

last were sharply distinguished from the lower elements (earth, air, fire and water), and assigned a circular, as opposed to a rectilinear, motion. So it puzzled Aristotle why the belt of elemental fire should also move in a circle, as he thought it did, on the evidence of comets and suchlike, which he took to be carried along in that belt. His own answer to the question seems to waver,[47] but Philoponus' answer is again that an impetus is implanted from the rotating heavens into the fire belt.[48]

So far, Philoponus' applications of impetus theory are only beginning. Aristotle had split dynamics into unconnected areas. Projectile motion was explained by the external pockets of air. The heavens were thought to be alive and their motion, like that of animals, was explained in psychological terms. The fall of rocks and rising of flames was explained non-psychologically by reference to an inner nature, while the rotation of elemental fire, we have seen, was a special case. Philoponus' next move has the effect of unifying dynamics. The context is the discussion of creation in the book of Genesis. This is the subject of his *de Opificio Mundi*, which has been dated to thirty or forty years later.[49] In a few lines,[50] he extends impetus theory, in one form or another, to all the remaining cases, and he is enabled to do so by the belief, for which he has argued in so many earlier works, that God created the universe. It is God who implants (*entheinai*) a motive force (*kinêtikê dunamis*) in the sun, moon and other heavenly bodies at the time of creation. It is God who then implants the downward inclination (*rhopê*) in earth and the upward inclination in fire. It is God who implants in animals the movements which come from the souls within them.

The impetus which God implants in heavenly bodies seems closely analogous to that which a thrower implants in a javelin, but in the other cases, the analogy is less close. The impetus implanted in the elements, earth, air, fire and water, must be a complex one. For in Philoponus' view, elements lose their weight or lightness, that is, their inclination to move down or up, once they reach their proper places.[51] Earth has weight, he thinks, only when lifted away from its resting position. What God implants in earth, then, is not an inclination to move down, but an inclination to move down, *if* dislodged. Similarly with animals, what God implants when he implants their souls is nothing as simple as the impetus in a javelin, and indeed at this point the analogy begins to be strained.

I have been insisting that, in so far as Philoponus unifies dynamics, it is his belief in a creator God which enables him to do so. It is true that belief in a creator does not distinguish him from pagan Neoplatonists who, in a different sense, accepted a creator. This point is made by Michael Wolff in Chapter 4, n 90. But it is no accident that they did not put their belief to the same use. For this, two things were needed. Philoponus first had to introduce the idea of impetus for projectiles, and this, for all the inspiration he got from

[47] Aristotle *Cael* 1.2, 269a9-18; a30-b2; *Meteor* 1.7, 344a11-13.

[48] Philoponus *in Phys* 384,11-385,11.

[49] See p 39 below for evidence on chronology.

[50] Philoponus *Opif* 28,20-29,9.

[51] Philoponus *contra Aristotelem*, ap Simplicium *in Cael* 66,8-74,26 = frr 37-46 Wildberg.

Proclus,[52] had not occurred to the pagan Neoplatonists, as Michael Wolff points out. Consequently, they could not take the second step of generalising the idea. It was Philoponus who did this and he did it by means of his belief in a creator, and in a book devoted to the biblical account of creation. In saying that his belief in a creator God *makes possible* his unification of dynamics, I am not saying that it *motivates* that unification. Concerning the claim of motivation, I agree with Wolff's strictures.[53]

In order to understand the situation, it is important to be clear what is meant by impetus. Modern historians and the thinkers they are commenting on use the term in two different ways. Impetus in the strict sense is equated with *vis impressa* – an internal force impressed from without. But there is also a looser use of the term to apply to a force internal to the moving body, whether impressed from without or not.[54] I believe that this ambiguity has led to confusion. If we drop the requirement of impression, we shall be inundated with examples of impetus. There will be no need to wait for Philoponus' *de Opificio Mundi*. Already in the *Physics* commentary he would ascribe the fall of rocks and the rotation of stars to internal forces. And he would have had many forerunners. Aristotle and the Stoics both ascribe many cases of motion to internal forces, and both have been cited as forerunners, as has the Aristotelian Alexander of Aphrodisias, and the author of the pseudo-Aristotelian *Mechanica*.[55] But not only do these thinkers refrain from generalising the idea of internal forces to all cases; more important, they do not treat their internal forces as *impressed*. Other forerunners have also been suggested, but I can leave the claims of Hipparchus and of Proclus largely to the searching criticism of Michael Wolff in Chapter 4.[56]

[52] Documented by Michael Wolff (1971) 92 and 94; and in Chapter 4 below, also by Jean Christensen de Groot (1983). What should be acknowledged, however (see n 56), is that Proclus does introduce something like an impetus impressed by God, to explain, not the circular motion of the heavens, but the infinite duration of their existence and circular motion.

[53] I believe that F. Krafft (1982) 60 has expressed the same view as myself, if he means that Philoponus' belief in a creator God makes possible, not that it motivates, the unification of dynamics.

[54] Walter Böhm (1967) 369, even suggests that the concept of impetus changed eventually from that of an impressed force to that of an inner soul-like principle.

[55] On Aristotle, see H. Carteron, *La Notion de force dans le système d'Aristote*, Paris 1923, 11-32, translated into English in Jonathan Barnes, Malcolm Schofield, Richard Sorabji, eds, *Articles on Aristotle* I, London 1975, 170, n 40; G.A. Seeck, 'Die Theorie des Wurfs', in G.A. Seeck, ed, *Die Naturphilosophie des Aristoteles*, Proceedings of the 4th Symposium Aristotelicum, Darmstadt 1975, 386. On the Stoics, see Walter Böhm op.cit. 346; 369; G.E.R. Lloyd, *Greek Science after Aristotle*, Cambridge 1973, 158ff; Michael Frede, 'The original notion of cause', in Malcolm Schofield, Myles Burnyeat, Jonathan Barnes, eds, *Doubt and Dogmatism*, Oxford 1980, 249. On Alexander of Aphrodisias, see the report on Girolamo Cardano in E. Wohlwill, 'Die Entdeckung des Beharrungsgesetzes' I, *Zeitschrift für Völkerpsychologie und Sprachwissenschaft* 14, 1883, 387. See also S. Pines (1961) for the view that Alexander has as good a claim as Philoponus to have provided the framework for impetus theory. On pseudo-Aristotle *Mechanica*, see H. Carteron, loc. cit.; F. Krafft op. cit. 60.

Stoic influence has indeed been found by J.E. McGuire (1985) in Philoponus' account of nature in the *Physics* commentary. But when Philoponus turns nature into a *vis impressa* thirty or forty years later in the *de Opificio Mundi*, he is departing from the Stoics.

[56] I would add only that the point Wolff makes about Hipparchus can also be made about Alexander: he still has the force that moves a projectile implanted in the air, rather than the

The fact that Philoponus' impetus is an *impressed* force bears also on the question of his influence. Shlomo Pines made a major contribution when he detected impetus theory in Avicenna, Barakāt, and in all subsequent theories of the Islamic East.[57] Subsequently he found it in an earlier source, which draws directly from Philoponus, and which Fritz Zimmermann dates in Chapter 5 (n 19) as early as the ninth century.[58] But even after that, Pines was modest about his own discovery, saying that there was no direct evidence for Philoponus, but that 'perhaps' the intrinsic movement he postulated in the case of light and heavy bodies 'facilitated a framework for' impetus theory.[59] At most he pointed out that Philoponus went beyond Alexander in extending the idea of internal force to projectiles,[60] but he did not draw attention to the fact that in Philoponus, unlike Alexander, the force was *impressed*.[61] Had he done so, he might have felt free to see Philoponus as taking the crucial step in relation to Islamic theories of projectile motion. For in the Islamic accounts, which Fritz Zimmermann describes in Chapter 5, the impetus in projectiles is also an impressed force. The idea of impression is applied, however, only to forced motion, not to the fall of rocks or the rotation of the heavens, and this is what we should expect. For Islamic writers had no interest in Philoponus' Christian tract, the *de Opificio Mundi*, where impression is extended to these other contexts.

In the mediaeval Latin West the tradition comes still closer to Philoponus. Buridan and his pupil Oresme seem astonishingly like him in their views, and Buridan even shares the idea of Philoponus' *de Opificio Mundi* that stellar movement is due to an impressed impetus, impressed by God at the time of the Creation.[62] This idea is not likely to have come via Islamic sources, nor is there a known Latin translation of the *de Opificio Mundi*. On the other hand, scholars have been too ready to follow Anneliese Maier's view that impetus theory was not transmitted to the Latin West at all, but was an independent development there.[63] The minimal idea of impetus as an internal force (*mayl*) would have been transmitted to the Latin West, so Zimmermann shows, when Ghazali's summary of Avicenna was translated into Latin in the second half of the twelfth century. The idea of impetus as a force not only internal, but also impressed, is harder to trace. Although present in Avicenna, it is represented only more obscurely (see Chapter 5, n 12) in Ghazali's

---

projectile, according to the Arabic text cited by S. Pines (1961) 30. On Proclus I would further refer to his views on the need for a repeated impression (*endidonai*) of finite amounts of power (*dunamis*) from an external and infinite power source, in order to keep the cosmos moving and existing for ever: *in Tim* 1.260,14-15; 1.267,16-268,6; 1.278,20-1; 1.279,11; 1.295,3-12; 1.473,25-7 (= schol.); 2.100,18-29; 2.131,3; 3.220,1-3; ap Philoponum *aet* 238,3-240,9; 297,21-300,2; 626,1-627,20, to be discussed in my *Matter, Space and Motion*, forthcoming, ch. 14.

[57] S. Pines (1938a, b).

[58] S. Pines (1953).

[59] S. Pines (1961) 54; Michael Wolff was equally resistant in (1978) 168.

[60] S. Pines (1961) 49; 51.

[61] ibid. 53.

[62] e.g. *Quaestiones Super Libros IV de Caelo et Mundo*, lib II, q 12, Latin quoted, with German translation, by Michael Wolff, *Geschichte der Impetustheorie* 226. See ch 7, 212-46 for Buridan and Oresme.

[63] A. Maier (1951) 127-33, who influences S. Pines (1961).

summary, in a reference to an internal force that is *violent*. The channel of transmission here could, then, have been Ghazali, but it might instead have been one of the many other people living in Spain who knew, or commented on, the work of Avicenna.

For Galileo, impetus theory is such a commonplace that he does not name any authorities for the view. On the other hand, in the strict sense according to which impetus is an *impressed* force, Galileo does not go as far as Philoponus or Buridan.[64] In a discussion in the *Dialogue Concerning the Two World Systems*,[65] Galileo makes Salviati express ignorance of whether the causes of fall and of celestial motion are external or internal. Galileo himself applies the notion of impressed force only to projectile motion and to a limited range of other cases. Philoponus' application of the notion was very much wider than that.

Impetus theory was not finally replaced until a certain conception of inertia came to be accepted. It is a matter of controversy whether the relevant conception of inertia is to be found in Galileo, or in Descartes, or in Newton.[66] On one view, Newton's third definition shows him to be in the tradition of Philoponus, treating inertia as itself a type of inner force (*vis insita*) like the innate inclination (*emphutos rhopē*) of Philoponus' *Physics* commentary, though not impressed like the implanted inclination of his *de Opificio Mundi*.[67] On another view, Newton's theory of inertia makes a force necessary only for limited purposes, for example, to *start* a javelin moving. It is needed to *divert* any body from its state of rest or of uniform rectilinear motion. But so long as a body continues to move at uniform speed in a straight line, this persistence requires no force at all. Instead, the body merely conforms to Newton's first law, which says that a body will persist in a state of rest, or of uniform rectilinear motion, unless a force acts on it to the contrary. In particular, the impressed force which brings a body into a state of uniform rectilinear motion does *not* persist in the body thereafter,[68] and in this it differs from the older conception of impressed impetus. Certainly, this conception of inertia came to prevail eventually.

The idea of inertia was unificatory, because it showed how rotation related to rectilinear motion. It was no different in principle from accelerated or decelerated rectilinear motion. All these equally required a force, since they were equally deviations from the motion now viewed as basic: uniform rectilinear motion.[69] The problem of how to accommodate in one system two such different motions as the rectilinear and rotatory had troubled Greek

---

[64] Walter Böhm is taking impetus theory in the loose sense when he ascribes it quite generally to Galileo (1967) 365; 369-70.

[65] Translated into English, S. Drake, 2nd ed 1967, 234. I thank Michael Wolff for the reference.

[66] On the claims of Galileo and Descartes see Richard S. Westfall, 'Circular motion in seventeenth-century mechanics', *Isis* 63, 1972, 184-9.

[67] Newton, *Opera Omnia*, vol.2, London 1979, p 2, Definition 3, cited by J.E. McGuire, comment on I.B. Cohen, in R. Palter, ed., *The Annus Mirabilis of Sir Isaac Newton, 1666-1966*, Cambridge, Mass., 1970, 186-91, Michael Wolff (1978) 328, and Walter Böhm (1967) 371.

[68] Newton op.cit., p 2f, Definition 4, cited and discussed by Michael Wolff (1978), 315-16, and Richard Westfall op.cit. 189.

[69] See Westfall op. cit.

philosophers such as Democritus from Presocratic times, as David Furley has shown.[70] Philoponus leaves us with the diverse intentions of the Creator in implanting this or that impetus, and in this respect his unification of dynamics does not achieve as much as Newton's.

I have treated the introduction of impetus theory as significant, but there is one significance which I do not think it had. It has been suggested that it had the effect of demolishing Aristotle's division of motion into natural and unnatural (or forced), and that it was intended to do so.[71] It is true that it violates Aristotle's criterion for classifying projectile motion as *unnatural*, for it leaves no external cause which (like Aristotle's pockets of air) maintains active contact with the projectile. But Philoponus would evidently wish to find an alternative criterion for distinguishing unnatural motion (and that would not be difficult), for he views the distinction between natural and unnatural motion as valid even after the introduction of impetus theory, and even in the very context where he extends impetus theory to the maximum.[72] In connection with motion other than that of projectiles, my own belief is that Philoponus' view is remarkably parallel to Aristotle's. There are even ironical echoes, conscious or unconscious, of Aristotle's view: in the case of the natural rise of steam, Aristotle allows the motion to be triggered off by an external agent or quasi-agent, and still to be natural, provided that the agent does not remain in active contact with the rising body. One of the permitted external agents of the rise of steam is the generator (*gennêsantos*),[73] who is presumably the man who boils the kettle and makes the steam. Philoponus retains Aristotle's generator when he explains the rise of light bodies, but transforms him from the humble boiler of a kettle into God who created the light elements in the first place.

Velocity in a vacuum

Philoponus' other main contribution to dynamics concerns velocity in a vacuum. Aristotle had connected vacuum and motion, for a large part of his attack on the possibility of a vacuum consisted in drawing out the supposed implications for motion. So far from being needed for motion, as his predecessors had said, a vacuum would make motion impossible. One argument claims in effect that, if resistance to motion is reduced to nothing, as it would be in a vacuum, speed would have to rise, absurdly, to infinity.[74] Epicurus in the generation after Aristotle seems to have thought he could get out of the difficulty.[75] But he does not diagnose what is wrong with

---

[70] David Furley, 'The Greek theory of the infinite universe', *Journal of the History of Ideas* 42, 1981, 571-85.

[71] So Michael Wolff (1971) 45-52; (1978) 68; and in a modified version below; Walter Böhm (1967) 18 and 339. I shall return to this subject in *Matter, Space and Motion*, forthcoming, ch. 14.

[72] Philoponus *contra Aristotelem* ap. Simplicium *in Cael* 34,9, and *Opif* 29,6.

[73] Aristotle *Phys* 8.4, 256a1. Similarly *Cael* 4.3, 310a32; 311a9-12.

[74] Aristotle *Phys* 4.8, 215a24-216a11.

[75] Epicurus describes his atoms as moving through the void only 'quick as thought' rather than infinitely fast. He might have filled out this idea in various ways. An atom progressing a

Aristotle's argument. Philoponus was the first to do that. All motion takes time, he replies. What you achieve by removing resistance is not the necessity for time, but the necessity for *extra* time spent in overcoming the resistance.[76]

The point about *extra* time is repeated by Galileo, who in his *de Motu* (*c*.1590) acknowledges Philoponus as a proponent of finite velocity in a vacuum.[77] The point was known to mediaeval Islamic thinkers, since it was made by Avempace (died 1139), who is also acknowledged by Galileo. Avempace had been credited by Averroes as the first to make the point, but the prior claim of Philoponus was recognised at the beginning of the sixteenth century by Pico della Mirandola, as Charles Schmitt has explained.[78]

We must not exaggerate Philoponus' contribution. It has been pointed out[79] that his positive account of velocity in a vacuum falls short of Galileo's on at least two counts. First, he thinks, wrongly, that there will be a marginal difference in speed of fall according to the weight of the body. For this he refers to experiments with dropping weights such as were subsequently credited to Galileo. Secondly, in talking of weight, he is thinking of gross weight, not of Galileo's concept of specific weight which takes into account the volume of a body and permits a direct mathematical comparison between the falling body and the medium through which it falls. None the less, it must be said that at the time of the *de Motu*, Galileo himself had not yet reached the idea of the equal velocity of freely falling bodies in a vacuum.

Vacuum and space

Philoponus' remarks on velocity form part of a larger discussion of the vacuum. Unlike Aristotle, he believed that vacuum, though never achieved, was in a tenuous sense possible, and David Sedley discusses what this sense is in Chapter 7.[80] Philoponus' idea of vacuum is linked to his idea of place or space as an extension (*diastêma*). Aristotle too had linked these ideas. He refused to think of place as an extension (*diastêma*) stretching between the walls of a container.[81] He insisted instead that it should be viewed, roughly speaking, as a thing's surroundings.[82] He then ruled out the idea of a vacuum *inter alia* because it involved the discredited idea of place as an extension

---

minimal unit of length does in a sense do so infinitely fast, if it disappears from its first position and reappears in the next at the next minimal unit of time. On the other hand, for lengthier journeys it would be perfectly feasible to propose a maximum finite speed of one new unit of length for each new unit of time.

[76] Philoponus *in Phys* 678,24-684,10, translated into English in Cohen and Drabkin (1958) 217-21.

[77] Drabkin's translation of *de Motu*, p 50 n 24.

[78] For details see Ernest Moody (1951) 163-93; 375-422; Edward Grant (1965) 79-95; Charles Schmitt (1967) 146-9; 154-5; cf Chapter 12 below.

[79] For these caveats, see E. Grant (1965).

[80] Sedley suggests that space is separable from body only in thought, but that in thought one needs to ignore the body with which it is inevitably associated, in order to understand what it is in itself.

[81] Aristotle *Phys* 4.4, 211b14-212a2.

[82] ibid. 212a2-21.

which might be empty.[83] Philoponus restores both ideas. Place or space should be viewed as an extension,[84] and this extension could, 'so far as depended on it', be void, or vacuous.[85] Philoponus' spatial extension should be viewed as a three-dimensional expanse which is immobile and contains body.

Philoponus' restoration of the ideas of extension and vacuum is not new. Within Aristotle's own school his immediate successor Theophrastus had collected the most worrying doubts about Aristotle's view of place as surroundings,[86] and this is one of the few points on which Philoponus and Simplicius actually agree that Aristotle is wrong.[87] Theophrastus' successor Strato returned to the idea of place as extension, void 'in its own nature', but always in fact filled with bodies.[88] He is even said to have gone further and allowed tiny interstices of actual vacuum within bodies.[89] Outside Aristotle's school, the view of place as extension was even more widespread, and many of these believers in extension allowed vacuum to exist as well.[90] What is surprising is that the mediaeval Latin West was less robust in rejecting Aristotle's account of place, and was prepared to go through many contortions to preserve it.[91] Because of this, Philoponus was in a position to influence the eventual break from Aristotle. His defence of spatial extension and of the possibility in some sense of vacuum attracted Pico della Mirandola in the early sixteenth century, and after Pico ideas of this kind became widespread. Philoponus was consequently credited by such thinkers as Galileo at the end

---

[83] ibid. 4.7, 214a20.

[84] Philoponus, Corollary on Place at *in Phys* 557,8-585,4, summarised by David Furley below.

[85] For the last qualification see e.g. 579,6-9. For vacuum, see the Corollary on Place and the Corollary on Void, at *in Phys* 675,12-695,5, both summarised by David Furley below and both due to be translated by him for publication.

[86] Theophrastus ap. Simplicium *in Phys* 604,5-11.

[87] See their Corollaries on Place. That by Simplicius in his *in Phys* is partly translated in S. Sambursky (1982), and is being translated by J.O. Urmson.

[88] Strato ap. Simplicium *in Phys* 601,24; 618,24; ap. Aëtium *Placita* 1.19 in Diels *Dox* 317.

[89] Strato ap. Simplicium *in Phys* 693,11-18. See David Furley, 'Strato's theory of the void', in J. Wiesner, ed., *Aristoteles Werk und Wirkung*, Berlin and N.Y., vol 1, 1985.

[90] For Epicurus, see e.g. *Letter to Herodotus* 39-40; Lucretius 1.419-444; Sextus *M* 10.2,; Themistius *in Phys* 113,11; Simplicius *in Phys* 571,24-5. For the Stoics, e.g. Sextus *M* 10.3; Themistius *in Phys* 113,11; Simplicius *in Phys* 571,24-5. For Galen, Themistius *in Phys* 114,7; Simplicius *in Phys* 573,19-32, and Philoponus himself at *in Phys* 576,13. Such a view is reported by Syrianus *in Metaph* 84,27-86, 7 (translated in S. Sambursky (1982) 57-61). It is ascribed not only to Strato but also to Platonists by Simplicius *in Phys* 601,24; 618,24. Simplicius himself says that place is not merely extension (*diastasis*), but extended space, and hence a substance, not a mere accident, *in Phys* 623,20. Of these believers in extension, the following allow vacuum to exist, some inside, some outside, the cosmos, and some in microscopic pockets: Epicurus, the Stoics, probably Galen, Strato and in addition Hero of Alexandria who has a passage corresponding verbally to one of Strato's (Hero *Pneumatics*, introduction, translated in part in Cohen and Drabkin, 248-54). Others again share Philoponus' view that extension could be void 'so far as depends on it', but never is. So probably Strato and the Platonists reported by Simplicius and the people reported by Syrianus. Philoponus himself refers to the force of vacuum as much discussed (*poluthrulêtos, in Phys* 570,17).

[91] See E. Grant, 'The medieval doctrine of place: some fundamental problems and solutions', in A. Maierù and A. Paravicini Bagliani, *Studi sul XIV secolo in memoria di Anneliese Maier*, Rome 1981.

of the sixteenth century and Gassendi in the seventeenth.[92]

Philoponus is once again influenced by his views on creation in the account he gives of place. For one thing, having denied the possibility of an infinite past, he cannot allow space to be infinite,[93] and in this his account differs from most of the Renaissance accounts that he helped to inspire. There is a further appeal to the creation when Philoponus argues for his conception of spatial extension. He cites the case of wine becoming gaseous: the newly created matter bursts the jar, which shows that newly created bodies need to be accommodated in a three-dimensional extension such as he is arguing for. The point can be extended to all bodies when we recall that bodies were originally created by God.[94]

Natural place unexplanatory

Another target of Philoponus' attack is Aristotle's appeal to natural places in his explanation of how the four elements move. In his geocentric cosmos, the natural place of fire is up above at the periphery with only the heavens above it, and of earth is down below at the centre. Air and water are assigned intermediate positions. Aristotle thinks that natural places help to explain the natural movement of the elements towards them, for he says that elemental motion shows that place has some power (*dunamis*).[95] The most likely explanatory role, although he never says this, is as a final cause or goal (not consciously sought) of motion. Aristotle complains that mere vacuum could not explain (be *aitia* of) the motion, for a vacuum contains no differences and hence not the differences of up and down.[96] If *per impossibile* the earth were dislodged from the central position which he assigns it, stray clods of earth would fall not to join the dislodged mass, but to their natural place at the cosmic centre.[97]

This type of explanation was already questioned by Aristotle's successor Theophrastus.[98] Place is not an entity in its own right, he suggested (*kath' hauton ousia tis*). An animal's limb has a place, because the animal has a nature and form which requires a certain arrangement (*taxis*) of the animal's parts. For this reason each limb seeks (but not consciously) its position in the arrangement. This explanation assigns no power to *place*, but appeals to the nature of the whole organism. Moreover, the case of animals and plants is

---

[92] See Charles Schmitt (1967) 140-3, 146-9, 154-6.

[93] *in Phys* 582,19-583,12.

[94] ibid. 573,22-574,1. This is a variant on the argument which Aristotle associates with Hesiod that a creation of the world would require a space to house it, for which see Aristotle *Cael* 3.2, 301b30-302a9; *Phys* 4.1, 208b27-209a2; Sextus *M* 10,11; Gregory of Nyssa *in Hex*, PG 44, 80B-C.

[95] Aristotle *Phys* 4.1, 208b11. The denial at 4.1, 209a20 that place can serve as any of the four causes, or four modes of explanation, is merely part of a puzzle or *aporia*.

[96] Aristotle *Phys* 4.8, 214b12; 214b32-215a1; 215a9-11.

[97] Aristotle *Cael* 4.3, 310b3.

[98] Theophrastus ap. Simplicium *in Phys* 639, 15-22. With Theophrastus' suggestion compare the one ascribed to Aristotle by Peter K. Machamer, 'Aristotle on natural place and natural motion', *Isis* 69, 1978, 377-87.

treated merely as one example of something more widespread.

Simplicius shows how the example of animals came to be generalised by his teacher Damascius and himself.[99] The cosmos as a whole is an organism, whose parts are earth, air, fire, water and the heavens. Moreover, the heavens and the earth each have parts of their own. All these parts are subject to a certain arrangement within a larger whole. Simplicius believes that he and his predecessors Iamblichus and Damascius are in the tradition of Theophrastus.[100] But in fact Iamblichus had introduced the very antithesis of Theophrastus' idea. For he restored to place an active power (*drastêrios dunamis*), and claimed that each body was actively held together and prevented from dispersing by its own proper place.[101] Simplicius, and Damascius as Simplicius interprets him, thought of this proper place as preserving the arrangement of the body's parts,[102] and as moving together with the body when it moved.[103] The wider, surrounding place might also move, and it too could be viewed in the same active way as holding together the surroundings.[104]

Against this background, Philoponus can be seen as reverting to Theophrastus' suggestion in so far as natural place is concerned.[105] He denies that place has any power (*dunamis*)[106] Place for Aristotle is a mere surface, the inner surface of one's surroundings, and it is laughable to suppose that the elements move to their proper places in search of a *surface*, or even in search of an (undifferentiated) extension. In an animal the head is on top of the body, because that is good for the animal, not because the head seeks the surface of the surrounding air at that point. The whole cosmos too is an organism, and it is good for the whole that the heavens should surround the other parts. So the heavens have a (non-conscious) impulse (*hormê*) to be so related to the other parts of the cosmos. Philoponus here adds to Theophrastus' account an idea taken from his belief in a Creator. What the four elements are seeking is not a surface or an extension, but the arrangement (*taxis*) which was originally alloted to them by God, for then they achieve their perfection. All this answers Aristotle's objection that in a vacuum there could be no natural motion. On the contrary, the elements would still seek their God-given arrangement.

Philoponus' account, like Theophrastus', is teleological or purposive. The goal of motion, the God-given order, is different from Aristotle's. But that

---

[99] For Damascius, see Simplicius *in Phys* 626,17-628,7; for Simplicius see his *in Phys* 628, 34-629,12; 637,25-30; *in Cat* 364,23-35. Texts and English translations are mostly available in S. Sambursky's valuable collection (1982); fresh translations will be available soon.

[100] Simplicius *in Phys* 639,12-15; 22-3; 639,36-640,1; 642,17-18.

[101] Iamblichus ap. Simplicium *in Cat* 361,7-362,33; *in Phys* 639,22-640,11.

[102] Simplicius *in Phys* 625,27-628,23, 629,3; 13-15; 19-20; 631,8; 14-15; 644,10-645,17; *in Cat* 364,18-35.

[103] Simplicius *in Phys* 629,8-12; 637,25-30.

[104] Simplicius *in Cat* 364,31-5.

[105] Philoponus *in Phys* 581,8-31; 632,4-634,2. For a different assessment of Philoponus' idea as *ad hoc*, see Michael Wolff below, p 96.

[106] Iamblichus' term *drastêrios dunamis* is reserved by Philoponus for such things as the impetus of impetus theory, *in Phys* 385,7.

difference does not qualify my earlier claims that Philoponus' treatment of motion is parallel to Aristotle's. There still the same types of ingredient in natural motion, at least in the later theory of the *de Opificio Mundi*: a final cause or goal of motion different from but parallel to Aristotle's goal, an external generator (God) different from but analogous to Aristotle's generator, and an inner nature or tendency to be moved thanks to the initial action of that generator.

We may ask whether Philoponus' removal of power (*dunamis*) from place lies in the direction in which scientific progress was to occur. Copernicus' overthrow of the geocentric cosmos, of course, disrupted Aristotle's scheme of natural places. Force is restored to space by certain interpretations of the General Theory of Relativity. For they invite us to consider gravity as an effect of the geometry of space. More recent interpretations, however, avoid talk of cause and effect, and treat gravity and spatial geometry as having a merely functional relation.[107]

Matter as extension

Philoponus gives a role to extension not only in connection with space, but also in connection with his concept of matter, which forms another part of his attack on Aristotelian ideas. He is talking of matter in a special sense. He does not mean body, but that *aspect* of a body which carries its properties. In a bronze statue, the bronze might be thought of as a subject which carries the properties of the statue, including its shape or form. But Philoponus is concerned with the *ultimate* subject of a body's properties. The characteristics of bronze might belong to its elemental ingredients, earth, air, fire and water, which would then be a more fundamental subject. The characteristics of earth, air, fire and water might in turn belong to some more fundamental aspect of the body which would serve as an ultimate subject. Such an ultimate subject was called first matter or prime matter, and the idea of it was ascribed to Aristotle.

The text which most influenced Philoponus, so I believe,[108] was Aristotle's *Metaphysics* 7.3, where Aristotle may be construed as thinking (whether in his own person or not) of the three dimensions, length, breadth and depth, as being the first properties to be imposed on prime matter.[109] This view is still reflected in Locke.[110] The danger in this idea of an aspect which

---

[107] For the older type of account, see Bertrand Russell, *The ABC of Relativity*, London 1925. As regards Copernicus, it has been said that he retains a kind of natural place for the fixed stars: William A. Donahue, *The Dissolution of the Celestial Spheres*, Cambridge Ph.D., 1973, in the Cambridge University Library.

[108] In this, I depart from Michael Wolff (1971) 112-19, but agree with Ian Mueller, who associates Philoponus' early text *in Cat* 83,13-19, with *Metaphysics* 7.3 ('Aristotle on geometrical objects', *Archiv für Geschichte der Philosophie* 52, 1970, 156-71, repr. in J. Barnes, M. Schofield, R. Sorabji, eds, *Articles on Aristotle* 3, London 1979). Philoponus also draws on Aristotle *Physics* 4.2, 209b6-11, which he takes to describe not ultimate matter, but matter already endowed with the three dimensions (*onkôtheisa, in Phys* 515,19).

[109] Aristotle *Metaph* 7.3, 1029a12-19.

[110] John Locke *An Essay Concerning Human Understanding*, 1690, 2.23.2.

serves as ultimate subject is that it will seem to be, as Locke was himself to confess, a mysterious 'something, I know not what'.[111] Philoponus in his earlier writings, including the *Physics* commentary, or its early version of 517, accepted the conventional view of prime matter. But in the *de Aeternitate Mundi contra Proclum* of 529 he had a new idea.[112] Why not treat length, breadth and depth, or three-dimensional extension, as the ultimate subject of properties, and dispense with the lower-level subject which he found in Aristotle's passage? The great advantage from our point of view of this manoeuvre, although it is not the reason that Philoponus himself gives, is that three-dimensional extension, or expanse, is something perfectly familiar. We are no longer left with a 'something, I know not what' as our ultimate subject of properties.

The shift in Philoponus' views is marked by his use of the phrase 'second subject'. In his earlier works, the *Summikta Theôrêmata* (probably) and the commentaries on the *Categories* and the *Physics*, Aristotelian matter is viewed as the first subject of properties, while matter endowed with the three dimensions, length, breadth and depth, is called the second subject.[113] It is this second subject which carries such further properties as colour. Just such a double set of subjects is found also in John Locke's passage, in a sentence which has puzzled modern commentators.[114] The change comes in Philoponus' *de Aeternitate Mundi*, where he most consciously promotes three-dimensional extension to *first* subject (*prôton hupokeimenon*),[115] and explicitly draws a contrast with being *second* subject.[116] I think it is a distraction to draw attention to an earlier, but rather different, use of the phrase 'second subject' in Porphyry,[117] which seems to me to be merely a divergent reflection of the same common source: Aristotle, *Metaphysics* 7.3.[118]

Philoponus' promotion of three-dimensional extension to *first* subject is accompanied by another promotion of it to being the form, differentia, essence or essential attribute of body.[119] His idea is that it performs two disparate but

---

[111] ibid.

[112] Philoponus *aet* 11.1-8, pp 405-45; some of the most significant passages are: 405,23-7; 424,4-11; 424,23-425,14; 428,7-10; 428,14-25; 440,6-8.

[113] *Summikta Theôrêmata*, judging from *in Phys* 156,16; *in Cat* 83,13-19; *in Phys* 578,32-579,18.

[114] Material substance is the subject for solidity and extension to inhere in, while the solid and extended parts are the subject for colour and weight to inhere in (Locke, loc. cit.). Jonathan Bennett thought Locke was confusedly introducing an irrelevant point about primary qualities (*Locke, Berkeley, Hume, Central Themes*, Oxford 1971, ch 3), while John Mackie was agnostic on Locke's meaning (J.L. Mackie, *Problems from Locke*, Oxford 1976, ch 3).

[115] Philoponus *aet* 406,10-11; 414,3; 425,11-12; 428,23-5; 433,4-5; 440,6-8.

[116] ibid. 426,22-3.

[117] Porphyry ap. Simplicium *in Cat* 48,11-33, used by Michael Wolff in his pioneering study (1971) 115.

[118] Porphyry applies the expression 'first subject' to matter, just as Aristotle himself explicitly does at *Metaph* 7.3, 1029a1-2. (I am grateful to David Sedley for this extra evidence that *Metaphysics* 7.3 is the relevant passage). But because Aristotle does not specify how the expression 'second subject' is to be used, Porphyry diverges from Philoponus, applying it not to Aristotle's level of length, breadth and depth (as 7.3, 1029a14-17 would suggest), but to the more complex level of complete physical things, like Socrates or bronze (as could be suggested by 7.3, 1029a23-4). For this reason I think his usage does not throw light on Philoponus.

[119] Philoponus *aet* 405,24-7; 423,14-424,11; 424,24; 425,5-6; 427,8; 435,21-2. Michael Wolff helpfully draws attention to this second idea in (1971) esp 118-19 and (1978) 151-2, but because

compatible functions: not only does it serve as the first subject of properties,
but three-dimensionality also actually defines body, as he repeatedly says in
the *de Aeternitate Mundi*.[120]

Philoponus' idea that three-dimensional extension on its own could define
body represents a lapse, I think, from his earlier views.[121] His idea that it
could serve as the ultimate subject of properties also seems startling, for we
ourselves would more readily think of extension as a *property* of bodies than as
a subject. None the less, the idea of extension as subject seems to me a
possible one, provided that certain difficulties can be met. One difficulty
arises when we try to say what kind of extension serves as the ultimate
subject. It is not *geometrical* extension, for this is something Philoponus
locates, along with other mathematical entities, in the mind.[122] It is
presumably not *spatial* extension,[123] for that is something static unlike the
matter of bodies. It is rather what he elsewhere calls *corporeal* extension, the
extension of bodies. An example of corporeal extension would be the volume
of a body, and Philoponus uses the word volume (*onkos*).[124] But we need to
think of a volume without thinking of its particular size in order to get at
Philoponus' idea of prime matter. For particular sizes belong not to the
concept of matter, but to the concept of properties imposed on matter. This is
not to say that the extension in question ever *exists* without having a size
(Philoponus' description of it as 'indefinite' as regards size[125] should not be
allowed to mislead). It is merely that one can *think* of it without thinking of
the size which it undoubtedly has. I say that it *has* size, not that it *is* size, for
in calling it indefinite, Philoponus does not mean that it is a determinable
property (viz. size). He means that one can think of it without thinking of its
particular size, or even of the fact that it has some size or other.

Philoponus may have created a difficulty for himself over how to
distinguish his corporeal extension from spatial extension. In earlier writings
he had distinguished it as having Aristotelian prime matter underlying it.[126]
But he has robbed himself of that criterion in the *de Aeternitate Mundi* by
abolishing Aristotelian prime matter, and he has not worked out a new way of
drawing the distinction. I think the distinction can be redrawn, but that the
task is not as straightforward as it may at first appear. We shall see that one
alternative taken by subsequent thinkers was to deny that the distinction was
a real one.

Philoponus does not specify whether his corporeal extension is universal or
particular. It will more easily serve as the ultimate subject of properties in a

he understands *hupokeimenon* as 'substance', not as 'subject', he omits to draw attention to the
first idea. For a different assessment of Wolff on this point, see Wildberg (1984).

[120] Philoponus *aet* 414,10-17; 418,25-6; 419,3; cf *Opif* 37,21.

[121] *in Phys* 505,8-9; 561,3-12; 22-3 recognised that three-dimensional extension could not be
the definition of body, or *place* would qualify as body.

[122] Philoponus *in Phys* 500,22; 503,17.

[123] I am grateful to David Sedley and Christian Wildberg for the point.

[124] *Onkos*: Philoponus *aet* 424,10; 424,16; 428,8; 434,4.

[125] ibid. 405,26; 424,10; 424,16; 424,24.

[126] Philoponus *in Phys* 561,11; 577,10-16; 687,31-5.

particular bronze statue, if it is itself particular, although it will no doubt have to derive its particularity from the particularity of the statue.[127] The particular extension of the statue will be a subject in that it has such and such dimensions as properties, and the extension endowed with those dimensions will be a subject in that it has such and such a shape, colour, weight and so on.

Part of the interest of Philoponus' account lies in the history of subsequent developments. His analysis of body as extension endowed with properties is a contribution to a debate which is still unfinished, and in which similar, and yet interestingly different, proposals have repeatedly been made. First, we should notice the irony of Philoponus' relation to his opponent Simplicius. They arrived at a similar destination by utterly different routes, for Simplicius believed that indefinite extension was what Aristotle had had in mind all along when he spoke of prime matter.[128] Philoponus believed that it is what he ought to have had in mind, but didn't. However, the convergence between the two thinkers is not total. For Simplicius denies that first matter can be thought of as a body, even as a body without qualities,[129] whereas that is precisely how Philoponus does think of his three-dimensional extension.[130]

An obvious point of comparison is provided by Descartes. When Descartes says in the *Principles of Philosophy* that quantity differs only in thought from extended substance, and that extension constitutes body,[131] this sounds so far like Philoponus' view that three-dimensional extension actually is qualityless body. When Descartes further claims that if we strip away what is not entailed by the nature of body, we will eventually get down to length, breadth and depth,[132] he reminds us of Philoponus' view that three-dimensional extension is essential to, and definitive of, body. But we must be careful, because there are differences. To mention nothing else,[133] Descartes has a quite different attitude to the distinction between corporeal and spatial extension which, like some of his predecessors,[134] he questions. It

[127] An undiscussed problem here is whether the particular extension underlying the water in a kettle can persist through the change, as prime matter is supposed to, when the water is transformed into steam. Philoponus would have to say that the same extension persists differently distributed and with different dimensions. No doubt he could make sense of saying this through any one change, even if not through a series of changes.

[128] Simplicius *in Phys* 229,6; 230,19-20; 230,26-7; 230,31; 232,24; 537,13; 623,18-19. I depart here from H.A. Wolfson (1929) 582.

[129] Simplicius *in Phys* 201,25-7; 227,23-230,33; 232,8-13.

[130] Philoponus *in Phys* 156,10-17; *aet* 405,11; 16; 19; 412,28; 413,2; 6-7; 414,16; 22; 415,2; 4; 7; 17-18; 417,22; 26; 418,7; 25; 419,3; 421,11; 20-1; 424,18-19; 426,21-2; 442,17.

[131] Descartes *Principles of Philosophy*, 1644, part 2, secs 8-11.

[132] sec 11.

[133] There is not either the same insistence in Descartes' passage that extension is the subject of properties, and indeed he would recognise only a very few properties such as motion, figure and size.

[134] John Buridan (c. 1295-1356) had refused to draw a distinction between space and corporeal extension, insisting that space was nothing but the dimension of a body: *Questions on the Physics*, book 4, question 10, fol 77v, col 1, and book 4, question 2, fol 68r, cols 1-2, in Johannes Buridanus, *Kommentar zur aristotelischen Physik*, Frankfurt 1964, facsimile reprint of Paris 1509: 'Spacium non est nisi dimensio corporis'. So E. Grant (1978) 554-7, and (1981) 15, who shows

is the *same* extension, he insists, which constitutes space and constitutes body.[135] The difference lies only in our way of thinking about it.[136] The absence of a distinct spatial extension in turn means that Descartes has no truck with the existence of an empty space, or vacuum.

Newton objects strongly to Descartes' analysis of body and insists that body is much more than extension. Yet he finishes up in his early *de Gravitatione* with a variation on the idea that body is extension endowed with properties. His explicit reason is his dislike of an unknowable prime matter, and it is to avoid this that he substitutes extension. His extension, however, differs from that of either Philoponus or Descartes, in that it is a static, spatial extension, which had existed without properties until God, at the time of creation, gave properties to it.[137]

Even today the idea of an extension endowed with properties is being invoked in a different form. At the sub-atomic level, physicists are increasingly dispensing with talk of particles and thinking in terms of fields of force. Space or space-time is viewed as the fundamental subject of properties, and these properties are manifested, often very briefly, at particular points. The following is a statement by Einstein:

> We may therefore regard matter as being constituted by the regions of space in which the field is extremely intense ... There is no place in this new kind of physics both for the field and matter, for the field is the only reality.[138]

In modern philosophy too, Strawson has raised the question whether, instead of thinking in terms of bodies, we could think of the world as a system of regions to which features were to be ascribed.[139] In these modern treatments, matter is analysed in terms of an extension endowed with features or properties. Of course the extension is not Philoponus' corporeal extension, but space or space-time, and the aim is to provide not an analysis of, but a substitute for, our everyday talk of bodies.

---

that Franciscus Toletus (1532-1596) also goes some way to blurring the distinction. Writing about Philoponus and his corporeal extension, he treats it as a kind of space, albeit an intrinsic space (*spatium intrinsecum*: in Toletus, *Commentaria una cum Quaestionibus in Octo Libros Aristotelis de Physica Auscultatione* fols 123r, col 2-123v, col 1, Venice 1580). In fact the idea of intrinsic place had been used not by Philoponus, but by Damascius and Simplicius, though in a rather different context (Simplicius *in Phys* 627,16-32; 628,21-629,12; 631,1-6; 631,32-633,18; 637,25-30; 638,23-639,9; *in Cat* 364,22-35).

[135] secs 10-11.

[136] secs 11-12.

[137] Newton *de Gravitatione*, tr. Rupert A. Hall and Marie Boas Hall, 139-41. I am indebted to the valuable discussion in J.E. McGuire, 'Space, infinity and indivisibility: Newton on the creation of matter', in Z. Bechler, ed, *Contemporary Newtonian Research*, Dordrecht 1982, 145-90.

[138] Einstein has more than one such statement. This one is cited in French without a reference in Louis de Broglie, *Nouvelles Perspectives en microphysique*. Reasons for preferring a field conception are given in Michael Redhead, 'Quantum field theory for philosophers', *Proceedings of the biennial meeting of the Philosophy of Science Association*, 1982.

[139] P.F. Strawson, *Individuals*, London 1959, ch. 6, part 2, and ch. 7. Rudolf Carnap had

A corollary of such views is that extension is not merely relative to bodies. In the modern accounts, bodies are even eliminated and space-time points are distinguished by reference not to bodies, but to properties.

The disruption of Aristotle's categories

Philoponus' idea that three-dimensional extension is the essence of body has damaging implications for Aristotle's scheme of categories. Extension would most naturally have been placed by Aristotle not in his first category, the category of substance, but in a subordinate category, the category of quantity. And that is where Philoponus places it in his *Physics* commentary. But now that the *de Aeternitate Mundi* makes three-dimensional extension the defining characteristic of body, he thinks that it belongs in the same category as bodies. He consequently transfers it right out of the category of quantity into the category of substance.[140] This completes a disruption of Aristotle's scheme of categories which had begun already in the *Physics* commentary. For there, although content to leave extension in the category of quantity, Philoponus concluded that quantity could not after all be made subordinate to substance. He gave as one ground that corporeal and spatial extension could, so far as depended on them, exist on their own without substance. (Spatial extension on its own would be vacuum.)

An analogous disruption of the categories is to be found in Philoponus' opponent Simplicius. Talking of place, Simplicius insists that it is not a mere accident of body – magnitude – which would put it in the category of quantity, but a substance (*ousia*).[141]

In the sixteenth century, Patrizi took the process of disruption further. Patrizi had translated into Latin what he took to be Philoponus' commentary on Aristotle's *Metaphysics*, and would certainly have known Philoponus' *de Aeternitate Mundi* and the *Physics* commentaries of Simplicius and Philoponus. Going further than Philoponus, and talking of spatial, not corporeal, extension, he put space outside the Aristotelian categories altogether, on the grounds that it exists independently of the physical world.[142] It is neither a quantity nor a substance, in the sense of the categories, although there is another sense in which it is substance. This particular idea about space seems to have been influential: it recurs in Gassendi, Charleton and Newton, who

---

earlier proposed space-time points, rather than regions: *The Logical Syntax of Language*, London 1937 (translated from the German), 12-13.

[140] *aet* 423,14-424,11, esp 424,5: it falls under the category of substance, like a differentia. It is (405,26; 424,9; 24; 425,5-6) the essence or substance (*ousia*) of body, and an essential or substantial quantity (*ousiôdes*), 405,24; 424,6.

[141] Simplicius *in Phys* 623,19-20.

[142] Francesco Patrizi *de Spacio Physico*, probably 1587, translated into English by Benjamin Brickman, *Journal of the History of Ideas* 4, 1943, 224-45. See further John Henry, 'Francesco Patrizi da Cherso's concept of space and its later influence', *Annals of Science* 36, 1979, 549-73. For the discussion of Aristotle's categories, see Brickman, 240-1.

applies it to all extension.[143] It has been thought to be original with Patrizi.[144] But it seems very possible that Patrizi was extrapolating from what he found in Simplicius or Philoponus.

I have not mentioned all the implications of Philoponus' treatment of prime matter as extension. He takes it to be relevant to Christology[145] – it is hard to see how, but Henry Chadwick makes a guess in Chapter 2. He also uses it to argue for the perishability of the heavens. Because the celestial element has three-dimensional extension as its matter, it must be viewed not as simple, but as a compound of form and matter. Moreover, its matter is the same as belongs to the perishable elements down here. On both grounds, the heavens must be declared perishable.[146] Simplicius, in a different but related context, had drawn the opposite conclusion. Speaking of those who had made qualityless body serve as prime matter, he complains that they will make the matter of things down here as imperishable as that of the stars.[147]

Rejection of the fifth element

Philoponus' assertion of the perishability of the heavens introduces another strand in his attack on the Aristotelian world view. Aristotle had introduced a fifth element, aether, as constituting the stars. Philoponus' first doubts about the fifth element were expressed in one of his treatises on creation and most fully expounded in another.[148] It is no accident that this was the context in which he came to reject the fifth element. For Aristotle had specially tailored his fifth element to be a stuff incapable of generation or destruction. It had to be ruled out, then, if Philoponus was to maintain that matter had a beginning.

The rejection of the fifth element is important also for impetus theory. For Aristotle had introduced the fifth element as the element with rotatory motion, and his main argument for it is that it is needed to explain the

[143] Details on Gassendi in E. Grant (1981) 199; 204-6; 209; John Henry op.cit. 568; Charles Schmitt (1967) ch 5, 143-4. On W. Charleton see: J.E. McGuire, 'Body and void and Newton's *de Mundi Systemate*: some new sources', *Archive for the History of Exact Sciences* 3, 1966, 233. On Newton see: E. Grant (1981) 242; 244; J.E. McGuire loc. cit. and 'Existence, actuality and necessity: Newton on space and time', *Annals of Science* 35, 1978, 463-509. In Newton (*de Gravitatione*, tr. A. Rupert Hall and Marie Boas Hall, 132), the un-Philoponan denial that extension is substance naturally gets an un-Philoponan justification.

[144] E. Grant (1981) 187 n 40; 194; 204-6.

[145] In the extracts from Philoponus *Tmêmata against Chalcedon* collected in the Chronicle of Michael the Syrian 8.13, translated into French by Chabot, 2.92, see p 108.

[146] Philoponus, fragmentary, work in Simplicius *in Phys* 1331,20-5; cf *contra Aristotelem* ap. Simplicium *in Cael* 89,22-5; 134,16-19. There may be influence from Alexander *Quaestiones* I 15, 26,29ff.

[147] Simplicius *in Phys* 232, 8-13.

[148] First clear doubts: *aet* 491,12-492,4; 492,20-493,5; 517,7-519,20. After that the extant fragments of the *contra Aristotelem* are devoted to an attack on Aristotle's eternal fifth element followed by an attack on the eternity which that fifth element would imply. Further attacks are found *passim* in the *Meteorology* commentary, which is placed before the *contra Aristotelem* by Evrard (1953), after by Christian Wildberg (1984). Contrast the earlier orthodoxy of Philoponus *in DA* 331, 33; *in Phys* 262,1; 340,31; 341,1.

rotation of the heavens.[149] So long as it does this job, there is no room for explaining celestial rotation as due to a divinely impressed impetus. Conversely, Philoponus is able to say that, since there is another explanation of the rotation, the fifth element will not be needed for explaining it.[150] In this passage, as it happens, he is not yet thinking of the rotation as due to impetus. But that is the explanation which he eventually reaches.[151]

I have already referred to other implications of Philoponus' rejection of aether. Its supposed divinity,[152] and the supposedly supernatural character of its motion,[153] came to seem to him unacceptable. Presumably he thought it detracted from the majesty of the Christian God. At any rate, Simplicius thinks that a motive for downgrading the heavens is to venerate God,[154] and complains that the result is precisely the opposite, an act of impiety.[155]

Aristotle's postulation of a fifth element created a further difficulty: it became a problem to explain the sun's heat. The fifth element could not, like the other four, possess such contrary characteristics as heat or cold which would inevitably, in Aristotle's view, make any given portion of it liable to destruction. The sun, being made of the fifth element, could not then really be hot, but must heat us through its motion creating friction in the belts of elemental air and fire down here.[156] Aristotle does his best to explain why the moon, which is closer, does not also set up friction, and why the sun sets up most friction at noon, while Alexander tries to explain why the effects of friction are not felt in the shadows.[157] But the task is too hard, and Philoponus abandons the theory of friction. At first he suggests that light plays the role of arousing the innate warmth of the air, a theory which at least explains why the shadows where no light penetrates are cool.[158] Later he is able to give a simpler account. For once he abandons the fifth element and declares that the heavens are composed of a mixture of the purest parts of the four elements, with fire predominating,[159] he can conclude that the sun

---

[149] Aristotle *Cael* 1.2.

[150] Philoponus *aet* 492,20-493,5. But see n 52 for Proclus' alternative method of introducing a divine impetus.

[151] In *de Opificio Mundi* 28-9. The *aet* passage still describes the rotatory motion as supernatural.

[152] Simplicius *in Cael* 370,29-371,4.

[153] It was not Aristotle, but the other pupils of Ammonius, who treated the motion of the heavens, or of the fire belt, as supernatural. So Damascius ap. Philoponum *in Meteor* 97,20-1; Simplicius *in Cael* 21,1-25; 51,22-6; 35,13; Olympiodorus *in Meteor* 2,19-33; 7,21-30. At first Philoponus agreed, treating as supernatural the motion of fire (*in Phys* 198,12-19; 378,21-31; *aet* 240,28-241,10; 259,27-260,2; 278,21-8), or of fire and the heavens (*aet* 492,20-493,4). Supernatural motion is, however, denied for the heavens at *aet* 278,21-8 and for fire at *in Meteor* 97,20-1.

[154] Simplicius *in Cael* 26,4-5.

[155] ibid. 70,17-18.

[156] Aristotle *Cael* 2.7; *Meteor* 1.3, 341a12-36.

[157] Aristotle loci cit.; Alexander *in Meteor* 19,13-19; cf 18,8-19,13 on how the sun can heat us when it is separated from the regions down here by three impassible spheres which it does not move. For Philoponus' reply see his *in Meteor* 42,32-43,25; 52,6-53,26.

[158] Philoponus *in DA* 331,33-332,22.

[159] Philoponus *aet* 518,14-18; *contra Aristotelem*, ap. Simplicium *in Cael* 84,15-22; *in Meteor* e.g. 53,2; 53,23.

simply possesses heat.[160]

Philoponus' rejection of a fifth element is not new. He sees himself as returning from Aristotle to Plato, and once again there was a precedent within Aristotle's own school. For Xenarchus in the first century B.C. had rejected the fifth element, and Simplicius accuses Philoponus of plagiarising Xenarchus' work (now lost).[161] Even Aristotle's immediate successor, Theophrastus, had expressed doubts,[162] and although he appears to have remained orthodox,[163] the next head, Strato, did not.[164] But even if not new, Philoponus' attack on the fifth element in his *contra Aristotelem* was massive and the treatise was to exert influence in every direction. In Islam it provoked a reply from al-Farabi,[165] among Jewish philosophers it influenced Gersonides,[166] in Byzantium Gemistos Plethon,[167] and in the Latin West Thomas Aquinas.[168] Nor did the controversy subside quickly. The revolution of Copernicus did not extend to the fixed stars, and so the fifth element was able to survive his theory.[169] In 1616 Cremonini was still defending the fifth element against Philoponus, and some have seen a relic of it in Newton's ether.[170]

Directionality of light

Not only on heat but also on light it has been said that Philoponus 'completely rejects' Aristotle, turning light from a static to a kinetic phenomenon better suited to the needs of geometrical optics, and changing the meaning of Aristotle's word *energeia* in the process. It is an important contribution to have drawn attention to Aristotle's innovation here,[171] but I am not sure that the innovation has been rightly understood.

We need to distinguish light from the action of colour. Each can be called an *energeia*. Light is the state in virtue of which a transparent medium can *actually* be seen through, whereas in the dark the medium is only *potentially* seeable-through. This is what Aristotle means, as Philoponus sees, when he

[160] Philoponus *contra Aristotelem*, ap. Simplicium *in Cael* 87,29-31; 88,8-10; 89,15-19; *in Meteor* e.g. 41,37; 42,31-2; 43,14-25; 49,29-34; 50,28-34; 52,13-18; 52,27-53,6.

[161] Simplicius *in Cael* 25,23; 42,20.

[162] Theophrastus *de Igne* 4-6, cited by Robert Sharples, 'Theophrastus on the heavens', in J. Wiesner, ed, *Aristoteles Werk und Wirkung*, Berlin and N.Y., vol 1, 1985.

[163] Theophrastus ap. Philoponum *aet* 520,18-21. P. Steinmetz seeks to discount the evidence of this passage: *Die Physik des Theophrastos von Eresos*, Bad Homburg 1964, 164.

[164] Strato is said to have made the heavens of fire, Stobaeus *Eclogae* 1.23.1, Aëtius 2.11.4 (Diels *Dox* 340).

[165] See M. Mahdi (1967) 233-60.

[166] Judging from unmistakable echoes in the account given of Gersonides in Seymour Feldman, 'Gersonides' Proofs for the creation of the universe', *American Academy for Jewish Research* 35, 1967, 113-37.

[167] See S. Pines (1938a) 22.

[168] Reported in work in progress by Christian Wildberg.

[169] William A. Donahue, op. cit.

[170] C. Cremonini (1616), and see, for Newton, Paul Moraux (1963) 1171-1263.

[171] The pioneer, as so often, is S. Sambursky in (1958); (1962) 110-17; (1970) 136. I will take this opportunity of saying how much I have learnt from the writings of both Sambursky and Wolff; points of disagreement merely reflect the fact that both are pioneers.

calls light the *actualised* state (*energeia, entelecheia*) of the transparent.[172]

There is also an *energeia* of colour. Philoponus uses the expression, unlike Aristotle,[173] for something that goes on in the medium between the observer and the thing seen. Aristotle agrees that colour acts on the medium,[174] but he prefers to call this action a *kinêsis* (change), rather than an *energeia* (activity). To this *kinêsis* he applies the language of motion.[175] His *de Generatione Animalium* speaks of the *kinêsis* as 'arriving', 'from the outside', 'from' a distant object, and as 'taking a straight course' or 'being scattered'.[176] The *de Sensu* talks of it as being 'via' the medium 'from' the sense object.[177] The *de Anima* uses the phrase 'in turn' of the action of air at a reflecting surface, and insists that the medium and observer are not affected 'together' by colours, as they are by tactile qualities.[178]

However, there need be no suggestion in these passages of Aristotle that the influence of colour travels, in the sense of requiring a time lag and reaching the half-way distance before the whole. The last passage, for example, in denying 'togetherness', is not postulating a time lag, but insisting that the medium plays an active role in stimulating vision, and is not merely a passive co-recipient of a colour's influence. Moreover in each case, despite language suggestive of travel, Aristotle need only be trying to convey the idea of directionality – of a direction of causal influence. It is hard to make the point that the influence comes from the colour and is exerted in certain directions, without suggesting that there is travel in an unwanted sense.

Whatever may be the case about colour, it is made emphatically clear by Aristotle that light does not travel in the sense of affecting one part of the medium before another.[179] Light should rather be thought of as a *state* in virtue of which transparent things are actually seeable-through, thanks to the presence in them of fire or a similar substance. Or it can be thought of as the *presence* of that fiery stuff.[180] Rather than allowing this state or presence to reach one point before another, Aristotle introduces the idea of what later came to be called discontinuous 'leaps', an idea whose history I have traced

[172] Aristotle *DA* 2.7, 418b9; 419a11; Philoponus *in DA* 324,31.

[173] Aristotle uses the expression *energeia of colour* for something that goes on not in the medium, but inside the beholder, the action of colour on his senses, an action in which colour achieves its highest level of actuality (*energeia*): *DA* 3.2, 425b26-426a26.

[174] e.g. Aristotle *DA* 2.7, 418a31-b1; colour is *kinêtikon* of the transparent.

[175] I am not thinking here of those uncharacteristic passages in *Cael* 2.8 and in *Meteor* 3.3ff, where Aristotle adopts the popular rival hypothesis that sight goes out from the eyes, rather than the influence of colour coming in towards them. The language of motion is used there too, but the theory is rejected at *Sens* 438a25-7; *Mem* 452b10-11; *DA* 435a5-10; and *Gen An* explains (780b35ff) that it makes no difference for his purpose there which theory is assumed. Philoponus *in DA* 333, 18-35, suggests that he resorts in the *Meteorologica* to the theory of sight going out, merely because it is easier to follow and meanwhile makes no difference to the principles there under discussion.

[176] *GA* 5.1, 780a29; 780b35-781a12.

[177] *Sens* 2, 438b4-5; 6, 446a21.

[178] *DA* 3.12, 435a5-10; 2.11, 423b12-17.

[179] *DA* 2.7, 418b18-26; *Sens* 6, 446b27-447a11.

[180] State: *DA* 3.5, 430a15. Actualised state: *DA* 2.7, 418b9; 419a11. Presence: *DA* 2.7, 418b16; b20; *Sens* 3, 439a20.

elsewhere,[181] although the word 'leap' is not the one used by Aristotle, and it may be too suggestive of travel. The light of a lamp can fill a whole transparent volume without having to reach the half-distance before the whole. At the end of his discussion of the discontinuous leap of light, he may well mean to extend the idea to the influence of colour. For having said that it happens with light, he adds:[182] 'and for the same reason it happens with seeing too, for light produces seeing.'

As with colour, so with light, Aristotle is himself impelled to use language suggesting travel. He talks of rays,[183] and of the reflection of rays or of light.[184] Once again these references should be construed as concerned with directionality, rather than as reimporting the forbidden idea of successive arrival. But this time Aristotle seems to be in difficulties even on the subject of direction. For if light is merely the presence of fiery stuff in a transparent volume, he has omitted to explain why we get shadow in oblique corners of that volume. The volume, corners and all, has fiery stuff present within it, and so, on his definition, should be illuminated throughout. Moreover, he seems regrettably unaware of his omission, when he seeks to convince people that light must be reflected since otherwise (but why, on his account?) light would not spread round corners.[185] He is unaware again when he gives the correct explanation of lunar eclipse, that the dark patch is the earth's shadow not an opaque obstacle:[186] why should there be a shadow, given his definition of light?

That Aristotle cannot easily explain the directionality of light is less serious for his account of vision than it would otherwise have been, in so far as he thinks of light not as what carries the message to our eyes, but only as what makes it possible for the influence of colour to bring us the message.

So much for Aristotle's theory. As regards Philoponus, I think that he is not so much making a complete break with Aristotle as revising Aristotle's treatment of light on the model of his treatment of colour.[187] He does so in a series of passages which take up problems, some peculiar to colour (why do we not see colour in any and every direction?), some to light (how does the burning glass work?).[188] He makes it explicit that he is treating them in parallel, and yet that he is aware that they are distinct.[189]

Philoponus entirely agrees with Aristotle in separating directionality from ordinary travel. Light does not reach one point before another,[190] and Philoponus applies to it the language of Aristotle's so-called leaps: 'all at once' (*athroos*), 'suddenly' (*exaiphnês*), 'without time lapse' (*achronôs*), 'simultaneously' (*hama*).[191] Aristotle had said that only qualities, not bodies, were

---

[181] Sorabji (1983) esp chs 5 and 25. Aristotle *Sens* 6, 446a20-447a11; *Phys* 8.3, 253b13-31.

[182] *Sens* 6, 447a10-11.

[183] *Meteor* 3.4, 374b4.

[184] *Meteor* 1.3, 340a28; *DA* 2.8, 419b29-33.

[185] *DA* 2.8, 419b29-31.

[186] e.g. *An Post* 2.8-10.

[187] Sambursky does not consider Aristotle's treatment of the action of colour.

[188] Philoponus *in DA* 329,14-341,9.

[189] ibid. 331,3-7.

[190] ibid. 325,1-330,28.

[191] ibid. 327,3-5; 328,34; 330,14-15; 330,26; 344,33-345,11.

capable of filling a volume in this discontinuous way. Philoponus agrees, and concludes that, since light behaves this way, it cannot be a body.[192] Certain philosophers of the Hellenistic period tried to extend Aristotle's discontinuous leaps to the motion of bodies.[193] Philoponus would not agree with them. If bodies performed discontinuous leaps, they would have infinite velocity, and Philoponus, we have seen, considers infinite velocity absurd.

So far Philoponus has remained close to Aristotle. He regards himself as still following Aristotle when he takes it that in Aristotle's view the whole air is filled with the things we see.[194] Aristotle did not in fact say this, but Philoponus treats it as a justifiable inference in a neighbouring passage. The action of a colour *must* be distributed throughout the air, just because observers at different angles are equally affected by it.[195] But now Philoponus has to confront two problems which Aristotle never considered. First, if the action of colours is distributed everywhere why do we not see distant things as clearly as near ones? Secondly, why can we not see everything regardless of the direction of our gaze? The first question can be answered by saying that the action of a colour weakens the further it is from its source, the second by saying that the air merely lets the action of a colour through; it is not itself changed in such a way as to contain visible images in every direction. Philoponus compares the effect familiar to us from stained glass windows, which can throw a pool of colour on to a distant wall without colouring the intervening air.[196]

We can now evaluate the interpretation reported at the beginning of this section. The suggestion was that the Aristotelian theory which distributes the action of colour throughout the air is a *static* theory and that Philoponus is demolishing it in favour of a *kinetic* theory.[197] This cannot be right, first because Philoponus describes himself as *defending* (*sunagônizomenos huper*) Aristotle's idea that colour influence spreads throughout the air,[198] and secondly because Philoponus and Aristotle are alike in accepting directionality, while rejecting travel. Directionality is presupposed by, and gives rise to, Aristotle's theory that the effect of colour is everywhere. What Aristotle is said to deny is that the action of colour travels (*phoitan*) to our eyes.[199] Philoponus, as Aristotle's defender here, must agree with him on directionality, and this will be what he is referring to when he talks of the

[192] ibid. 327,2-7; cf 330,14-15.
[193] Sorabji (1983) 53 for leaps of variable length; the leaps of atomic length (18-19, 347-8, 369-71) have a different provenance.
[194] Philoponus *in DA* 334,38.
[195] ibid. 330,33-5.
[196] ibid. 334,40-335,7; 335,7-30. Earlier Aristotelians had already made some of these points: if the air between the observer and the thing observed were coloured, observers whose gazes intersected could be forced to see contrary colours, Alexander *Mantissa* 147,16-25; *DA* 62,5-13. In fact there are many examples in which a medium remains unaffected, Alexander *in Meteor* 18,8-28. The whole air transmits colours, sometimes the same portion transmitting contrary ones to differently situated observers, so it cannot be coloured by the colours it transmits, Themistius *in DA* 59,24-6.
[197] Sambursky loci cit.
[198] Philoponus *in DA* 335,13-14.
[199] ibid. 334,38.

action of colour weakening as it 'progresses' (*proienai*).[200] The talk of 'progressing' need not imply a time-taking process of travel any more than does similar talk when it is put into Aristotle's own mouth. For example, Philoponus elsewhere ascribes to Aristotle the view that the action of colour 'arrives' (*aphikneisthai*) at our eyes, in some non-time-taking sense. He does so in his later commentary on the *Physics*, when he wants to make the idea of impetus acceptable in dynamics by comparing it to the action of colour.[201]

What I conclude is that Philoponus does indeed change Aristotle's theory of light to make it directional in the way it needs to be. On the other hand, he does not introduce travel in the sense of a time-taking process. Nor does he overthrow Aristotle's theory of the action of colour on the medium. Instead, he gives to light the same directionality as was already to be found in Aristotle's account of the action of colour.

The attack on Aristotle in retrospect

I have reported Thomas Kuhn as thinking that just one of the items credited above to Philoponus constituted a scientific revolution, the introduction of impetus theory. What I have been emphasising is that this idea is only one strand in a far broader attack on Aristotelian science. Not all the ideas were new,[202] but even the few that were not were argued with detail and thoroughness, and were often to prove more influential in Philoponus' version than in those of his predecessors. Also striking is the interconnection of Philoponus' views. Creation holds a central position. It permits the expansion of impetus theory, it supports the attack on natural places, it provides one argument for space as extension, and the infinity arguments for it necessitate the finitude of space. In turn it is buttressed by the abolition of the fifth element and by the ascription to the stars of extension as matter.

Philoponus' other ideas are interconnected too. With the abolition of the fifth element, a Christian sense of reverence can be satisfied, the sun's heat can be explained and impetus theory can be applied to the heavens. Impetus theory can in turn be supported by the analogy of colour theory. Meanwhile, instantaneous change of place, banned for bodies in a vacuum, can be exploited to prove light incorporeal. In several of these cases, theology influences scientific theory to an extent paralleled in antiquity only perhaps in Plato's *Timaeus*.

It is equally true that Philoponus applies his metaphysics to theology. The idea of matter as extension is applied to questions of Christology, and the location of mathematical entities in the mind is extended to universals and eventually to the Trinity. Philoponus' approach to questions of Christian doctrine, as we shall see, is that of a logician – a logician who has already shown himself reluctant to postulate such unnecessary entities as Aristotelian matter or separate mathematical space. He reduces still further the number of

[200] ibid. 335,7.

[201] Philoponus *in Phys* 642,3.

[202] Not the belief in vacuum as a possibility, in space or matter as extension, in the dispensability of the fifth element, or in the unexplanatoriness of natural place.

entities to be accepted in discussing the nature of Christ and of the Trinity.

Christian doctrine: Christ and the Trinity

It was to such subjects that Philoponus increasingly turned in later life. This led to controversy, to the suppression of his ideas in the Christian world and to the delay of his eventual influence. The shift occurred around 553,[203] when he was probably in his fifties and most of the work so far discussed was already behind him. That the transition was not altogether abrupt is made clear *inter alia* by the finding of Christian Wildberg (see Chapter 11) that the earlier work *contra Aristotelem* already contained a large component of Christian doctrine. But 553 was the date of the Fifth Ecumenical Council held in Constantinople, and this provided a trigger for a fresh concentration of energy. Over the next twenty years Philoponus delivered a sequence of three blows. The story has been made much clearer through the recent translation of additional Syriac fragments by A. van Roey.[204]

Philoponus first published the *Diaetêtês* or *Arbiter*, following it up with a series of further defences of the monophysite view that Christ had one nature, not two natures, human and divine.[205] Although that view was predominant in Philoponus' part of the world, it had been rejected at the Council of Chalcedon in 451,[206] and was to be rejected again at the new Council of 553 under the influence of the Emperor Justinian. Some time before his own death in 565, Justinian summoned Philoponus to Byzantium to explain his position. Philoponus' letter survives in which he excuses himself from the journey on grounds of old age.

Philoponus had to develop his own interpretation of, and arguments for, the monophysite position. Among other things, he argues that his opponents themselves allow that Christ is only one *hypostasis*, distinct from the other two in the Trinity. But if there is only one *hypostasis*, there should be only one nature.[207]

Philoponus' second blow divided the monophysites. In *On the Trinity*, also known as *On Theology*, published late in 567, and also in *Against Themistius* and *Letter to a Partisan*, Philoponus apparently committed himself to tritheism, in regard to the persons of the Trinity. At any rate he declared that each of these three hypostases was God,[208] that there were three Gods,[209] and that they were a plurality of substances.[210] As regards the Trinity, it is a universal,

---

[203] For chronology, see below, p 40.

[204] See Bibliography s.v. Roey, A. van.

[205] See below and Bibliography for the list of monophysite writings.

[206] For an irreverent account of the proceedings at Chalcedon, see Geoffrey de Ste Croix's papers, forthcoming.

[207] Philoponus *Four Tmêmata against Chalcedon*, in the Chronicle of Michael the Syrian, vol 2, p 103; vol 4, p 225.

[208] *Against Themistius*, tr. in Ebied, Van Roey, Wickham (1981) 51,13-52,5.

[209] Two texts of uncertain origin, cited in the Chronicle of Michael the Syrian, translated in Ebied et al. (1981) 31-2.

[210] *On Theology*, translated into Latin, van Roey (1980) 161, fr 17; fragment of uncertain origin, cited in the Chronicle of Michael the Syrian, translated in Ebied et al. (1981) 31.

and so exists in our minds. There is a single God only in thought;[211] if the
Trinity were a single God, it would be a fourth one.[212] We have already
noticed Philoponus locating mathematical objects and mathematical place in
the mind. Here he treats universals the same way and draws his shocking
conclusion about the Trinity.

The insistence on three substances was a natural development, given that
the *Arbiter* had already equated nature with substance,[213] while assigning
one nature to Christ. A later tritheist text of uncertain origin talks
interchangeably of three substances and three natures.[214] In each case, the
effect of Philoponus' view is to reduce the number of independently existing
entities. Hypostases, natures and substances are not independent of each
other, and universals are not independent of us.

Resurrection and soul

Philoponus' third blow split the tritheists. In *On the Resurrection,* written before
575, and in *Against the Letter of Dositheus,* Philoponus declared that in the
resurrection we should receive not our old bodies, but new ones.[215] The new
body is said to be immortal or eternal,[216] but another passage insists that
immortal and eternal bodies will be quite unlike our old bodies, different not
only numerically, but also in kind. Moreover, if those who are resurrected are
given immortality, they will be of a different substance and nature from
ourselves, for man is by definition mortal.[217]

If these reports and quotations are accurate, Philoponus' idea is very
startling. Christians have always wanted to be sure that it would be *we* who
were resurrected. Some modern philosophers have argued that this would
require the same body to be resuscitated,[218] and Thomas Aquinas, who
thinks that at least some of the same matter would be required, valiantly
considers whether there will be enough matter to go round for the
resurrection of cannibal communities.[219] Yet here is Philoponus denying not

---

[211] *Against Themistius,* tr. in Ebied et al. (1981) 51,5-9; *On Theology* translated into Latin, van
Roey (1980) 148; the two texts of uncertain origin, tr. in Ebied et al. (1981) 31-2.

[212] *Against Themistius,* tr. in Ebied et al. (1981) 33 and 52,3-5.

[213] See Ebied et al. (1981) 25-6.

[214] Cited in the Chronicle of Michael the Syrian, translated in Ebied et al. (1981) 31.

[215] Timotheus of Constantinople *de Receptione Haereticorum* PG 86, 44A; 61C; Nicephorus
Callistus *Ecclesiastica Historia* book 18, ch 47, PG 147, 424D; Paul of Antioch, in J.-B. Chabot,
'Documenta ad origines monophysitarum illustrandas', *Corpus Scriptorum Christianorum Orientalium*
17, Paris 1908, 330 (103, Louvain 1933, 230); John of Ephesus *Historiae Ecclesiasticae* Part III
Corpus Scriptorum Christianorum Orientalium 2.51, p 85,26-35; 3.17, p 106,12-16; cf 5.5, p 194,3 and 9,
English translation by Payne Smith.

[216] So Philoponus' near-contemporary Timotheus of Constantinople three times, op. cit. 61C.
Nicephorus Callistus, copying Timotheus in the fourteenth century, says so twice, although in
Migne's text he describes it as mortal on the third occasion, op. cit. 425A.

[217] Fragment 32, in Syriac with French translation in A. van Roey (1984), where all the
fragments are collected.

[218] e.g. Bernard Williams, 'Personal identity and individuation', *Proceedings of the Aristotelian
Society* 57, 1956-7, 229-52; 'Bodily continuity and personal identity: a reply', *Analysis* 21, 1960,
43-8.

[219] Aquinas *Summa Theologiae* III supplement, qq 69-86 (treatise on the resurrection).

only sameness of body, but sameness of substance and nature, if the sources can be believed.

His view on individual humans goes with a view about the world as a whole. This too, according to the *contra Aristotelem*, will be changed into another world which is more divine. Simplicius, who reports this, implies that Philoponus views the transformation as a destruction of the world,[220] although not, as we learn from Wildberg's new Syriac fragment, as a dissolution into nothing.[221]

The resurrection is further described as the uniting of our rational souls to the new, immortal body.[222] Philoponus had from his early writings viewed our rational soul as immortal, and this view is evidently retained even after he describes the soul as created.[223] But it seems less likely that he keeps to his early view that our rational soul has an eternal luminous body eternally attached to it.[224] Since Philoponus now believes that there will be a new heaven and a new earth, it would be appropriate for him to hold that at that time the resurrection body will replace the luminous vehicle.

Influence of Philoponus

I have spoken of the influence of Philoponus' views about creation, about impetus theory, about motion in a vacuum, about vacuum and space, about the position of extension in the categories and about the fifth element. I have also referred to the unpopularity of his views on Christian doctrine and to the anathema of 680, which deterred Christians from mentioning him explicitly. But the Arab conquests proved favourable to his ideas. A particularly well documented example is provided by his arguments for a beginning of the universe, which were repeated again and again, with elaborations, by Islamic and Jewish thinkers.[225] However, when Bonaventure propounded them in Latin in the thirteenth century, with Philoponus' own examples, he did not mention Philoponus and may well not have known of his authorship. Because of this, it has been thought that Bonaventure invented the arguments.[226] Only the *de Anima* commentary is known to have been translated into Latin in the thirteenth century, and it has been doubted whether more than a small

[220] Philoponus *contra Aristotelem*, ap Simplicium *in Phys* 1178,2-5.

[221] Philoponus *contra Aristotelem*, fr 133 in Wildberg, from Brit. Mus. Add 17214, fols 72b-73a.

[222] Timotheus op. cit. 61C; Nicephorus Callistus op. cit. 426A says mortal.

[223] Immortal: *in DA* 12,15-17; 16,2-26; 241,27-8; 242,16-19. Created (no objection: *aet* 468,26-469,5; affirmed *Opif* 23,21-27,5; 276,19-280,10, but see p 183, n 14 for another interpretation). The idea of something created but immortal is shown in Chapter 10 below to depend, in Philoponus' view, on God overriding Nature. For another example, see Preface, p xi.

[224] ibid. 18,24-8 (cf. 138,8-9) and Latin translation of commentary *in DA* 3 24,60-5. The luminous body appears to be abandoned at *Opif* 26,8-9.

[225] See the classic account by H.A. Davidson (1969). But my account must be qualified by reference to Zimmermann's remarks in Chapter 10 below: Philoponus was indeed well known for his arguments in favour of a beginning, but his impetus theory was appropriated without adequate acknowledgment.

[226] See E. Gilson, *La Philosophie de Saint Bonaventure*, Paris 1924, 184-8; John Murdoch, 'William of Ockham and the logic of infinity and continuity', in Norman Kretzmann, ed, *Infinity and Continuity in Ancient and Medieval Thought*, Ithaca N.Y. 1982, 166; G.J. Whitrow, 'On the impossibility of an infinite past', *British Journal for the Philosophy of Science* 29, 1978, 40 n 1.

part of that was translated.[227] Admittedly, as we have seen, many of Philoponus' other ideas filtered through, but the main work of translation into Latin was postponed until the sixteenth century.

Philoponus then came into his own in the West. William Wallace has estimated that in his early writings, Galileo mentions Philoponus more often than Plato, Albert or Scotus.[228] The arguments for creation are mentioned in Galileo's early notebooks, and the acceptance of finite velocity in a vacuum in his later work, *de Motu*. The former helped to make Philoponus respectable in the sixteenth century, as Charles Schmitt points out in Chapter 12, and so eased the way for his ideas on matter, space, vacuum and motion.

Antecedents

Philoponus' ideas, of course, had antecedents; he was steeped in the history of Greek philosophy. Vitelli estimated that there are six hundred citations of Themistius, though not by name, in the *Physics* commentary alone.[229] I think it is a mistake, however, to see Philoponus as supporting one school of thought rather than another among his predecessors. The Stoics and Plato have been picked out. It has been said that Philoponus was a Christian Stoic, and that he abandoned Aristotelian physics to expound and defend Stoic theories, as their most brilliant propagator.[230] As regards Plato, it is true that on the subject of the fifth element Philoponus particularly wants to contradict Aristotle, and to go back to Plato's simpler scheme of four elements, a fact about which Simplicius complains (while criticising his understanding of Plato).[231] But the preference for Plato on this particular subject is a special case.

I have already warned against the view that impetus theory comes from the Stoics. I will now take as a specimen Philoponus' discussion of first matter in the *de Aeternitate Mundi*. The case is instructive because superficially the treatment of matter as extension may seem so reminiscent of Plato's discussion of space in the *Timaeus*, while the equation of it with 'the three-dimensional' is, at least verbally, in accord with the Stoics, as he explicitly acknowledges.[232] The Stoic term 'qualityless body' is also applied

[227] William of Moerbeke translated *in de Anima* 3,4-8 and some fragments found in the margin of a Themistius commentary. Gennadius Scholarius tells a strange story in the fifteenth century that the extant commentary by Thomas Aquinas on the *de Anima* which he (Gennadius) translated into Greek was substantially identical with one by Philoponus. There is, however, no trace of such a commentary by Philoponus, much less of a Latin translation of it used by Thomas. For discussion see Jugie (1930), Schissel von Fleschenberg (1932) and Verbeke (1966) lxxi-lxxxii.

[228] Charles Schmitt warns in Chapter 12 below (n 77) that Wallace's calculations vary slightly. See William A. Wallace, *Prelude to Galileo, Essays on Medieval and Sixteenth Century Sources of Galileo's Thought*, Dordrecht 1981, 136 (contrast 196-7).

[229] *CAG* vol 17, index s.v. Themistius.

[230] Pierre Duhem (1913) vol 1, 313; 321; Gustave Bardy (1924) col 834; similarly Sheldon Williams in (1967). Of course, there is some Stoic influence, see e.g. J.E. McGuire (1985).

[231] Simplicius *in Cael* 66,33-67,5; *in Phys* 1331,7-16.

[232] Philoponus *aet* 414,3-5. I am grateful to Gisela Striker and Christian Wildberg for Stoic references, and for raising the question of the relation to Stoicism.

to the three-dimensional.[233] And Zabarella takes Philoponus' doctrine to be the same as that of the Stoics.[234] None the less, I think that the apparent similarities are misleading.

In writing earlier than the *de Aeternitate Mundi*, Philoponus is actually opposed to the Stoics. For while they had said that matter was three-dimensional and was qualityless body, Philoponus still recognised, beneath the level of three-dimensional qualityless body, an Aristotelian matter, which was not body at all.[235] It is true that the *de Aeternitate Mundi* removes this point of difference by eliminating Aristotelian matter, and transferring the name of 'matter' to the three-dimensional. But, first, this convergence with Stoic views results from dissatisfaction with Aristotle rather than from love of the Stoics, and, secondly, it calls into question another point of agreement with the Stoics. For how can he now justify saying that the level of the three-dimensional still deserves to be called '*body*'? He had earlier appealed to the underlying Aristotelian matter,[236] when he wanted to justify the slightly different claim that the extension in question was corporeal (i.e. an extension of bodies), rather than spatial. But, as we have seen, that justification disappeared. Nor is Philoponus attracted by the justification offered to the Stoics for describing matter as body: that matter has resistance (*antitupia*).[237] Plotinus had attacked this justification, saying that matter would then no longer be qualityless, and no longer simple but a compound,[238] and Philoponus suggests no answer to this objection. In other words, his agreement with the Stoics that the three-dimensional is *body* looks like a hangover, which he knew how to justify only so long as he disagreed with the Stoics on another point, and assigned to the three-dimensional a thoroughly non-Stoic incorporeal matter as its substratum.

His distance from the Stoics becomes clear again when we consider the different motives for thinking of matter as body. The Stoics believed that matter was something real and something acted on, that acting or being acted on was the criterion for being fully real, and that only body could satisfy this criterion. It is doubtful that Philoponus would accept any of this. He is at great pains to insist elsewhere that light, colour, heat and impetus can act, even though they are incorporeal. Nor is it clear that he would agree that matter can be acted on. Certainly, it can receive qualities, but he is keen to protest that it does not undergo change in the process.[239]

Simplicius gives three further arguments, whether or not they were used by the Stoics, for the Stoic view that matter is body,[240] but none of these is used by Philoponus in *aet* XI 1-8. Instead, he appears to have two reasons for

[233] *in Phys* 156,10-17; *aet* 405,11; 413,6-7; 414,22; 415,2; 4; 426,21-2; 442,17.
[234] Giacomo Zabarella, *de Rebus Naturalibus Libri XXX*, Frankfurt 1607 (first published 1590) *de prima rerum materia*, liber secundus, col 211.
[235] e.g. *in Cat* 83, 14-17; 'prime matter which is without body, form or shape before being given volume (*exonkôtheisa*)'.
[236] Philoponus *in Phys* 561,11; 577,10-16; 687,31-5.
[237] Plotinus 6.1.28 (18-20).
[238] id. 6.1.26 (17-23).
[239] Philoponus *aet* 412,15-28; 413,24-414,5; 414,16-20.
[240] Simplicius *in Phys* 227,26-228,17.

describing matter as body. First, if matter were incorporeal, bodies would be composed wholly of the incorporeal, since their other constituent is incorporeal form.[241] Secondly, the three-dimensional (which is now viewed as matter) constitutes the actual definition of body, and so cannot but be body.[242] This second reason is completely un-Stoic in spirit. For one thing, it is not so much a ground for applying the word 'body' to matter, as a ground for applying it to the three-dimensional, which, for reasons unconnected with the Stoics has come to be viewed as matter. More decisively, the argument trades on an account of 'body' which the Stoics would probably reject, because it fails to insist on bodies having resistance (*antitupia*).[243]

A final contrast comes in Philoponus' description of qualityless body as an extension (*diastêma*).[244] The Stoics think of it as essentially extended, but not, so far as I know, as an extension.

The other bogus source for Philoponus' views on matter is Plato's *Timaeus*. Aristotle identified Plato's concept of space with his own concept of matter.[245] Hermodorus, another contemporary, agreed,[246] and from then on so did most ancient writers.[247] This might, then, seem to be the source for the view of Philoponus' *de Aeternitate Mundi* that matter is three-dimensional extension. But it cannot be. For one thing, Philoponus interpreted Plato in a non-literal way, as meaning that matter was analogous to space, not identical with it.[248] Secondly, matter as Philoponus conceived it in the *de Aeternitate Mundi* was a mobile, corporeal extension, not a static, spatial one.

In rejecting the claims of Plato and the Stoics, I am not denying that there were antecedents for Philoponus' view of matter as extension. On the contrary, Moderatus of Gades, a Middle Platonist of the early first century A.D., is said to have ascribed to Plato and the Pythagoreans the same view, that matter is indefinite extension.[249] And there is a precedent closer still:[250] Plotinus reports an opponent who asks why we should give to Aristotelian prime matter which (as Philoponus later says)[251] lacks magnitude, the role of receiving properties. What receives properties is magnitude (*megethos*) and volume (*onkos*). Why not, then, dispense with Aristotelian matter and postulate just magnitude and the various qualities?

[241] Philoponus *aet* 443,6-13; 22-3. This is the third of three arguments to show why Aristotle's incorporeal matter is impossible, but the others (428,26-436,16; 436,16-443,6) trade on that matter lacking size, form or extension rather than on its not being body.

[242] ibid. 414,10-17; 418,25-6; 419,3.

[243] Stoic body is three-dimensional *with resistance*: Galen *de Qualitatibus Incorporeis* 10; Plotinus 6.1.26 (20); matter also has resistance (Plotinus 6.1.28 (18-20) = SVF II, 381; 501; 315; 318), though for attempts to discount these passages, see Margaret Reesor, 'The Stoic concept of quality', *American Journal of Philology* 75, 1954, 56-67 and now more persuasively Eric Lewis, Ph.D. diss., forthcoming.

[244] Philoponus *in Phys* 577,13; 687, 30-3; 688,30.

[245] Aristotle *Phys* 4.2, 209b11-13; *GC* 2.1, 329a14-24.

[246] Hermodorus ap. Simplicium *in Phys* 247,30-248,19; 256,35-257,4.

[247] For some of the references, see J.C.M. van Winden, *Calcidius on Matter*, Leiden 1959; Willie Charlton, *Aristotle, Physics Books I and II*, Oxford 1970, 141-5.

[248] Philoponus *in Phys* 516,5-16; 521,22-5.

[249] Moderatus ap. Simplicium *in Phys* 230,34; 231,17-20.

[250] Plotinus 2.4.11 (1-14).

[251] Philoponus *aet* 430,16; 430,25; 436,17; *Opif* 37,18-27; 39,14-18.

This shows that Philoponus' line of reasoning in the *de Aeternitate Mundi* had been around for at least three hundred years.

Other writings

I have not by any means mentioned the full range of Philoponus' works. Among what survives is the oldest extant treatise in Greek on the astrolabe and two books of grammar concerned with accentuation,[252] although the standard of all these has been judged very low.[253] Arabic writers ascribe medical works to Philoponus,[254] and there are two Greek *mss* bearing his name: *On Fevers* and *On Pulses*,[255] although the medical attributions have all been questioned.[256] He wrote a commentary on Plato, which has now been lost,[257] and another, which is extant, on a mathematical treatise.[258]

Another area in which Philoponus worked, but was less successful, was logic, which he treated extensively in his commentaries on Aristotle's logical works. Some modern logicians would agree with Simplicius' charge of incompetence.[259] But Philoponus' commentaries often record interesting views on logic which are not preserved, or not fully, by his predecessors.[260]

Chronology of Philoponus' writings

I shall conclude by considering what is known about the chronology of Philoponus' writings, although some of the findings of the next two paragraphs will need to be re-evaluated in the light of Verrycken's recent work.[260a] Among the philosophical writings before the works on Christian doctrine, there are two fixed dates. The commentary on Aristotle's *Physics* is dated to 517 by a reference at 703, 16-17, while the *de Aeternitate Mundi contra Proclum* is dated by a reference at 579,14 to 529. Various writings can then be placed before 517, for example, the *Summikta Theôrêmata* to which there are certain or probable references back at *in Phys* 55,26; 156,17, and *in An Post* 179,11; 265,6. In the first of these references, Philoponus refers to earlier arguments by him against the

---

[252] See Bibliography.
[253] So A.P. Segonds in the introduction to his French translation of the treatise on the astrolabe, and Lloyd W. Daly, in his edition of one of the grammatical works, in agreement with A. Gudeman and W. Kroll in their encyclopaedia article on Philoponus.
[254] Listed by M. Steinschneider, 'Al-Farabi-Alpharabius-des arabischen Philosophen Leben und Schriften', *Mémoires de l'académie impériale des sciences de St Petersbourg*, série 7, XIII 4, 1869, 163-5. Philoponus is called a doctor in a florilegium in Syriac, newly edited and translated into French by A. van Roey (1984), fragment 33.
[255] See Bibliography.
[256] They are rejected by M. Meyerhof (1930) and (1931). But a more hospitable view is taken, for example, by O. Temkin (1962) 105 n 58.
[257] The commentary on Plato's *Phaedo* is referred to at *in An Post* 215,5.
[258] On Nicomachus' *Introduction to Arithmetic*.
[259] Simplicius *in Cael* 28,14-30,26; 30,16; 31,1-6; 166,12-13. See e.g. Tae-Soo Lee (1984) 43.
[260] See A. Bäck (1986).
[260a] Since this volume went to press, Verrycken (1985) has complicated the situation by splitting up the *Physics* commentary and assigning only the earlier stratum to 517. He aligns *in DA*, *in GC* and *in Cat* with this earlier stratum; *in Meteor* and some of *in an Post* with the later, which he places *after* the *aet contra Proclum*. It is indeed the *aet contra Proclum*, he suggests, not the *Summikta Theôrêmata*, to which the *Physics* commentary refers back at 55,24-6.

eternity of the world. If he has in mind the *Summikta Theôrêmata*,[261] there is room to conjecture that works which leave the eternity of the world unchallenged at relevant points may be earlier than that work and than the *Physics* commentary. This would be true of *in DA* 18,24-8; 138,31; 324,15-16 and *in GC* 2,10 and 11.[262]

There are other works which might be placed before 517. For example, the *Categories* commentary could be thought earlier than the *Physics* commentary, on the grounds that it does not yet express the doubts referred to above on the priority of the category of substance to the category of quantity. It is also earlier than the commentary on the *Prior Analytics*, which refers back to it.[263]

By way of contrast, the *contra Aristotelem* is relatively late, having been written after 529, since it refers back to the *de Aeternitate Mundi contra Proclum*,[264] while the *de Aeternitate Mundi contra Proclum* refers forward to it.[265]

The *Meteorology* commentary is also late, and belongs to the same period. Not only does it refer back to the *Physics* commentary,[266] but its doctrines constitute an advance on the *de Aeternitate Mundi contra Proclum*. Like the *contra Aristotelem*, and much more decisively than the *de Aeternitate Mundi contra Proclum*, it rejects Aristotle's idea that the heavenly bodies are made of a fifth element, which is neither hot nor cold. On the contrary, the sun is predominantly made of fire, and so its warming of us ceases to be a mystery. These developments of doctrine have been traced by Evrard,[267] who adds that the *Meteorology* commentary also recants the earlier view that the rotation of the fire belt is supernatural.

As for the relative order of the last two closely related works, Evrard takes the *Meteorology* commentary to be referring forward to the *contra Aristotelem* at 16,31, but in Chapter 11 Christian Wildberg argues that it refers back to the *contra Aristotelem* at 24,38-25,2; 91,18-20, and 97,16.

The *contra Aristotelem* is by no means the last of Philoponus' works dealing with the eternity of the world. S. Pines has translated an Arabic summary of a lost work by Philoponus, arguing that the world was created in time.[268] It starts off, after an allusion to the Trinity, with references back to two earlier works on the subject, the *de Aeternitate Mundi contra Proclum* and the *contra Aristotelem*. It goes on to say that it seemed necessary to compose, after the books refuting the arguments of the eternalists (Proclus and Aristotle), a book specially devoted to improving the proofs for the temporal creation of the world. As explained in the Bibliography, it is not clear if there was one, or more than one, of these 'non-polemical' works on the subject, i.e. works not directed against a particular individual.[269]

[261] As É. Evrard (1953) 340, suggests.

[262] The absence of a challenge on eternity is pointed out by A. Gudeman-W. Kroll (1916), É. Evrard (1953), R.B. Todd (1980), and H.J. Blumenthal, in his book in preparation on the interpretation of Aristotle in late antiquity.

[263] Philoponus *in An Pr* pp 1; 40; 273.

[264] *contra Aristotelem* ap. Simplicium *in Cael* 135,27-8.

[265] *aet* 134,17; 258,22-6; 396,24; 399,23; 483,20.

[266] *in Meteor* 35,18.

[267] É. Evrard (1953).

[268] S. Pines (1972).

[269] Polemical here carries no special implication of animosity.

A still later work, probably the last, on the eternity of the world is the *de Opificio Mundi*, which treats the biblical account of creation in Genesis. Evrard has pointed out that it refers back at 118,3-4 to the *contra Aristotelem*,[270] and that its opening, 1,2ff, refers back on the one hand to Philoponus' past discussions of the eternalists' arguments and on the other hand to his own arguments for a creation in time. The latter may be a reference to Philoponus' non-polemical writing on the subject.[271] It is hard, however, to decide between the date of 557-60 for the *de Opificio Mundi*, the date most fully supported by Evrard, and that of 546-9, subsequently reargued by Wolska.[272]

On Evrard's dating, the *de Opificio Mundi* falls well within the period in which Philoponus concentrated on matters of Christian doctrine.[273] This period, we have seen, started in, or just before, 553 with the monophysite treatise *Arbiter* or *Diaetêtês*. Chadwick suggests that Philoponus was writing just before, rather than, as Šanda says, just after, the Fifth Ecumenical Council which Justinian held at Constantinople in that year. His aim will then have been to influence the debate there on the nature or natures of Christ, for which Justinian had announced the general programme back in 551. Philoponus' *Epitome* of the *Arbiter* will possibly, and his two *Apologies* defending it will certainly, have been written after the Council of 553. The *Apologies* refer back to the Council. The *Four Tmêmata against Chalcedon* constitute a much more outspoken attack on the Fifth and the Fourth (Chalcedonian) Council for their views on the natures of Christ. Chadwick suggests a date between 553 and 555. Another monophysite work, *On Difference, Number and Division*, should perhaps be dated to after 556-7, since it disclaims Tritheism, which did not become an issue until then,[274] and which Philoponus did not espouse until later. It is hard to date the *Letter to Justinian*, in which Philoponus excuses himself from going to Constantinople to explain his monophysite views. Because he pleads old age, and refers to Justinian's old age and anticipated arrival in heaven, a later date has been suggested. although one before Justinian's death in 565. Chadwick proposes around 560, Šanda somewhat later.

By that time at least, Philoponus will have written the *de Opificio Mundi*, which on the later dating belongs to 557-60. The *de Paschate*, which discusses whether the Last Supper was the passover meal, may have been influenced by the *de Opificio Mundi*, which displays similar interests in book II.

The anti-Arian treatise *Against Andrew the Arian* should probably be dated to before Philoponus' espousal of Tritheism in 567. For van Roey has translated a Syriac fragment of an anti-Arian work by Philoponus which is explicitly said to have been composed before then, and all the anti-Arian

---

[270] Evrard (1953) 338.

[271] Evrard (1943).

[272] Evrard (1953) 299-300, following E. Stein, *Histoire du Bas-Empire*, vol 2, 1949, 627 n 2 and 701 n 1; W. Wolska (1962) 163-5.

[273] For the dating of works in this period, see Henry Chadwick in Chapter 2 below, A. Šanda (1930), H. Martin (1962), A. van Roey (1979, 1980, 1984), Ebied et al. (1981).

[274] For the origins of Tritheism, see Martin (1962), Ebied et al. (1981).

Richard Sorabji

fragments are viewed by van Roey as coming from the same work.[275]

Martin has offered a very precise date for *On the Trinity*, which he regards as Philoponus' first Tritheist work.[276] He locates it between 1 September 567 (the earliest date for a discourse on the Trinity by the Chalcedonian Patriarch of Constantinople to whom Philoponus replies) and 3 January 568 (when Philoponus' reply was condemned at the monastery of Mar Bassus at Bitabō by the oriental Archimandrites). At least Philoponus' work will not have become known before 17 May 567, when the same Archimandrites condemned Tritheism in general terms without mentioning Philoponus.

Finally, Philoponus' novel work *On the Resurrection* began to fragment the Tritheists in 574, and the first recorded replies date from 575.[277] Philoponus' writings thus span at least sixty years from well before 517 to as late as 574. The timetable could be set out as follows:

before *Summikta Theôrêmata*	in *DA*
	in *GC*
before 517	in *Cat*
	Summikta Theôrêmata
517	in *Physica*
529	de *Aeternitate Mundi contra Proclum*
after 529	*contra Aristotelem*
	in *Meteorologica*
after *contra Ar* but before *Opif*	Non-polemical work or works against the eternity of the universe.
546-9 (or 557-60)	de *Opificio Mundi*
about 553	*Arbiter* or *Diaetêtês*
553 or after	*Epitome* of *Arbiter*
	Two *Apologies* for *Arbiter*
after 553	Four *Tmêmata against Chalcedon*
after 556-7	*On Difference, Number and Division*
557-60 (or 546-9)	de *Opificio Mundi*
after *Opif*	de *Paschate*
before 565	*Letter to Justinian*
before 567	*Against Andrew the Arian*
567	*On the Trinity*
about 574	*On the Resurrection*

---

[275] A. van Roey (1979). The fragment is translated into Latin on p 241.

[276] H. Martin (1962).

[277] See Ebied et al. (1981) 22; A. van Roey (1984).

CHAPTER TWO

Philoponus the Christian Theologian

Henry Chadwick

In his role as a Christian theologian John Philoponus presents several faces. On the one hand, within a Neoplatonic framework his Christian beliefs, and especially his monotheism, lay at the root of the impulse which led him to question the validity and coherence of Aristotle's ideas about the celestial bodies, to join the Platonists in challenging 'quintessence', and to say explicitly that while Aristotle was obviously a clever man and a master of logic, nothing is to be accepted as true merely on his authority.[1] If his arguments seem good, then one should accept; not otherwise. Philoponus' Christian beliefs also impelled him to challenge Proclus on the eternity of the world, and so to subject to fresh scrutiny the concepts of time and infinity. Likewise his book *de Aeternitate Mundi contra Proclum* repeatedly insists that when all form is abstracted from the underlying matter, matter does not lose its three-dimensionality. A passing observation in one of his late and highly theological writings shows how conscious he was that this was a new discovery.[2] Moreover, he even thought his discovery had some bearing on the logical problems of Christology.

The young Philoponus does not appear a man obviously interested in theology. He does not, like Boethius in his first commentary on Porphyry's *Isagoge*, encourage his readers with the thought that dialectical studies will set one on a ladder up to God. Occasional passages in Philoponus' early commentaries on Aristotle, *de Anima*, the *Physics* and finally the *Meteorologica*, can be seen to show a gradually mounting interest in concerns of special importance to Christians. If the thunderflash in Damascius' *Vita Isidori*, where he refers to Ammonius as a man who compromised his pagan loyalty by concessions to the Church, may be stretched to imply that Ammonius' submission to Christianity had gone a long way, then perhaps Philoponus' teacher Ammonius could already have helped him along his path. The attack on Aristotle, however, is altogether an 'insider' critique. And even the vast onslaught on Proclus in the lengthy book on the eternity of the world includes one handsome, perhaps politically significant, acknowledgment of how excellent Plato is on the idea of God – *if only* he had not been afraid of the

[1] The commentary on the *de Anima* is illuminating on this point.
[2] Michael Syr. *Chron* VIII 13. See Chabot's translation, p 108.

Athenian mob and had had the courage of his inner convictions, how different subsequent history would have been.[3] It looks as if Philoponus is getting at somebody.

The commentaries on Aristotle were written in the golden age for Monophysite Christianity: the reign of the great emperor Anastasius, 491-518, the man who laid the foundations for the amazing achievements of sixth-century Byzantium under Justinian, but so gave heart to the critics of the council of Chalcedon (451) that he bequeathed an insoluble social and political legacy in church divisions such as even the subtle intricacies of Justinian's mind could not solve. The age of Justinian (527-65) was dominated by the agonising debates of theologians over questions of rarefied complexity which, nevertheless, had profound bearing on religious belief and practice at a devotional level. After 529 and Justinian's closure of the Platonic school at Athens, Philoponus may well have felt that amid the endless verbal confusions and mutual misrepresentations there was room for a professional logician to lend a hand in support of his own party. The date of his *de Aeternitate Mundi contra Proclum* (529) invites the suggestion that Philoponus saw the Athens affair as an opportunity and a challenge, whether he wrote in order to attract Justinian's favour by an attack on the principal architect of late Neoplatonic dogmatics or to avert unwelcome attention from the Alexandrian philosophers by demonstrating that not all of them were motivated by a cold hatred of Christianity as Proclus was.

Philoponus' earliest intervention in theology was almost certainly his essay entitled *Arbiter, Diaetêtês*. The work survives complete in a Syriac version (manuscripts in the British Library and the Vatican), and was edited in 1930 by A. Šanda together with a Latin version. Two Greek excerpts, one of some length, are preserved through the late seventh-century florilegium edited by F. Diekamp, *Doctrina Patrum* (1907), whence they strayed to become intruded into two manuscripts of the catalogue of heresies compiled by John of Damascus in the middle of the eighth century. In the *Arbiter*, Philoponus offers his services as a trained logician who thinks it may tidy up the ecclesiastical garden if the confusing terminology of the Christological debate is analysed and sorted out. His posture is somewhat akin to that of the role assigned to the dialectician in Syrianus' commentary on the *Metaphysics*: 'The philosopher aims at the salvation of his hearers, the sophist at their bamboozlement. The logician is a tester of what you are saying, not someone laying down truth for you to accept.'[4]

Let us look back for a moment on the controversy. In 451 the council of Chalcedon's Christological definition had bequeathed a legacy of broken ecclesiastical communion and consequent civil disruption, with dangerously mounting tension between the Chalcedonian West and the Greek East where Monophysite or anti-Chalcedonian pressure on the government was too strong to be resolutely resisted, even if the emperors had been (as some were not) perfectly convinced that Chalcedon had got things right. Yet the

---

[3] *aet* p 331 Rabe.
[4] Syrianus, *in Metaph* 63, 21ff.

definition of 451 was expressly intended to bring peace and was drafted with no small finesse as a statement of consensus between the two main warring schools of theology which, for convenience, we label Antioch and Alexandria. Theodore of Mopsuestia (d. 427) had given a striking lead to the Antiochene school in his theology of redemption by the perfect self-offering of Jesus, model to humanity in faith, obedience, holiness and divine Sonship: he is the 'pioneer', the captain leading his people to salvation. Only one who is all that we are in our essential humanity can be our redeemer. By faith, through obedience to his word and through participation in him by baptism and being joined with him in the eucharistic memorial of his sacrifice of love, those who follow him are brought to the beatific vision of God in the transcendent higher world. Man was created to be the link or linchpin between the created realms of spirit and matter. So Jesus is the second Adam, the sign of the renewal of creation, and the supreme exemplar of what the Creator intended humanity to be.

The Alexandrian theological tradition, running through Athanasius, Apollinaris and Cyril, adopted a less cheerful estimate of the finite created order. Redemption is there seen as being achieved by the sovereign power of the Creator; because humanity cannot take itself by the hair and pull itself out of the mire, the redeemer must transcend our mortality, ignorance, and finitude if he is to lift us up to the realm whence he himself comes. The redeemer of the world cannot simply be part of the world. He who once for all suffered in the flesh remains unchanging in what he eternally is, 'one of the Trinity', and the mother of the incarnate Christ is not merely the physical agent or channel of his humanity. She is *Theotókos*, Mother of God.

The Alexandrian incarnational doctrine is threatened by the intensity of its own power. It can too easily slip into regarding the humanity of the redeemer as an incidental, secondary, merely accidental tool to the real work of redemption achieved by the divine presence within the veil. That would be to leave us with a myth of God Incarnate, and well known awkwardnesses begin to beset us. The Antiochene doctrine, on the other side of the house, is threatened by its own reasonableness and accessibility which can begin to pass into a reducing or minimising of the significance of Jesus. It can slip into treating the redeemer as an exceptionally inspired person, full of rare wisdom, a model of virtuous living to a degree seldom achieved in the story of our wretched race; one to whom virgin birth and resurrection appear like luxury trappings added to impress the simple but with the disadvantage of raising an intellectual hurdle for the educated. Perhaps indeed on this most minimising view the importance of Jesus is hardly found in anything he really said or did so much as in the substantial continuing existence of an ethical community seeking to follow the example of loving self-sacrifice which the stories about Jesus symbolise.

Ancient men had two principal frames of language for speaking about divine presence in and to human life: either incarnation or inspiration. Both categories antedate Christianity. But in Christian history the co-operative complementarity between them has tended to become a fierce rivalry.

The Alexandrian tradition was never so eloquently and cogently formulated as by Cyril, bishop of Alexandria from 412 to 444. His conflicts with the 'inspired man' Christology of Theodore and his pupil Nestorius gave him something approximating to hero-status for monastic Egypt and for many ascetics in Syria and Palestine. The central issue in the debate lay in the manner in which Christian theologians ought to express the unity of the person in whom both God and man are present for our redemption. Nestorius followed Theodore in seeing the union as one of will: the moral grandeur of Jesus as man is to have had a will one with God's will. The incarnate Lord is a kind of sublime partnership. Cyril abrasively rejected this explanation. For him the immutable eternal Word of God has descended to make his own a particular soul and body, thereby rendering the significance of the incarnation one of universal consequence for our race. Cyril's favourite natural analogy for the union is the coming together of body and soul to constitute the human person, a single person, one nature, one *hypostasis*. One can distinguish the two natures out of which the one Christ comes only by mental abstraction or *theoria*. Just as soul and body produce one person, so in Christ there is a similar union, a single nature constituted by the bonding together or 'synthesis' of divine and human. Cyril bequeathed to his successors not only a technical vocabulary of nature, person and *hypostasis*, but also an awareness that these terms already had a background of usage in Neoplatonic logic. Cyril was less than a professional logician, but he saw that many of the axioms and arguments of Porphyry had their application in the problems of Christological language. The discussion in Plato's *Parmenides* of identity and difference fascinated the Neoplatonists. If it was being said that two identical things are nevertheless distinct, the late Platonists (like Proclus) liked to add qualifying adverbs such as 'inseparably and indivisibly'. If it was affirmed that distinguishable things are nevertheless *au fond* identical or come to return into an identity, then they like to qualify that statement by such adverbs as 'unconfusedly and immutably', i.e. without ceasing to be what they are.

The Christological definition produced by the council of Chalcedon in 451 was a brilliantly constructed piece, in which the central contentions of the school of Antioch were protected but set within a qualifying framework of Alexandrian language. Indeed, the clauses protecting the 'two natures' tradition of Antioch were derived from a mosaic of phrases taken out of Cyril himself, turning his concessives into substantial statements (e.g. 'the difference of natures is not destroyed by the union'). There was, however, one crucial point where Chalcedon departed from Cyril, and that was in a preposition. Cyril had insisted that the one Christ is the product of (*ek*) two natures.

The first draft of the definition laid before the Council used exactly this language, and it would have saved an infinity of trouble and division had it been possible for that to be approved. But the exalted secular bureaucrats presiding over the Council on behalf of Marcian and Pulcheria had instructions to see that whatever the formula of faith contained, it must be in conformity with the requirements of Pope Leo. The new emperor Marcian had not yet gained recognition from his western colleague Valentinian, and the palace was sharply aware that western dissatisfaction with the orthodoxy of the eastern emperor would greatly reduce the political influence of Byzantium in the western half of the Mediterranean, already rapidly passing under barbarian control. The situation in 451 strikingly anticipated that prevailing at the accession of Justin in 518 when Pope Hormisdas was able to enforce submission to Rome as arbiter of orthodoxy and failed to realise that he was being hugged now only that his successors might be the better squeezed later. In 449 Leo had sent to Constantinople his famous Tome setting out the western understanding of the Christological question and using at one point the formula 'in either nature'. At Chalcedon the Roman legates pressed for the preposition 'of' to be replaced by 'in'. They were with reason angry with Dioscorus, the courageous but highly imprudent bishop who had succeeded Cyril at Alexandria, because Dioscorus had wished to set aside Leo's Tome. Moreover, the disadvantage of the preposition 'of' was enhanced by an unwise intervention in the Council by Dioscorus, declaring that 'of two' was acceptable to him, but not simply 'two'.[5] So Rome would surely welcome a formula that Dioscorus would find it hard to accept without fatal loss of face. But it was awkward and politically disastrous for the future that Cyril had never said 'in' two natures. The revised draft of the definition included 'in', but then qualified this by affirming not only 'one person' but also 'one *hypostasis*', language that ought to have satisfied the most ultra-Cyrilline divine.

Neither the radicals on the Nestorian side nor the zealots of the Alexandrian or Cyrilline side thought it comprehensible to affirm two natures but only one *hypostasis*. The Monophysites got their convenient but resented nickname from whose who accepted Chalcedon, because they could not abide the Chalcedonian clause 'known in two natures' through which Nestorians could merrily drive a coach and four. To the Monophysites, if Christ is a real union of God and man like soul and body making one person, he is one nature, composite, not indeed simple, but a single end-product. On the other side, the Nestorians did not really think 'known in two natures' was easily compatible with the assertion of one *hypostasis*. As a Greek metropolitan sadly remarked in a letter to Rome in 512, the two warring factions of the Nestorians and the Monophysites had coherent and incisive formulas with logical bite, whereas the orthodox were left in the difficult position of pursuing a via media between the two which ended in a series of unhappy negations. The Greek metropolitan pleaded with the Pope to give some authoritative guidance on how the dilemma could be solved. We do not know that he received a reply.[6]

In short, Justinian's empire was racked by theological disagreements which deeply affected social and political life. The emperor himself was a firm Chalcedonian who longed to reconcile the alienated Monophysites to the formula 'in two natures' and was ready to accept anything Monophysite

---

[5] *Act Chalc* i, 332.

[6] I analyse the letter and seek to place it in its setting in my *Boethius* (Oxford 1981). The text is printed among the letters of Pope Symmachus in A. Thiel's edition of the papal letters of this period.

divines might propose to him if only they would swallow that. His wife
Theodora had received wise pastoral care, at one point in the distress of her
turbulent youth, from an Alexandrian priest of Monophysite allegiance, and
was well known to provide a refuge within the royal palace at Constantinople
for numerous Monophysite bishops extruded from their sees by their inability
to subscribe to the Chalcedonian formula being enforced by her husband.
But on 28 June 548 cancer removed her from the scene, and Justinian
decided to attempt a grand reconciliation of the rival parties by getting the
Pope, Vigilius,[7] to agree to a series of formulas designed to silence
Monophysite criticism of Chalcedon by demonstrating that the definition of
451 was truly in line with Cyril and in no sense made room for the radical
two-nature doctrines of Theodore of Mopsuestia and his disciples.

The most probable context for Philoponus' initial intervention in the
Christological debate is the immediate run-up to the Council of
Constantinople of 553 at which Justinian obtained everything he wanted not
only from the Greek bishops but, after painful vacillations, even from Vigilius,
who was in effect tortured into submission. (It is among Justinian's more
remarkable achievements that, side by side with his monuments in
architecture and legal codification, he succeeded in enraging both the pagan
philosophers and the Pope, not to mention the principal historian of his
military campaigns and buildings, Procopius.)

Philoponus' *Arbiter* is a cool analysis by a man who presents himself as a
detached dealer in clear and incisive language. According to a report in
Nicephorus Callistus, the book was dedicated to Sergius the patriarch of
Constantinople. He must mean Sergius, monophysite patriarch of Antioch
(558-61) who was in fact resident at Constantinople, and at the time of
writing was still in presbyteral orders. For the internal content of the *Arbiter*
strongly points to a date shortly before 553, but after Justinian's declaration
of 551 delineating the dogmatic pattern of the forthcoming council's decisions
– a declaration to which Philoponus may refer. The initial standpoint adopted
by Philoponus is closely akin to that of Justinian himself. He starts by
remarking that most of those in dispute hold remarkably similar theological
positions and are in disagreement only in words. Both the principal parties
entirely concur in rejecting extreme and absurd positions. So there is room, in
this mutual misunderstanding exacerbated by misrepresentation, for a
formula of peace and reconciliation. Philoponus starts from Cyril's analogy of
the unity of body and soul. He rejects the analogy which compares the unity
of Christ's person to matter as a substrate acquiring the accident of whiteness
or heat. Nestorianism comes down to seeing the union as a mere uniting of
accidents, not of natures (cap. 8). On the other hand, the unity is not a simple
unity, but composite, a *synthesis*.

When the logician is confronted, however, by Chalcedon's formula, one
hypostasis and two natures, he shudders to a standstill, wondering what the
terms could mean. By *hypostasis* one understands an individual existent, and
it is through a plurality of such individuals that genera and species have

---

[7] A clear account of Vigilius is given by A. Lippold's article in PW.

being. Universals exist only in concrete reality through the individual *hypostases* constituting them. *Hypostasis* is therefore a narrower term than nature. Nature is that which many *hypostases* or individual existents share. One could at least find it comprehensible if Chalcedon had affirmed two *hypostases* sharing in a single nature. Even though such a formula could not be true christologically, it would at least not look like nonsense. Moreover, in the doctrine of the Trinity, orthodox tradition speaks of three *hypostases* sharing in a single nature (cap. 28).

Towards the end (45) Philoponus turns to consider the prepositions 'of' and 'in' which lay at the hottest point in the furnace of the controversy. Why, the Chalcedonians asked Philoponus and his party, do you not say 'in?' A triangle consists *in* three straight lines, a house *in* wood and stone. 'Surely (argue the Neochalcedonians) we may say both in and of two natures, as long as we exclude a Nestorian notion of a moral union of wills. Can you not accept Chalcedon's formula and simultaneously affirm one composite nature? To refuse is mere pigheadedness.'

We see here how Philoponus' tradition was being put under pressure by the Neochalcedonian move to assert both 'in' and 'of'. Philoponus had to find arguments for holding that 'in' can never be acceptable. So he reasons: if the component parts mutually coinhere or pervade each other, natural speech would say 'of', implying that there is an end-product which is a single entity. The preposition raises a question about the relation of whole and parts. We say a whole is 'in' the parts when the parts are spatially separate and distinct, like the parts of the human body. But we say a man is 'of' soul and body, not that he is 'in' soul and body. Nevertheless the *Arbiter* makes a momentous concession: on condition of the affirmation of 'one composite nature', we may allow 'in', even though it is vastly less appropriate than 'of'. Unless the 'in' is qualified by the affirmation of 'one nature of the incarnate Lord', it is unacceptable; and that was an addition which Chalcedon failed to provide.

Writing at Alexandria where the mass of the population and their clergy had decisively rejected the Chalcedonian council, Philoponus was naturally anxious to reinforce the defences of the Monophysite position. Yet it is striking to find him willing to recognise the force of the Chalcedonians' fear that to speak of one nature in Christ might be taken to imply that the union of God and Man produced a tertium quid, neither fully divine nor fully human. Polemical writers of the Chalcedonian party seized on the *Arbiter*'s concession that 'in' might be acceptable if glossed, and in embarrassment he had to withdraw it, saying that the concession was merely an unreal hypothetical condition, suggested for rhetorical purposes.[8] So the established misrepresentations continued, the Monophysites insisting that Chalcedon provided cover for Nestorianism, the Chalcedonians regarding all rejection of the Definition as committing one to the extremist position of Eutyches (abhorrent to all moderate Monophysites) that in the incarnate Lord there is only one nature and that is divine.

From 553 onwards Philoponus found himself being attacked on both sides

---

[8] See *Solutio duplex* 18 and 23 (ed Šanda).

because his explanations of the Monophysite position were regarded by the hard-liners as making concessions, welcome or unwelcome according to one's viewpoint. By this stage of the controversy the sense of group rivalry was so powerful that no one wanted to be told the fearful truth that the main parties believed the same things in everything that really mattered. The hard-line men nursed the deepest suspicions of all attempts at mutual agreement or comprehension. It was axiomatic for them that those who imagined they had reached agreement could not have begun to see what 'the real issue' was; that if a formula was proposed which one side could accept, the other side could not be sincere in saying that they could also agree to it, or, alternatively that the agreement must conceal hidden ambiguities in which the same words were being understood in different senses. In short, the rivalry had produced the deadlock all too familiar in modern ecumenical discussions, where the extremists on either side were not willing to recognise their faith in any terminology other than that with which they were familiar, and felt that any statement which the other side could conscientiously accept must, for that reason alone, be inadequate to protect the truth. Those who talked of agreement across the divide were regarded as either diabolically clever or unbelievably stupid.

Although Justinian himself disavowed the intention, many contemporaries understood the policy of his council of Constantinople in 553 to be that of reconciling the Monophysites to Chalcedon by censuring the three *bêtes noires* of the Monophysite demonology – Theodore of Mopsuestia, Theodoret, and Ibas of Edessa – and by glossing Chalcedon's 'in two natures' with almost every formula that its critics used to oppose it. Like many intelligent observers of the time, especially in the West, Philoponus thought it ridiculous to affirm Chalcedon and then condemn Theodoret whose orthodoxy had been accepted at Chalcedon and who had been allowed to take his seat among the bishops in synod. The council of 553 seemed to be reaffirming two natures and then subjecting the formula to death by a thousand lethal qualifications. For that council accepted 'one nature of the incarnate Word', 'one composite *hypostasis* of Christ', and virtually everything else in the Monophysite armoury except the unqualified 'one nature'.[9] 'Now they are saying with us one composite nature' (remarked Philoponus);[10] but a century earlier Chalcedon had used no such language, and was it not rather late in the day to be rectifying the error? Even if the Chalcedonians had lately come to see something of the truth, the stream remained incurably polluted by a disastrous century of diphysite heresy, and a true Monophysite like Jacob Baradai saw in Justinian's policy nothing but a terrifying threat to what he understood to be authentic orthodoxy. Hence the catastrophic decision to establish a rival hierarchy. Jacob seems to have understood that even if dogmatic formulas can be harmonised, mutual recognition of rival ministries presents so vast a challenge to human vanity that no such proposal is likely to succeed, above all if one side is denying the validity of the other's sacraments,

---

[9] See *Acta concil. oecumenicorum* IV/1, p.242,15 ed Straub; Philoponus *Apology for the Arbiter* 6, p 108 Šanda.

[10] *Apology for the Arbiter* 10, p 111.

and if the denial of validity is coming from the embattled minority.

Not long after the council of 553 Philoponus felt it necessary to publish a full frontal attack on Chalcedon and on Justinian's attempts to make 'heresy' palatable by the censure of the chapters drawn from the three Antiochene theologians. This book, *Against the Fourth Council* or *Four Tmêmata (Divisions) against Chalcedon*, apart from an uninformative chapter in Photius' *Bibliotheca* 55, survives only through a summary included in the history of Michael the Syrian (VIII 13). To judge from this summary, the book was a tough piece of polemic, ruthlessly exposing the incoherence of Justinian and his council. Moreover, 'whatever orthodox formulas heretics may use, they remain heretics in the inward intention with which they use them, and the presence of sound language is merely a trap for the unwary'. What legends the Chalcedonians had propagated to surround their fateful council of 451 with a bogus aura or nimbus! They had inflated the numbers present to 630 (or did Philoponus mean 636 to give double that of the 318 fathers of Nicaea?), whereas a simple count would show that at the censure of Dioscorus only about 200 were there, and of them several bishops were represented by presbyters or deacons or other proxies – not exactly evidence of a highly responsible decision in so weighty a matter. And under what canon or imperial sanction did Leo proceed when he wrote debarring Dioscorus from his place in the council? Merely by a usurped authority. Philoponus sees the papacy as acting with a new and extraordinary arrogance towards the eastern churches, most strikingly illustrated by Vigilius' behaviour at the council of 553 (behaviour which had not greatly pleased Justinian and the Greek bishops, it is fair to add). For although residing in Constantinople, he had not deigned to sit with the bishops in synod, but kept to himself and afterwards confirmed the synodical condemnation of the Three Chapters in writing. In short, he acted as if he were something other than a bishop.

Philoponus next launches an onslaught on the logic of the formula 'one *hypostasis*, two natures'. 'One composite *hypostasis*' (accepted in 553) must mean 'one composite nature'. A letter of Theodoret damaging to its author is quoted, in which Theodoret interpreted the one *hypostasis* of Chalcedon by claiming that in Scripture *hypostasis* often stands for a plurality, e.g. Deuteronomy 1:12 'how can I endure your *hypostasis*?' (i.e. seditious assembly), or 1 Kingdoms 14:4 where Jonathan attacked the *hypostasis* of the Philistines. Evidently Theodoret was ready to understand the term to cover a multitude of individuals, not just a Nestorian partnership of two. Philoponus' argument was picking up a current complaint; for canon 5 of Constantinople (553) condemns this exegesis.

It would have been better if the council had said 'one composite nature, *phusis*'. 'I do not say the composite entity is exactly as the things of which it is composed. But it belongs to no different genus from simple entities. A discourse is not a sentence, a sentence is not a syllable, a syllable not a letter. But all are of the genus Words.' So a composite nature is tightly coherent, not a loose amalgam.

Michael the Syrian or his source found the logical parts of Philoponus' argument rather taxing on the intellect. He touchingly ends by beseeching any reader skilled in logic and rhetoric, for love of the crucified Jesus, to pray for the poor excerptor who has done his best. As I have mentioned, at one point Philoponus drags in his little demonstration that even unformed qualityless matter retains three-dimensionality. The argument appears to serve the point that all synthesis is of particular and specific entities, not of abstractions or universals. Philoponus rejects the notion that the divine and human united in Christ can be the universal substance (*ousia*) of divinity and universal humanity: all universals are mental abstractions which have no existence outside the mind.

The Syriac tradition preserves a letter from Philoponus addressed to Justinian himself, in which the incoherence of Chalcedon is contrasted with the clarity of Monophysite Christology, and also a short tract for his friend Sergius dealing with the relation of a whole and its parts. Can the parts be said to be *in* the whole either actually or potentially? The theological relevance of the argument emerges when Philoponus observes that the whole does not consist in the parts but is a product of them; and that is 'of', not 'in'.

A tract 'On Difference, Number and Division' is printed in Šanda's edition, and he doubted its authenticity principally because it begins with regrets that some are now teaching the Trinity to consist of a plurality of essences. Tritheism is a doctrine with which Philoponus' name was soon to be associated. The tract, however, is concerned with Christology, not with Trinitarian questions. The argument is against the notion of some moderate Chalcedonians and probably some Monophysites as well, that one can grant the differences of natures to continue after the union, provided that one at once denies that there is either separation or the possibility of numbering them. Philoponus thinks this formula is confusing realities by a smokescreen of words. In one composite nature no real division is possible. The difference between the natures is discerned, as Cyril had said, exclusively by mental abstraction. The concrete reality is one composite nature.

Here the right wing of the Monophysite party (I do not feel sure that one can really call a 'movement' a group which showed less and less inclination to move) seems to be more the target than the ostensible Chalcedonian opposition. Philoponus devotes several paragraphs to justifying the possibility of using plural terms about a unitary reality, e.g. because its definition needs many words to encompass its significance; or because it is so large and intricate that one aspect of it is insufficient to give understanding. One should be on one's guard, he warns, against a plurality which results from everyday usage and which may not be exact in logic. Like Severus of Antioch, he allows that the one composite nature of the incarnate Word has a plurality of properties, some divine, others human (ch 34). But properties can be plural when the entity possessing them is only one. A man as animal is both rational and mortal (15).

The tract ends (37) by confessing that the incarnation is a sublime mystery beyond human reason. But that does not excuse anyone using slipshod or confused terms.

Four or five years later Philoponus' friend Sergius had been elevated to become Jacobite patriarch of Antioch, 557-60; a titular office since he resided in Constantinople. To this Sergius, Philoponus dedicated a major essay on the Mosaic cosmogony, *de Opificio Mundi*. Most of this work is of more interest to the historian of theology and exegesis than to the historian of science and philosophy. The work is a sustained polemic against the opinions of an unnamed opponent. The opinions are identical with those found in the text of Cosmas, a Nestorian merchant of Alexandria who traded south of the Red Sea and was nicknamed Indicopleustes. Cosmas' theological hero was Theodore of Mopsuestia, and he wanted to treat the first chapter of Genesis as an authoritative guide to creation-science. This landed him in such delightful paradoxes as the observation that although the Bible shows paradise to be located in the East, westward migration is somehow the providential order. (The remark is perhaps neglected evidence that contemporary trading conditions in the East were not as good as, say, in Theoderic's Italy or even Visigothic Spain, and that there had been some movement of the population away from the historic centres like Antioch-on-the-Orontes, catastrophically damaged by earthquake in 526.)

Philoponus more than once insists that Moses never intended to provide a scientific cosmogony, but aimed to teach the knowledge of God to benighted Egyptians superstitiously worshipping the sun, moon and stars (i,1; iv,17). Sunk in idolatry they needed to raise their minds beyond visible fiery matter such as the sun and stars. (We meet here the thesis against which Simplicius directed substantial parts of his commentary on the *de Caelo*, in refutation of Philoponus' book *de Aeternitate Mundi contra Aristotelem*.) Philoponus thinks it foolish to quote Ecclesiastes 1 (nothing new under the sun) as if Solomon were teaching us science rather than ethics (iii,10). The fact of God's creation is revealed, but not how it all came about (ii,13).

Theodore of Mopsuestia had disliked the Platonising spirit of St Basil's *Hexaemeron*, and because of his distrust of allegory had ended by taking the Bible with a prosy literalism.[11] When Theodore's Latin contemporary Augustine composed his *Literal Commentary on Genesis*, he took it for granted that the Bible is here teaching no natural science, that trying to reconcile Genesis with the Ptolemaic cosmogony was not sensible, and that Christians who tried to use scripture in that way merely made their faith look ridiculous. One recalls how in the *Confessions* Augustine records that a major undermining of his confidence in Mani resulted from his discovery that the Manichee myth explaining eclipses was at variance with the findings of professional astronomers.

Philoponus thought it utterly absurd of Theodore to suppose that the sun, moon and stars move because they are propelled by angels. Do they push or pull, he asks? Being an admiring reader of Ptolemy, Philoponus believed in a ninth starless sphere beyond the planets and the visible stars. The stars themselves differ in size, position, order and colour because they burn different kinds of matter as fuel; and the planets differ vastly in their velocity

---

[11] Theodore even allowed himself to speak of the chaotic darkness ordered by God as a substance (*ousia*), which to Philoponus seemed Manichee language. Philoponus abominated the notion that when Christians speak of creation out of nothing, 'nothing' is a name for the matter of which the world is made (Simplicius *in Cael* 136,18ff).

(iii,4). But incorporeal things would need no three-dimensional space (i,16).

In *de Opificio Mundi* Philoponus does not restate his argument against Proclus that the world is a contingent non-necessary entity created out of nothing. God created by his will, and gave the laws of nature under which it operates. Miracle is allowed some restricted possibility within the context of the gospel history, as (for example) the three-hour eclipse of the sun at the Pascha when it was full moon. The divine glory of the creation consists in its order. Only when free choices are made by animate beings are we faced by disorder and evil. Their inflexible constancy is a ground for denying souls to the celestial bodies. Wherever we find souls, we find inconstancy (vi,2), something unreliable and indeterminate. In an age when Origen was a subject of heated controversy, Philoponus sharply denies that souls become embodied in matter in consequence of a precosmic fall (vii,2f), and refers his readers back to *de Aeternitate Mundi contra Proclum*, written thirty years earlier, for a refutation of the Platonic doctrine of Anamnesis.

Philoponus' works contain occasional comments on the culture of his time. For example, in the commentary on the *Meteorologica* he remarks that music and sculpture are at present in decline, but 'I think they will have a revival one day' (17,30). But he deplores nothing so much in contemporary society as the ineradicable passion for divination and astrology. He ends book iv of *de Opificio* with a round declaration that for him the principal vindication of the truth of Christianity lies in its requiring the renunciation of astrology. To Porphyry's vegetarianism he can be sympathetic (vii,5); but Porphyry's book on oracles he thinks a sadly decadent piece (iv,20). Pagans like Porphyry sceptical of the Mosaic cosmogony should ask themselves why everybody divides time into periods of seven days (vii,13). Pagans give the days planetary names, but no one knows exactly why a particular planet is assigned to a particular day. (Sixth-century Alexandrians had not the advantage of F.H. Colson's masterly little book, *The Week*, 1926, which would have told them.)

Man is a moral being and his life is assessed by his use or misuse of what he is given. Nothing in this material world is inherently evil. Adultery is evil, but not marriage or sexuality. Iron is wholly beneficent in agriculture or surgery, but is misused for weapons of death (xii,12). One must add that, unlike the great majority of ancient Christian fathers, Philoponus is willing to tolerate the necessity of capital punishment (303,6).

With astrology, Philoponus dismisses the myth of eternal return and the cycle of unending time (cf *in Phys* 456,17ff). The material cosmos is in continual change. No individual once perished can ever come to live again (vii,3 p 287,1). But a perishable thing is succeeded by something of the same sort. Quench a fire and you can never recover that fire. You can only start another one like it, and it is fire, but not the same fire.

The principle that nothing material can ever return once it has perished has an apparent bearing on the Christian hope of resurrection; that is, that survival after death is not a spooky animistic belief in ghosts, but is based on faith in the God who created the material as well as the spiritual world. Philoponus expounded the language of St Paul, that the body now 'is not sown as the body which shall be', to mean that resurrection is not mere resuscitation. It means that in the life of the world to come the soul is provided by the Creator with whatever new vehicle will then be appropriate to its new environment, created *ex nihilo* as the Creator wills.

This exegesis made some readers anxious, especially Monophysite friends and colleagues who felt that their entire position was weakened if Philoponus was successfully dismissed as a heretic on eschatology. Photius reports that Philoponus allowed himself to use mocking language about the naivety of respected figures (*Bibl* 21). So far as we can reconstruct the argument from the surviving fragments, Philoponus' doctrine seems to have been devised as a critique of Origen. Origen had said that the concept of a resurrection body ought to cause no difficulty to any philosophical Greek who thinks that matter is a qualityless continuum to which the Creator can give different qualities as and when he pleases. (See *contra Celsum* iv,57 and especially vii,32 which stands close to Philoponus.) But Philoponus disliked the disjunction of matter and form. For him it was a principle that God has created out of nothing both form and matter; that matter is irreducibly three-dimensional and is itself a kind of form. So 'everything in this sensible and visible world was brought into being out of nothing, is corruptible in principle and will perish in fact in both form and matter. Therefore the Creator will hereafter replace these material bodies by other and superior vehicles for the soul, incorruptible and eternal.' So the citation in Nicephorus Callistus' *Church History* VIII 47 (PG 147, 424D).

Not only Monophysite friends were distressed. Philoponus provoked a rival opinion from Eutychius the Chalcedonian patriarch of Constantinople in the 580s. He denied that the resurrection body will be in any sense palpable, and that led to protests from Gregory (the Great) at the time when he was papal 'nuncio' (apocrisiary) at the Byzantine court. So strongly did Gregory feel about the matter that he devoted a passage of his *Moralia* on Job to the topic (xiv,72-74).

De Opificio Mundi has a few passages in which Philoponus discusses central questions of Christology and Trinitarian dogma. He offers an annihilating critique of Theodore's notion that Christ is our redeemer because he is the realisation of perfection in humanity (vi,9-14). He is particularly insistent on the separate being of Father, Son and Spirit.

How should this plurality or independence be formulated by a dialectician? In the Christological controversy the Monophysites had stood firm that *hypostasis* is the concrete instantiation of nature: to speak of one *hypostasis* is to require one nature. In this context nature and *hypostasis* are virtually the same. Indeed some declared that the one *hypostasis* in Christ is identical in word and thing with the *hypostasis* of the Word of God in the Trinity of one God in three *hypostases*. 'One of the Trinity was crucified for us' was well known to be a liturgical acclamation which, like litmus paper, revealed the presence of Nestorians who turned pink when they heard such words.

In the *Arbiter* Philoponus treats substance (*ousia*) and nature (*phusis*) as terms which may apply either generically or specifically. There is such a thing as *idikotatê phusis*. Each individual human being is an individual and has his own reality. The Neoplatonic exegetes of Aristotle accepted that universals exist in the mind, not outside it (as we have seen Philoponus himself saying). So if in the Trinity the three *hypostases* equally share the divine nature, that divine nature is a universal which has existence only as found in the concrete realities of Father, Son and Holy Spirit.

Against Sabellian or modalist notions being advanced at Alexandria by a dissident Monophysite named Themistius (a deacon who led a group called Agnoetae and who opposed the patriarch Theodosius), and at Constantinople by the Chalcedonian patriarch John Scholasticus, Philoponus wrote his book *On the Trinity* to argue his case. The nature shared in common has no reality apart from the existents or *hypostases*. We must anathematise three deities, three natures, but also deny that there is an actual generic Godhead distinguishable even in thought from Father, Son and Spirit. Father, Son and Spirit are consubstantial in nature and substance, but not in their properties; there they are distinct. We do not say that the Father or the Spirit became incarnate.

Philoponus does not argue on religious grounds, e.g. that the doctrine of the Trinity is rooted in the idea of a salvation-history in which the one God discloses himself to humanity in the threefold process of redemption. His reasoning seems essentially nominalist: Divine unity is an intellectual abstraction, and the Trinity consists of three substances, three natures, considered in an individual rather than generic sense. Indeed the great John Chrysostom himself had written in his fourth homily on St John (PG 59,47) that the Logos is a substance (*ousia*) proper to the Word (*enupostatos*).

Philoponus was not the originator of the doctrine labelled Tritheism by its critics. The credit for origination lay with a Syrian Monophysite of Apamea named John Askoutzanges, 'with bottle-shaped boots'. He had studied philosophy of Constantinople and about 557 began to teach that in God there are consubstantial substances (*ousiai*), 'no doubt three if you press me, but let us leave the number indefinite, for God is indefinable'. His doctrines caused a rumpus, and called forth formal censure from the exiled Alexandrian patriarch Theodosius, in great senectitude still resident at Constantinople. His view of the matter may be read in Chabot's *Documenta Monophysitica*, 26-55.

Philoponus found Theodosius' terms of censure so unsatisfactory that he denied the authenticity of the document (Roey 1980, fr 26). How can one affirm consubstantiality unless there is a plurality of *ousiai* to share the one *ousia*?

The Tritheist controversy led to a split in the anti-Chalcedonian camp, already tending to fall apart into a multitude of precisionist sects in search of ever more exact definitions and defensive formulas. Part of Philoponus' tragedy as a theologian was that he belonged to a body which, by refusing to accept an ecumenical council (whatever exegesis its decisions might be given), became separated from the main Christian body and thereafter itself became more and more fissiparous. He himself contributed to that disintegration by his attempt (surprising in a mind so drenched in Neoplatonism) to explain a divine mystery with concepts originating in the created and finite order.

More successful as a philosopher and scientist than as a theologian, he was placed under formal anathema a century after his death at the anti-monothelete council of Constantinople, 680-81.[12] Yet in the fourteenth century Nicephorus Callistus thought him worth a complete chapter of his Church history, and remarks that, although his theology was not good, his commentaries on Aristotle were masterful in their lucidity and were still regularly studied.

Philoponus' Theologica

c. 552	*Arbiter* (*Diaetêtês*)
after 553	*Apology for the Arbiter*
	On the Whole and its Parts
c. 553-55	*Four Tmêmata against Chalcedon*
after 556-7	*On Difference, Number and Division*
557-60	*De Opificio Mundi* dedicated to Sergius, patriarch of Antioch, resident at Constantinople. An attack on the biblicist cosmology of Cosmas Indicopleustes (who is never mentioned)
c. 560	Letter to Justinian
before 567	*Against Andrew the Arian* (perhaps disclaiming an extreme Tritheism?)
567	*De Trinitate* against John Scholasticus of Constantinople
before 574	*On the Resurrection*

Of uncertain date

not before 567	*Against Themistius* Themistius was an Alexandrian deacon, separated from the Monophysite majority led, from Constantinople, by the patriarch Theodosius; he held that Christ really did not know the day of the Last Judgment, and that his ignorance in asking 'Where is Lazarus?' was not *oikonomia*. His followers were nicknamed *Agnoetae*
	De Paschate Philoponus discusses whether the Last Supper was the passover meal. Kindred interests appear in *de Opificio Mundi* ii

On the Tritheist controversies the principal source is the *Ecclesiastical History* of John of Ephesus, English translation by Payne Smith; Latin in CSCO by E.W. Brooks. Other references in Photius' *Bibliotheca*, cf codd 24, 230, 232.

The best modern account, with new texts, is by R.Y. Ebied, A. van Roey and L.R. Wickham (1981). Many old errors were corrected by E. Honigmann, *Evêques et evêchés monophysites d'Asie antérieure au VIe siècle*

---

[12] Sophronius of Jerusalem's Synodical letter to Sergius of Constantinople written in 634 includes John the Grammarian named Philoponus in an immense list of heretics: PG 87/3, 3192C.

(Louvain, 1951). See also R.Y. Ebied, 'Peter of Callinicum and Damian of Alexandria: the tritheist controversy of the sixth century', *Colloquium* 15/1 (October 1982) 17-22.

CHAPTER THREE

Simplicius' Polemics

*Some aspects of Simplicius' polemical writings
against John Philoponus: from invective to a reaffirmation
of the transcendency of the heavens.*

Philippe Hoffmann

I am not entirely comfortable at finding myself introducing a discordant note
into a collection intended to celebrate the refreshing originality of
Philoponus' ideas. I shall, however, be speaking for Simplicius, vindictive
pagan that he was, and shall hope to be an effective counterweight to what is
said in other chapters.

I shall be talking within the framework of a general interpretation of
Simplicius' commentary on Aristotle's *de Caelo*.[1] The commentary is an
exegetical work undertaken as a paean to the Creator or 'Demiurge'. Its basic
theory on the physical structure of celestial matter is that this matter is a
combination of the superior parts (*akrotêtes*) of the four elements, dominated
by the purely luminous superior part of fire. My aim will be to show how this
theory can be seen as a reaction to the theories of John Philoponus.
Philoponus had turned to the *Timaeus* for support in his *contra Aristotelem*, and
had attacked the Aristotelian doctrine that the heavens are made of a fifth
element and that the world is eternal. Well before Copernicus, Philoponus
denied that there was any substantial difference between the heavens and the
sublunary world. In his reply to the *contra Aristotelem*,[2] Simplicius reaffirms the
divinity, the transcendency, and the eternal nature of the heavens. His
exegesis aims to connect, rather than contrast, Plato's *Timaeus* and
Aristotle's *de Caelo*. It is, moreover, a religious act, a spiritual exercise
designed to turn the soul (both Simplicius' and his reader's) towards the
Demiurge. This conversion is our initiation into the grandeur of the universe
and of the heavens, and his description of the physical nature of the heavens is

---

[1] J.L. Heiberg, ed, *Simplicii in Aristotelis de Caelo Commentaria* (= *CAG* VII) Berlin 1894. The
references given in this chapter refer to the page and line of the Heiberg edition.

[2] Simplicius only quotes twice Philoponus' earlier work, *de Aeternitate Mundi contra Proclum*
(135,27-31 and 136,17), and he says that he has not read it. The criticism of the *contra Aristotelem*
is found not only in Simplicius' commentary on the *de Caelo* (Book 1), but also in his commentary
on the eighth book of the *Physics* (Diels edition, *CAG* X Berlin 1885).

57

one of the most valuable aspects of the revelation. Those readers still under Philoponus' spell cannot achieve this revelation until they have undergone a preliminary act of purification, which is the refutation of the arguments of Philoponus' *contra Aristotelem*. In this way, Simplicius' attack is directed at a target that is simultaneously philosophical and religious. A correct reading and interpretation of Aristotle's *de Caelo* leads not only to the acquisition of intellectual knowledge but also and above all to our elevation through thought (a thought that we 'live') to the whole universe and to the Demiurge. It is a form of prayer addressed to them. The sacrilegious blasphemy of the Christian Philoponus is countered by the Neoplatonist liturgy, a rightful celebration of their God.

I. Vocabulary, themes and images of the invective[3]

The historical background of the polemics is well-known, thanks principally to the studies of H.-D. Saffrey[4] and I. Hadot,[5] while certain doctrinal aspects have been covered by W. Wieland.[6] But no one has yet examined the 'connective tissue' of the polemics, that is, the transitions, the sentences which Simplicius uses to introduce his quotations, in short everything in the 'technical' philosophical discussion that is not itself explicitly philosophical but that is born of oratory or penmanship.[7] Now these literary bricks form a thematic structure that has strong philosophical overtones, and the very flow of the invective interacts with the doctrinal basis of the refutation. Gradually a picture takes shape before us of Philoponus doomed to sorrowful isolation by a series of oratorical manoeuvres that Simplicius deploys most brilliantly.[8]

In order to classify the vocabulary, themes and imagery of the invective, we must start from the target of the descriptions: Philoponus the *interpreter* and the *Christian*; Simplicius could only cope with this duality in Philoponus through the framework of two groups of representations that were specifically Neoplatonist.

---

[3] The study of the vocabulary of polemics which I am presenting in these few pages makes no claim to be exhaustive. Other texts and other expressions would have to be cited, and the subject would require more extensive development, especially as regards the literary tradition which gave Simplicius his materials. I am here only outlining an inquiry which I plan to take further eventually.

[4] H.-D. Saffrey (1954) 396-410.

[5] I. Hadot, *Le Problème du néoplatonisme alexandrin: Hiéroclès et Simplicius*, Paris 1978, 20-32 (and especially 26-9). The commentary on the *de Caelo* was written after 529, the commentaries on the *Physics* and the *Categories* after 532 or 538.

[6] W. Wieland (1960) 206-19. See also É. Evrard (1953) 299-357; and S. Sambursky (1962) (which brings out well the importance of Philoponus); A.-Ph. Segonds (1981) 12 and notes on 41-2. A more complete bibliography on Philoponus is given by R. Sorabji (1983) 428-9. See also L. Taran (1984) 104-15, esp. p.106 and n.71, p.112 and n.88.

[7] There is unfortunately no reference to Simplicius in S. Koster's work, *Die Invektive in der griechischen und römischen Literatur* (= *Beiträge zur klassischen Philologie*, 99) Meisenheim am Glan 1980, nor in the stimulating article by G.E.L. Owen, 'Philosophical Invective', in *OSAP* I, 1983, 1-25 (reprinted in his *Logic, Science and Dialectic*, London 1986, 347-64).

[8] We have only to remind ourselves of an ancient epigram praising Simplicius both as a philosopher and as a rhetorician (see I. Hadot, *Le Problème du néoplatonisme alexandrin*, 31-2).

Simplicius describes the requirements for a good interpreter of Aristotle in the prologue of his commentary on the *Categories*.[9] A good interpreter must, first of all, not be far removed from Aristotle's intellectual greatness (*megalonoia*). He must next be absolutely familiar with the whole Aristotelian corpus, must know all the passages of Aristotle's works and his linguistic peculiarities. He must also be intellectually honest, with integrity of judgment: he should not reject assertions that are correct, by understanding them 'indolently' (*kakoscholôs*), and, on the other hand, if a point needs examination, he should not insist (*philoneikein*) on justifying it at any price 'as if he had become a member of the Philosopher's sect'. Finally, and this is a prime requirement, his exegesis must bring out the most profound harmony between Plato's and Aristotle's philosophies. By distinguishing between the letter (*lexis*) and the spirit (*nous*) of the texts,[10] the interpreter will not abide by the former and condemn the (apparent) dissension between the two philosophers (*diaphônia*), but must 'consider the spirit (general or underlying: *nous*) and track down (follow the trail: *anichneuein*) the agreement between the philosophers on the majority of points (*tên en tois pleistois sumphônian*)'.

As for the listener (*akroatês*), that is, the student who is following a course of Aristotelian philosophy,[11] he must be good (*kalos*) and virtuous (*spoudaios*).[12] He must make a habit of examining and reflecting on Aristotle's principal concepts, either alone or in the company of others who are as 'enamoured of knowledge' (*philomathôn*) as himself.[13] He must beware of that eristic chatter (*eristikê phluaria*) which is the downfall of so many students who are bad readers of Aristotle. For 'whereas the Philosopher takes pains to prove everything with the aid of irrefutable scientific definitions, those who claim to be great scholars get into the habit of contradicting the evidence itself, thus blinding the eye of their soul' (*to omma tês heautôn psuchês apotuphlountes*). And punishment is threatened for the most inveterate of these sophists.

Since Simplicius produced the commentary on the *Categories* after those on the *de Caelo* and the *Physics*,[14] we may legitimately wonder whether this daunting array of qualities required both for the good interpreter and for the good student is a 'theoretical' invention drafted after the event, the fruit of an afterthought that resulted from Philoponus' manifold shortcomings, or whether it already existed complete in Simplicius' mind (and therefore in his

---

[9] K. Kalbfleisch, ed, *Simplicii in Aristotelis Categorias Commentarium* (= *CAG* VIII) Berlin 1907, 7,23-32.

[10] This distinction seems to correspond in exegetical practice to the distinction between *theôria* (a general explanation of an author's thought in a text) and *lexis* (an explanation of the words themselves). See A.J. Festugière, 'Modes de composition des commentaires de Proclus', *Museum Helveticum* 20, 1963, 77-100 (reprinted in *Études de philosophie grecque*, Paris 1971, 551-74).

[11] *in Cat* 7,33-8,8 Kalbfleisch.

[12] The student must already have received preliminary ethical instruction, not 'scientific' (he is not yet capable of this), but 'in keeping with upright opinion', before embarking on a study of Aristotle, which begins with the *Categories* and all the logical treatises (*in Cat* 5,3-6,5 Kalbfleisch; cf I. Hadot, *Le Problème du néoplatonisme alexandrin*, 160-4).

[13] This epithet sums up one of the basic requirements made of true philosophy since Plato (*Republic* 2.376b-c).

[14] I. Hadot, *Le Problème du néoplatonisme alexandrin*, 27-32.

culture).[15] In any case, this page of the commentary on the *Categories* is central: Philoponus figures simultaneously as a negative photograph of the good interpreter and of the good *akroatês* in Aristotle's works.[16]

We should also be guided by a second thematic framework, namely the Neoplatonist picture of Christians as H.-D. Saffrey has reconstructed it from the works of Proclus and Marinus.[17] Let us remind ourselves of its main points:

1. Christianity is a doctrine for the multitude and vulgar people (*hoi polloi*);
2. Christians are not Greeks, but foreigners;
3. They are ignorant, uncultivated men (*hoi anepistêmones, agnoia, anepistêmosunê*);
4. They possess an impious effrontery (*tolma, asebeia*); they are atheists who overturn any divine commandments and who touch what should be inviolate (*kinein ta akinêta*);[18]
5. They bring with them the shock of innovation (*kainotomia*) and they instigate dreadful confusion (*sunchusis*) and disorder (*paranomia*);
6. They are typhoons and gigantic vultures; evil (*mochthêroi*), their only aim is to harm (*epibouleuein*); they are intemperate (*akolastoi*), like neighbours who do not remain sober (*tines mê nêphontes geitones*).

Most of these themes will reappear in Simplicius' invective.

Philoponus, first of all, is far removed from Aristotle's intellectual pre-eminence, as is suggested in a short quotation from the second *Olympic* of Pindar (lines 87-88): 'But here is our young raven (*nearos ... korax*), or, rather, our jackdaw (*koloios*),[19] who *crows in vain against the divine bird of Zeus*, in Pindar's splendid phrase' (42,17-18). The very fact that he quotes Pindar is significant; the great lyric poet is one of the major figures of the Hellenic tradition that Simplicius, who elsewhere quotes Alcaeus,[20] claims to share. The history of Pindar's text at the end of antiquity shows that the selection of the *Epinicia* made in the second century displays some preferential treatment

---

[15] Some of the demands imposed by Simplicius reappear in parallel texts of the commentaries on the *Categories* produced by Ammonius, Philoponus himself, Olympiodorus and Elias. A comparison between these different prologues to the *Categories* will introduce a translation of Simplicius' prologue in a work now in preparation by I. and P. Hadot, and Ph. Hoffmann.

[16] Simplicius, in a tactical manoeuvre, sees Philoponus now as a misleading commentator, now as a pedantic and pretentious 'bad pupil' who should be rebuked.

[17] H.-D. Saffrey, 'Allusions antichrétiennes chez Proclus le diadoque platonicien', in *Revue des sciences philosophiques et théologiques* 59, 1975, 553-63. Cf. also Cl. Zintzen, *Damascii Vitae Isidori reliquiae*, Hildesheim 1967, 333 (s.v. *Christiani*).

[18] H.-D. Saffrey, in *Proclus. Théologie Platonicienne. Livre I*, Paris 1968, p.xxii, n.5 and p.xxiii, id., 'Allusions antichrétiennes', 560. The proverb is also quoted by Plutarch, *de Genio Socratis* 585F.

[19] Cf *in Phys* 1140,10 Diels. Like the jackdaw, Philoponus wears borrowed plumage; here (*in Cael* 42, 19-20ff) he adopts an objection by Xenarchus of Seleucia.

[20] *in Cael* 156,25-28: 'But since *the sow excites (us) again a little*, as the lyric poet Alcaeus has it (fr 99 Bergk), we must look again at this Grammarian, who betrays not only mindless folly but also great perversity in his arguments'.

of the two first books, in particular the *Olympian Odes*.[21] This may throw some light on the reaction to the quotation: the most cultured of Simplicius' readers would refer it back to the context of the great triumphal ode. The aim of this *Ode* is to celebrate the victory of Thero of Agrigentum. It begins with an invocation to Zeus (patron of the Olympic Games) and to Heracles (the hero who initiated these games). In the poem, the 'divine bird of Zeus' (the eagle) is Pindar himself, and the ravens that caw against him are doubtless (according to the scholiasts) his rivals Simonides and Bacchylides.[22] The transposition is clear: the eagle is no longer Pindar but has become Aristotle, and Philoponus is the raven. Further, in the eyes of the Neoplatonists, Zeus is none other than the demiurgic intellect[23] and Aristotle is his vicar. As for the figure of Heracles, hovering in the wings of the quotation, it will reappear on stage in the metaphor of the Labours, recalled later by Simplicius, who compares his refutation with the cleaning of the Augean stables.

In Simplicius' view, Philoponus is a mere greenhorn, a novice (*nearos*). Although both men are evidently contemporaries, Simplicius must already have been a philosophy student at Alexandria while Philoponus was still studying grammar.[24] This contemptuous epithet, reiterated as it is,[25] enshrines a play on words: Philoponus' novice chatter is also modernist chatter (*nea phluaria*, 201,7), that is, the dangerous novelty of Christianity. This youthfulness and novelty are, fortunately, doomed to an early death; their fate will be that of the Gardens of Adonis.

Philoponus' ignorance is boundless. He is unacquainted both with Aristotle's works and with the exegetic tradition,[26] nor does he have at his command the propaedeutic disciplines which, in the scholar's curriculum, normally precede the study of philosophy: for instance grammar[27] and astronomy.[28] He is woefully ignorant of logic and has no knowledge of syllogisms.[29]

---

[21] In the sixth century only the *Epinicia* were known and read, and choice was sometimes limited to the *Olympian Odes* alone. On this problem, see J. Irigoin, *Histoire du texte de Pindare*, Paris 1952, 93-121.

[22] Pindar contrasts the vast knowledge which the wise (*sophos*) poet has of nature with 'acquisitive' (expressed by the verb *mathein*) knowledge, superficial and restricted to tricks of the trade. This distinction of his will figure in Simplicius as one between the lover of knowledge (*philomathês*) and the late to learn (*opsimathês* – Philoponus). The context of the quotation suggests other ideas – vehemence (*labroi*) and garrulity (*panglôssia*) – which will be used to describe Philoponus.

[23] H.-D. Saffrey and L.G. Westerink, *Proclus. Théologie Platonicienne. Livre I*, Paris 1968, p lxvi; H.-D. Saffrey, 'La Théologie Platonicienne de Proclus, fruit de l'exégèse du Parménide', *Revue de théologie et de philosophie*, 116, 1984, 9.

[24] H.-D. Saffrey (1954) 402 n 4; I. Hadot, *Le Problème du néoplatonisme alexandrin*, 25.

[25] *in Cael* 26,11; 67,23; 90,21 (*neanieuesthai*); *in Phys* 1117,16 (*neanikon*) and 1169,8-9 (*neanieuma*).

[26] *in Cael* 126,6-7; 179,22-23 and 28-9. The verb 'not to know' (*agnoeô*) is often used by Simplicius (cf 82,30; 83,5 etc).

[27] *in Cael* 49,10-12; 74,5-7.

[28] ibid. 32,34-33,1; 36,27-33; 71,17-19.

[29] ibid. 30,16; 31,4-5; 166,12-13. See also pp 28-29 Heiberg. Since the command of the syllogism was the key to the demonstrative method, Philoponus' remarks are devoid of scientific value.

In these circumstances, he can hardly be a philosopher. And his Platonism is merely apparent.[30] Thus, he fails to understand that the fire of which the heavens are made, according to *Timaeus* 40a, is not the weightless fire animated by a vertical upward motion, i.e. the lightest of the sublunary bodies, but is the light referred to in *Timaeus* 58c. His state of mind is delineated in a few strokes:

> But since (I cannot tell why) Plato's doctrines seem to appeal to our man, although he does not subscribe to that school, as they say, and although he does not try to find (*ezêtêkota*) the real meaning of Plato's thought (*noun*) with a true love of knowledge (*philomathôs*), and although for this reason he thinks at one moment that Plato's doctrines tally with his own fantasies (*phantasiais*), at another that they clash (*enantiousthai*) with Aristotle's ... (84,11-14).

Whereas the lover of knowledge, that is, the reader who is genuinely a philosopher, seeks[31] the deeper meaning of Plato's writings, which accords with Aristotle's philosophy, Philoponus relies falsely on Plato in order to contradict Aristotle. In this way he condemns himself to misunderstanding both, and misses the truth. Simplicius takes the picture of the lover of knowledge (*philomathês*), who is nurtured from childhood on philosophical discussion, who is blessed with a wide acquaintance with the texts, and in whom the desire for truth takes the form of a longing for agreement (*sumphônia*) between Plato and Aristotle, and contrasts it with the picture of Philoponus, 'late to learn' (*opsimathês*): the *opsimatheis*, 'because their gaze encompasses few things (*eis oliga blepontes*), are struck by the apparent disagreement (*dokousa diaphônia*) [between Plato and Aristotle], and tend as chance may have it to one of two available theses, while rejecting the other' (159,2-9).

Another important theme figures in the text I quoted above (84,13-14). Philoponus' soul is dominated not by reason but by the passions and the imaginings (*phantasiai*) that arise from them. He is governed by his irrational soul. Yet the necessary condition for correct exegesis is a soul that is pure and rational, which will ensure a correct judgment.[32]

Like those pretentious sophists that Simplicius describes in the prologue to the commentary on the *Categories*,[33] Philoponus drowns in a flood of eristic chatter and logorrhoea.[34] His discussion is a straight line that suffers from

---

[30] ibid. 66,33-67,5; *in Phys* 1331,8 Diels.

[31] Simplicius' opponent cannot be *philoponos* (a lover of work), since he is in no way a man who seeks (*zêtêtikos anêr*), but an eristic. See below, p 68 and n 93.

[32] Which Simplicius calls *krisis adekastos* (*in Cat* 7,26 Kalbfleisch). Philoponus has not minded his manners or improved his character, which is a preliminary stage in the curriculum (see above, n 12). There is no doubt that he is only a beginner.

[33] *in Cat* 8,1-8 Kalbfleisch.

[34] One is struck by the richness of Simplicius' vocabulary: *phluaria* and the verbs of the same family (*in Cael* 49,25; 131,30; 178,11; 186,15; 190,16; 193,8; 199,19); *lêrein* (30,29; 136,14) and *lêrôdia* (*in Phys* 1147,31 Diels); *adoleschia* (*in Phys* 1141,8; 1159,26); *dapanan biblion* or (*pollous*) *logous* (*in Cael* 80,28; 81,22-3; 134, 10-11; 157,2; 172,25; 183,21-22; 189,28); *katateinein makrous/pollous logous* (58,17-18; 67,5-6; 71,19-20; 179,29-30; 186,30; 190,19-20). Simplicius also uses a technical term when he says that Philoponus' books have 'many lines': *polusticha biblia* (25,29; cf. *in Phys*

flowing on indefinitely.[35] In the inordinate length of his books, so verbose that no one can read them, his main aim is to impress fools[36] and the weak-minded. It is in any case for the common crowd that he is writing (*tois apo triodou*).[37] As we might have guessed, Simplicius prefers not the 'natural' audience of the ignorant doctrines inspired by Christianity, but the reader who is attentive, intelligent and cultivated (*hoi pepaideumenoi, hoi kathariôteroi, hoi enteuxomenoi epimelôs, hoi philomatheis*).[38] It is to these readers, worthy of it as they are (*axioi*), that he will reveal the mysteries of the universe by offering a true interpretation. There is no need to point out the ineptitude of Philoponus' theories to these enlightened minds, and Simplicius begs their forgiveness for the time he devotes to a refutation of the *contra Aristotelem*.[39] It is directed rather at the intermediate category of readers, who are neither hopeless idiots nor complete philosophers, and who have been led astray and deceived (46,14; 165,6) by the formidable sophist, either because they do not possess enough critical acumen,[40] or because they have read him too superficially (*epipolaiôs*),[41] or because they are impressionable enough to have been swayed by the mere prolixity of his output.[42] Simplicius wants to 'give help' (*boêthein*) to all these people.

This didactic conscientiousness, the fruit of a real desire to regain lost ideological ground,[43] takes the form, as we know, of long quotations followed by detailed refutations. 'I, too, find it necessary to chatter now,' says Simplicius (30,28-29). Thanks to him, the *contra Aristotelem* is in part known to us. Why did he put forward his adversary's writings so carefully? Leaving aside both the fact that long quotations were common practice in a scholastic tradition that was founded on exegesis, and also the conjecture that Simplicius was making an anthology of his opponent's least powerful texts, we can discover three answers in Simplicius' attitude. First, he is impelled by intellectual honesty, thus disarming any accusations of calumny or sycophancy (163,11-12); he must therefore quote the text he is refuting. Secondly, this 'honesty' may be a useful strategy: Philoponus' writings make

---

1118,2 and 1130,5), *polloi stichoi* (124,18) (on these terms, see B. Atsalos, *La Terminologie du livre-manuscrit à l'époque byzantine. Première partie: termes désignant le livre-manuscrit et l'écriture* (= *Hellênika. Parartêma* 21), Thessaloniki 1971, 158 and 279). Philoponus' garrulity is a collection of foolishness: *phlênaphoi* (25,31-32; 135,31). He offers us only empty speeches: *logous ... ekphusâi kenous* (*in Cael* 187,7; *in Phys* 1160,40-1161,1).

[35] *in Cael* 46,33-47,1 and 184, 21-22 (*chudên*).

[36] ibid. 25,30; 131,29-31 and 136,7-9; *in Phys* 1130,5-6.

[37] *in Cael* 131,28.

[38] ibid. 25,31; 26,2-3; 102,16-17; 166,16; 180,23; 201,8.

[39] ibid. 180,23-27; 184,27-185,2.

[40] ibid. 180,25.

[41] ibid. 184,31.

[42] ibid. 122,33-123,4. See also above, n 36.

[43] As I. Hadot has shown (*Le Problème du néoplatonisme alexandrin*, 26-7), Simplicius never went either to Alexandria or to Athens after the return from Persia, but did perhaps go to a Greek town in Asia Minor; at any rate somewhere near a big library. It also seems likely that he was writing for a cultivated public, able, in his view, to be led back to pagan orthodoxy. In any case, Simplicius seems to have felt that he was not writing for a mere handful of philosophers, but for other readers as well whom he is seeking to convert.

clear 'how poor is his aptitude for philosophical discussion' (*hopoian hexin echei peri logous*) (166,13-14). Lastly, in a move full of rhetorical cunning, Simplicius reiterates his intention of struggling against the incredulity of readers who refuse to believe that any one could write such foolish things.[44] This explains the need to reproduce all the relevant evidence.

While a serious philosophical approach involves the pursuit of being, *hê tou ontos thêra* (Plato *Phaedo* 66c),[45] Philoponus, for his part, is pursuing glory: he is *doxês thêratês* (25,23). This is his highest aspiration, and it can be summed up in one word: *kenodoxia*, vainglory (157,9). Simplicius distinguishes it carefully from the true glory achieved by the great philosophers – Plato, Aristotle, and their worthy interpreters: [Philoponus] 'believes that he is creating a glory for himself, vain and empty in reality (*kenên ontôs kai mataion doxan*), by contradicting those philosophers who are truly illustrious (*kleinous*)' (200, 27-29). The desire for glory leads to eristics and eristic chatter, and Philoponus is really on the warpath against Aristotle's proofs;[46] in this, he is acting from vainglory (*kenodoxôs*) and not from love of knowledge (*philomathôs*) (131,32). Two phrases sum up this contentious search for glory: *philoneikos kenodoxia* (143,14-15) and *kenodoxos philoneikia* (90,13-14). A list of the occurrences of the word *philoneikia* and of the verb *philoneikein* would certainly reveal their frequency as very high.[47]

Simplicius excels in identifying in Philoponus a patchwork of intellectual and moral shortcomings; not only, as we have seen, a lack of both culture and education,[48] but also superficiality,[49] opaque reasoning,[50] mindless folly,[51] inattention,[52] haste,[53] knavery,[54] and perversity.[55] He is indeed in a sorry state, being drunk, mad, and crazed.

In fact, he confuses the heavens and the earth and turns the world order topsy-turvy; how could anyone of sober judgment (*para tôn nêphontôn*) fail to see him as under the influence of drink (*methuein*)?[56] This uncultivated

[44] *in Cael* 48,22-25; 75,12-16; 82,11-14; 122,1-2; 163,11-12; 179,34-35; 180,26-27; *in Phys* 1159,4-7.

[45] The expression is picked up by Proclus, *Théologie Platonicienne* I. 5,14-15 Saffrey-Westerink.

[46] His only thought is to contradict Aristotle (*enantiologia*). Cf. for example *in Cael* 78,12-17 and 157,9 (*machetai*).

[47] Other words used: *duscherainein* (119,9; 135,26) and *aganaktein* (90,8-9 – against those who prove that the substance of heavenly beings transcends sublunary beings).

[48] For example *in Cael* 71,7 (*apaideuton*); 82,12 (*anagôgos*); 90,10 and 11.

[49] ibid. 135,14-15; 196,34-35.

[50] ibid. 135,1-2 (*pachutês logismou*).

[51] The key word is *anoia* (133,19-21; 156,27; 163,35; 182,18). Philoponus writes crazy things, *anoêta* (122,2; see also 75,13; 157,14-15; 164,23-24), and also unintelligent ones, *asuneta* (37,8; see also 75,14; 136,27; 178,28; 180,7; 186,30; 195,18). That is why he only convinces madmen, *anoêtoi* (25, 30 and 35).

[52] *in Cael* 71,7 (*anepistaton*); 157,11-12; 163,35 (*anepistasia*); 183,34;191,9-10.

[53] ibid. 56,26 (*propeteia*); 90,10; 158,32-33 (*propetôs* contrasted with *zêtêtikôs*); 173,30; 182,18.

[54] ibid. 133,19-21 (*panourgia*).

[55] ibid. 156,27 (*kakotropia*).

[56] ibid. 134,28-29.

madman[57] yaps like a dog,[58] he is a dissenter who is prevented by mania (*lutta*)[59] from turning his gaze towards truth. His readiness to contradict both Aristotle and tradition plunges him into darkness (*eskotômenos*)[60] and the eyes of his soul (*ta ommata tês psuchês*) are blinded (*ektuphloun*).[61] And what is blinding Philoponus is his aim, his *skopos* (67,15), whose content is explicitly stated: to deny, against Aristotle's advice (33,20-22), any substantial difference between the heavens and the sublunary world, in order to assert that they are 'of the same nature' (*homophuê*), and thus to show that the Heavens and the World are created and destructible and not eternal.[62] The fifth element was introduced by Aristotle as the eternal and indestructible matter of the heavens. The struggle which Philoponus puts up against it is intended to show that the heavens are 'of the same nature as the four elements taken as wholes' (*homophuê tais tôn tessarôn stoicheiôn holotêsin*) (59,15-18).

This aim (*skopos*) is impious (70,17-18), and, right from the beginning of his refutation, Simplicius condemns 'those who imagine that they are venerating the deity (*tous theosebein oiomenous*) if they believe that the heavens, destined to be at men's service, as they say, have no transcendency over the sublunary world, and if they reckon that the heavens are destructible just as it is' (26,4-7).

An untranslatable Greek proverb, *mia Mukonos*, expresses the idea that all inhabitants of the island of Mykonos were bald. Simplicius uses it to describe the dreadful confusion that Philoponus is trying to claim between things that are divine and things that are human: *ho ta theia kai ta anthrôpina eis mian Mukonon sunkukôn* (135,9-10). The realisation of the difference between the divine and the human (according to Homer's old dictum) is the very blessing that Simplicius is requesting from the 'Lord, Father, and Guide of reason in us' at the end of the commentary on Epictetus' *Manual*.[63] Philoponus is even doubly impious, for he is attacking both the 'blessed eternity of the Heavens' and 'the unchanging kindness of the Demiurge', 'who makes the heavens subsist after having created them'.[64] The blasphemous[65] and boastful[66]

---

[57] ibid. 88,28 (*maniôdôs*); 89,7-9 (Philoponus is not in his senses, cannot *sôphronein*); *in Phys* 1334,18-19 (*manikon, paraphron*).

[58] *in Phys* 1182,38 (*apaideutos prosulaktêsis*).

[59] *in Cael* 58,15.

[60] *in Phys* 1135,28-29.

[61] *in Cael* 74,4-5 and 141,19-21.

[62] ibid. 34,5-7; 35,31-33; 59,15-18; 73,5; 80,23-26; 81,10-11.

[63] I. Hadot, *Le Problème du néoplatonisme alexandrin*, 35 n 9. This realisation is also what Proclus is asking for in his *Hymn to all the gods* (see A.J. Festugière, 'Proclus et la religion traditionnelle', in *Mélanges Piganiol*, vol.3, Paris 1966, p.1586; reprinted in *Études de philosophie grecque*, Paris 1971, 580).

[64] *in Cael* 88,28-31 and 184, 29-30; *in Phys* 1182,28-32.

[65] *in Cael* 88,29 (*blasphêmein*); 137,20-21 (*blasphêmiai*).

[66] Pride full of boastfulness and vanity. *in Phys* 1140,5 (*megalauchoumenos*); 1160,9-10 (*megalauchia*); 1171,30 (*authadizomenos*); 1334,37 (*apauthadizetai*); *in Cael* 136,1 (*authadês*); 26,28 (*brenthuetai*); 130,14 (*brenthuomenos*). Philoponus speaks shamelessly, *gumnêi kephalêi* (*in Cael* 135,4: cf Plato *Phaedrus* 243b) or *anedên* (88,28). He is an insolent slanderer (185,4 *epêreazôn*). Cf also the use of *neanieuesthai* (see above, n 25).

boldness[67] which impels Philoponus to attack Aristotle, the vicar of Zeus, is also directed against the Demiurge and against the finest of his creations. This aim (*skopos*) is not only impious but irrational (*alogistos*),[68] for it emanates not from a search for truth[69] but from a passionate desire (*epithumia*).[70] This desire, as we know, has as its final object glory (*kenodoxia*), and it dominates Philoponus' untrammelled imagination,[71] Philoponus whose 'soul's eyes' are blind.

The final telling thrust of the polemic is provided by three images borrowed from ancient mythology and history.

A. *The gardens of Adonis*

Simplicius emphasises from the start the ephemeral nature of the doctrines he is setting out to refute: 'As for me, I am well aware that these liberties (*tolmêmata*) are like the so-called gardens of Adonis; to the minds of fools they seem to blossom, and yet in a few days they are faded' (25,34-36). The simile of those miniature gardens, artificially brought on and doomed soon to perish, symbols of the premature demise of Aphrodite's lover,[72] had been used by Plato (*Phaedrus* 276b-d) to discredit the lack of seriousness (*paidia*) of written speech.[73] Philoponus' attacks on Aristotle are not serious, and are not the result of *spoudê*. They are, in Simplicius' view, nothing more than an episode, irritating but short-lived, and his pagan faith is unshakable: Philoponus spoke in vain (*matên*)[74] because the cosmological and theological truth revealed by Plato and Aristotle is proof against all attacks. Simplicius' aim is to see to it that 'Aristotle's treatise *de Caelo* and the notion of the universe that reveres god (*theosebês*) should remain irrefutable in their ancient glory (*epi tês palaias eukleias*)' (26,12-14).

In spite of the serious nature of the steps Justinian took in 529 and of his own precarious position as pagan philosopher in the Byzantine empire,[75] Simplicius takes the same basic view as Proclus:[76] he does not believe that Christianity will survive, nor that pagan beliefs will pass away. The ubiquity and the ever-increasing inroads made by Christianity are a merely temporary phenomenon, regrettable though it may be. Besides, what is surprising for a Platonist is the idea that at a given moment the truth should be apparent to a

---

[67] *in Cael* 25,34; 26,10; 48,25; 90,7; 120,13-14; 134,26-27; 180,26-27 (*tolman*); 188,2 (*etharrêsen*); *in Phys* 1159,4-7; 1167,18-21; 1175,32.

[68] *in Cael* 35,34.

[69] ibid. 67,20-21.

[70] ibid. 70,15; 200,33.

[71] ibid. 84,13-14, and 134,28-29 (*phantasia, phantazesthai*); *in Phys* 1334, 32-7.

[72] See M. Detienne's book, *Les Jardins d'Adonis. La mythologie des aromates en Grèce*, Paris 1972, 187-226.

[73] On this text, see P. Hadot, 'Physique et poésie dans le Timée de Platon', *Revue de théologie et de philosophie* 115, 1983, 124-5.

[74] *in Cael* 176,13; 179,22; 184,7; 199,19-20.

[75] I. Hadot, *Le Problème du néoplatonisme alexandrin*, 23-4.

[76] H.-D. Saffrey, 'Allusions antichrétiennes chez Proclus', 561-2.

tiny cluster of the friends of knowledge, while the rabble, greedy for novelty, feeds on passing fancies? Present circumstances fall within a certain natural order of things. We know that when the Neoplatonist philosophers had to leave Athens in the oppression of 529, they did so in the conviction that in the end, the climate would change and that they would return. The care with which the statues that graced the 'philosophers' house' were hidden in a well confirms that the refugees nourished this hope.[77] But it was not fulfilled, and the public teaching of pagan philosophy had come to an end.[78]

B. *Herostratus*

Philoponus, whose ruling passion is *kenodoxia*, is compared by Simplicius to Herostratus, an obscure Ephesian who, to achieve fame, set fire to the temple of Artemis at Ephesus in 356 B.C. on the very night when Alexander the Great was born.[79] At the close of his refutation, Simplicius pronounces on his opponent the indictment that his whole attitude deserves:[80]

> Let this individual (*houtos*) rest with the fish and swim with them in the sea of irrationality (*en tôi tês alogias pontôi*); may he also rest with that infamous nonentity who, they say, because he was an outcast and wanted to make a name for himself by whatever means (*hopôsoun onomaston genesthai*), set fire to the temple of Artemis at Ephesus. For the man who sets out to prove at all costs that the heavens are destructible, or even who desires to (*epithumôn*), so as to make a name for himself, would, if he had the resources, have tried to destroy the heavens themselves to achieve this aim (200,29-201,1).

Philoponus' *skopos* is in itself a sacrilege comparable to the firing of the Artemesion. Must we, in this connection, remind ourselves of the importance and significance of Artemis, in particular of the Artemis of Ephesus, in later Neoplatonism?[81] Should we think of that other sacrilege: the transformation

[77] A. Frantz, 'Pagan philosophers in Christian Athens', *Proceedings of the American Philosophical Society* 119, 1975, 29-38 (in particular 36-7).

[78] In contrast to A. Cameron's claims in 'The last days of the Academy at Athens', *Proceedings of the Cambridge Philological Society* 195, 1969, 7-29; 'La fin de l'Académie', *Le Néoplatonisme*, Paris 1971, 281-90.

[79] See Bürchner's article, 'Ephesos', in *PW* V.2, Stuttgart 1905, cols 2810-2811, and Plaumann's 'Herostratos 2' in *PW* VIII.1, Stuttgart 1913, cols 1145-1146. In recent times, Herostratos has inspired J.-P. Sartre to write a short story (in the collection *Le Mur*): the Ephesian seems to him the paradigm of the 'black hero', as opposed to the 'white hero' (the aviator Lindberg).

[80] It is the 'correction' (*kolasis*), earmarked for the garrulous student who is seeking a sophistic kind of glory (*in Cat* 8,6-8 Kalbfleisch: cf *Topics* 1.11, 105a4-7). It is also the humiliating punishment which the heavens inflict on the impious (*in Cael* 84,29-30).

[81] H.-D. Saffrey, 'Quelques aspects de la spiritualité des philosophes néoplatoniciens de Jamblique à Proclus et Damascius', *Revue des sciences philosophiques et théologiques* 68, 1984, 177. On the Artemis of Ephesus as a symbol of nature, see P. Hadot, 'Zur Idee der Naturgeheimnisse. Beim Betrachten des Widmungsblattes in den Humboldtschen *Ideen zu einer Geographie der Pflanzen*', *Akademie der Wissenschaften und der Literatur, Mainz. Abhandlungen der geistes – und sozialwissenschaftlichen Klasse* 8, 1982, 3-33, and *Annuaire du Collège de France*, 1982-1983, 471. On the place of the 'vivifying' Artemis in Proclus' system, see H.-D. Saffrey and L.G. Westerink, *Proclus. Théologie Platonicienne*. Livre I, pp lxvi-lxvii. On the mentality of later Neoplatonism and the

of pagan temples into churches?[82] It would probably be more relevant to comment that Simplicius' text leaves Herostratus anonymous, in obedience to the verdict then passed on him by the Ephesians: they handed him over to the torturers and, since he had wanted to become *onomastos*,[83] forbade anyone to speak his name. This interdiction was respected by most of the authors who retailed the event, and that includes Simplicius. This is why, under parallel circumstances, he condemns his adversary Philoponus to an anonymity that has often been commented on, calling him 'the Grammarian'[84] or 'this individual' (*houtos*),[85] or else using various sarcastic phrases: 'man of noble race' (*gennadas*),[86] 'man of value' (*ho chrêstos houtos*),[87] or 'excellent man' (*beltiste*).[88] We also come across the phrase: 'this individual, whoever he may be' (*houtos hostis pote estin*).[89] Right from the beginning of his refutation, Simplicius makes it clear that he is free from passion and bears no personal grudge (*philoneikia*) nor resentment against Philoponus as an individual, whom he has in any case never met;[90] this passage certainly indicates historically that the two men never met in Alexandria,[91] but we must remember that Simplicius is trying to conceal his adversary behind a veil of anonymity, and this is also a means to oblivion, the natural fate of the gardens of Adonis. Further, as the epithet *philoponos* probably meant, in this case, less a person who was a 'militant layman'[92] than one who was a genuine philosopher,[93] it was out of the question that Simplicius should credit his opponent with such a description.

---

link it makes between the 'rational' philosophical approach and the genuinely pious attitude towards the gods adopted by traditional religion, see A.J. Festugière's two articles, 'Proclus et la religion traditionnelle', and 'Contemplation philosophique et art théurgique chez Proclus', reprinted in *Etudes de philosophie grecque*, Paris 1971, 575-84 and 585-96.

[82] Cf A. Frantz, 'From Paganism to Christianity in the Temples of Athens', *Dumbarton Oaks Papers* 19, 1965, 187-205 (with illustrations).

[83] Philoponus can be considered as perhaps the other side of the coin from Isidorus' friend, the 'pagan monk' Sarapion, who was so indifferent 'to human glory that his name was not even known' in Alexandria (cf *Vita Isidori* fr. 287 Zintzen p 231, lines 8-9; A.J. Festugière, 'Proclus et la religion traditionnelle', 584).

[84] See the indices of the Heiberg and Diels editions. Philoponus is also called 'Telchin', that is 'a grammarian who splits hairs about books' (*in Cael* 66,10 and *in Phys* 1117,15-16), and the term seems to have joined the traditional arsenal of epigrammatic insults against the Grammarians; cf *Palatine Anthology* XI.321 (and also poems 138, 322, 347, 399). A note by R. Aubreton (CUF edition, vol X, Paris 1972, 278), also draws our attention to the *Reply to the Telchines*, in which Callimachus, in line 27, describes his detractors as the 'deadly brood of slander' (*baskaniês oloon genos*); Simplicius, also, speaks of the *baskania* manifested by Philoponus towards the heavens and the universe (*in Phys* 1117,16-18).

[85] See *in Cael* 72,11 and 14; 119,7; 121,8 and *passim*.

[86] ibid. 48,14.

[87] ibid. 45,27; 83,25; 170,11; 176,13; 188,2.

[88] ibid. 58,4; 78,9. See also 132,3, where Simplicius speaks ironically of the wonderful conceptions (*thaumasias epibolas*) of Philoponus.

[89] ibid. 49,24-25; 90,12.

[90] ibid. 26,19.

[91] H.-D. Saffrey (1954) 402 and n 4; I. Hadot, *Le Problème du néoplatonisme alexandrin*, 25.

[92] H.-D. Saffrey (1954) 403-5.

[93] A.-Ph. Segonds (1981) 10-11 and 40 n 4.

C. The Augean stables

Philoponus' treatise *contra Aristotelem* is a mere pile of ordure. Simplicius calls on Heracles for help, and imitates his heroic achievement in cleaning the Augean stables:

> But since this individual who gives himself the title of Grammarian clearly seeks once again to persuade his peers to think of the world as destructible and as created at a certain moment of time; since he flies up against those who show that the heavens are uncreated and indestructible; since he releases a great mud-bath of arguments (*polun anakinei borboron logôn*) against the claims of Aristotle – come, let us call the mighty Heracles to our aid, and let us get down to cleansing the filth which is contained in the arguments of our adversary (... *epi tên katharsin tês koprou* ...) (*in Cael* 119,7-13 Heiberg).

> Now, I do not know how, when my intention was to clarify (*saphênisai*) Aristotle's treatise *On the Heavens*, I have tumbled into the Augean stables (*eis tên Augeou kopron empeptôka*) (*in Cael* 135,31-136,1 Heiberg).

> But since this famous Grammarian has amassed – not against Aristotle's demonstration, but against the fools (*kata tôn anoêtôn anthrôpôn*) – a bed full of dung (*polun surpheton*), let us call Alpheus to our aid along with Heracles and purify (*ekkatharômen*), so far as we can, the souls which have admitted this filth (*in Phys* 1129,29-1130,3 Diels).

The first and third quotations each mark a new stage in the polemic. Both have an opening which is solemn in tone and elegant in style; the syntax is not unique in the commentary on *de Caelo*.[94] Simplicius' invocation of Heracles is not a mere stylistic ornament. It rings out like a prayer; the philosopher is asking for help from a hero hallowed in tradition as the paragon of victorious moral and intellectual courage.[95] This god of the gymnasia, mythical founder of the Olympic games, whose statue graced the 'philosophers' house' at Athens,[96] accomplished, in cleaning the Augean stables, his fifth labour (*ponos*); Simplicius, following suit, shows, we might say, his love of labour (*philoponia*).

In order to purify the stables of the king of Elis, the hero diverted the Alpheus. The soft waters of the river are doubtless linked thematically with those of Simplicius' arguments (*potimoi logoi*): he quotes Plato (*Phaedrus* 243d) and wishes to 'wash away (*apoklusasthai*) the corrosive salt of the words he has heard' (*halmuran akoên*).[97]

But the word used to denote the filth of the Augean stables, dung (*kopros*), is echoed in a text in which Simplicius takes violent issue with atheism and

---

[94] Cf *in Cael* 66,8-10; 156,25-28 (a quotation from Alcaeus: see above n 20).

[95] See for example J. Pépin, *Mythe et allégorie. Les Origines grecques et les contestations judéo-chrétiennes*, Paris 1976, 103,399.

[96] A. Frantz, 'Pagan Philosophers in Christian Athens', p 34 and fig 16, p 37; H.-D. Saffrey, 'Allusions antichrétiennes', 561-2.

[97] Purification preceding a truthful speech (about love or about the heavens). See *in Cael* 201,1-2.

with the insolence (*hubris*) of the Christians:

> That it is innate in human souls to believe in the divinity of things celestial is
> shown most of all by those who slander things celestial[98] under the sway of
> their atheistic preconceptions. In fact, these individuals themselves affirm that
> the heavens are the abode of the divine and its throne,[99] and that the heavens
> alone are capable of revealing to those worthy of it the glory and pre-eminence of
> God. Can there be a more august conception? Yet, as if they had forgotten these
> thoughts, they think that remains worse than filth (*ta koprión ekblêtotera*)[100] are
> more honourable than the heavens (*tou ouranou timiôtera*), and they persist in
> dishonouring the heavens (*atimazein philoneikousin*) as if they (the heavens) were
> created simply to give free passage to their insolence (*hubris*) (370,29-371,4).

This remarkable extract reflects all that pagan piety saw as most revolting
and shocking in the veneration of Christ's dead body and of the relics of
martyrs, remains endowed with more value than things celestial. Philoponus'
doctrines are inspired by this *hubris*.

The case of the Christians is a proof by paradox of the axiom which states
that all men naturally harbour the concept of the divinity of heaven. They
are, in fact, out of key with themselves; the voice of truth speaks in their
rational souls, which proclaim that heaven is the seat and the throne of the
divine. That is what Matthew does, and the prophet David also says, at the
beginning of Psalm 18: 'The heavens declare the glory of God; and the
firmament sheweth his handywork.'[101] But the Christians forget truth,
stifled as it is in them by ignorance, atheistic prejudice, and passions: fury
(*philoneikia*) and insolence (*hubris*), which impel them to slander Heaven.
Their irrational soul, seat of passion and unrest, comes into conflict with their
rational soul, which it conquers and annihilates. Thus Neoplatonism, the
system of completeness, is able to give an account, philosophically speaking, of
the doctrine it is attacking, working from its own description of the human
soul and of its levels. From an anthropological point of view, Christianity as a
dominant *doxa* and as an historical phenomenon is readily explained as the
triumph of a doctrine which appeals to the unseeing multitude by its irrational
passions.

This is what is clearly explained in a fine passage from Proclus'
commentary on the *Alcibiades*:[102]

---

[98] In 370,30-31 Heiberg's edition of the text (*diablepomenoi*) must be corrected to *diabeblêmenoi* (according to K. Praechter, *Hermes* 59, 1924, 118).

[99] Cf Matthew 5:35.

[100] Cf Heraclitus, fr 96 Diels-Kranz (K. Praechter, in *Hermes* 59, 1924, 118-119 quotes, in this connection, two texts of the emperor Julian against the Christians). On pagan contempt for places belonging to the Christian cult, often built around the remains of martyrs, see also the texts cited by A.J. Festugière, *Antioche païenne et chrétienne: Libanius, Chrysostome et les moines de Syrie*, Paris 1959 (= *Bibliothèque des Ecoles Françaises d'Athènes et de Rome* 194), 81 and n 1; 83; 235 and n 1 (Julian, Libanius).

[101] *in Cael* 90,15-17.

[102] *in Alcibiadem* 264,5-18 Westerink. Cf H.-D. Saffrey, 'Allusions antichrétiennes chez Proclus', 557-8.

At the present time, the populace (*hoi polloi*) agrees that gods do not exist, and this has happened to it through a want of scientific education. To answer this difficulty, it must first be said that it is impossible that the wicked man should agree with himself; indeed, since he is bad, he must be at odds with his own way of life (*zôê*), and on the one hand, because of his rational nature, he must see truth in a certain way, and on the other, because of his passions and imaginings which are embedded in matter, he must be misled into ignorance and self-conflict; proof of which is his remorse once his passions have abated, and once the discord that he previously suffered, without realising it, has disappeared. So the atheist and the intemperate man can say things that are temperate and inspired by the gods in respect of their rational faculty, which is naturally related to the divine and belongs to the good part (of the soul), but in respect of their desires and their activities of imagination and of invention of forms, they are in an atheistic and intemperate state, and on the whole, because of their irrational soul, they inflict war and disorder of all kinds on themselves.

Dominated by his passions, led by his imaginings or fantasies (*phantasiai*), Philoponus is inveigled into the crassest absurdities; he even goes so far as to assert that the light of the heavens is in no way different from the light of a glow-worm or of fish-scales[103] His sensory faculty itself is affected, and he fails to notice that he is contradicting the teaching of the Psalm of David: according to the verse quoted above, what declares the glory of God is the heavens, not glow-worms and fish-scales![104] The contradiction that Simplicius detects between Philoponus and David is the reflection, or the manifestation, of the struggle that is rending the Christian soul apart, evil and unfortunate as it is.

Simplicius is therefore embarking on a systematic disparagement of Philoponus' assertions and arguments in his *contra Aristotelem*. The rhetorical apparatus that surrounds and introduces the quotations and their refutations is intended to convince the reader that he should resist the ruses of fallacious writing. Simplicius strips Philoponus' treatise of any intellectual validity and reduces it to brazen impiety, born of false opinions and vain imaginings (*phantasiai*). In the last analysis, the *contra Aristotelem* is no more than a passing symptom of the sickness of a soul prey to passions. Philoponus' pointless struggles are only the intemperance (*akolasia*) of a young man who should really look to his manners and who, totally uncultured as he is, is not even at the first stage of the curriculum, from a scholarly point of view. On a deeper level, it is the misfortune of the Christian soul, in all its impiety, ignorance and irrationality that is apparent in Philoponus' work.

In the face of this catastrophe, Simplicius' commentary sets out a correct exegesis, under the guidance of a soul governed by reason and free from passion and materialist imaginings. His final goal is the end, the *telos*, of Aristotle's philosophy, that is, the perfection of human happiness. And one of the stages of this happiness is the joy resulting from the contemplation of the

---

[103] *in Cael* 89,4-9; 90,4-5, 12-13, 17-18; 135,4-5.
[104] ibid. 90,15-18.

great things of the universe, which is celebrated in a liturgy dedicated to the Demiurge.

II. The religious significance of Simplicius' commentary

Simplicius' commentary on the *de Caelo*, like his commentaries on the *Categories* and the *Physics*, is a piece of scientific research that approaches Aristotle's text by means of 'a huge mass of historical documentation and long discussions with previous interpreters'.[105] We know that it is not keyed to an actual course of instruction; it is a book written to be read: an encyclopaedic work produced by a great mind for his 'peers'.[106] None the less, too much emphasis on this aspect – the most obvious – and too much concentration on the purely intellectual side of the *in de Caelo* would overlook the vital aspect, which is spiritual and religious.

The climax of the commentary is a personal prayer, written in prose, that Simplicius addresses to the Demiurge.

> These reflections, O Lord, Creator of the whole universe and of the simple bodies within it, I offer to you as a hymn (*eis humnon prospherô*), to you and to the beings you have produced, I who have ardently desired to contemplate (*epopteusai*) the greatness of your works and to reveal it to those who are worthy of it (*tois axiois ekphênai*), so that, conceiving (*logizomenoi*) nothing mean (*euteles*) or human (*anthrôpinon*) about you, we may adore you (*proskunômen*) in accordance with your transcendency (*huperochê*) in relation to all the things you have created (731,25-29).

The sober tone of this prayer recalls the restraint of the final prayers of the commentary on Epictetus' *Manual*[107] and of the commentary on the *Categories*;[108] one of them entreats the 'Lord, Father and Guide of Reason in us' to guarantee and ratify the victory of the rational soul over the irrational passions, and the other prays the 'Guardians of discourses' to allow the knowledge of categories, the basic elements of logic, to enable the soul, free at last from the vicissitudes of life, to attain higher knowledge. These three prayers share the general atmosphere of Neoplatonist spirituality,[109] and they set the commentaries, of which they are both a resumé and the crowning glory, within the general framework of the Neoplatonist curriculum.[110] This,

[105] I. Hadot, *Le Problème du néoplatonisme alexandrin*, 194.

[106] A. Cameron, 'La fin de l'Académie', in *Le Néoplatonisme*, Paris 1971, 289. But we have seen that the commentary is not intended only for Simplicius' 'peers'.

[107] See Simplicius *in Ench. Epict.* 138,22ff. Dübner; I. Hadot, *Le Problème du néoplatonisme alexandrin*, 35 n 9.

[108] See Simplicius *in Cat* 438,33-6 Kalbfleisch; I. Hadot, ibid. 36 (continued from 35 n 9.)

[109] H.-D. Saffrey, 'Quelques aspects de la spiritualité des philosophes néoplatoniciens', 169-79 (esp 174-5).

[110] A.J. Festugière, 'L'ordre de lecture des dialogues de Platon aux Ve/VIe siècles', in *Museum Helveticum* 26,1969,281-96, reprinted in *Études de philosophie grecque*, Paris 1971, 535-50; I. Hadot, *Le Problème du néoplatonisme alexandrin*, 148-9 and 160-4; P. Hadot, 'Les divisions des parties de la philosophie dans l'Antiquité', *Museum Helveticum* 36, 1979, 201-23 (see 221); H.J. Blumenthal, 'Marinus' Life of Proclus: Neoplatonist biography', *Byzantion* 54, 1984, 475 (with

as we know, was a huge programme, spiritually based, a means of gradual conversion towards beings ever more exalted, which was finally to lead to a return to the first principle.[111] The final stage in this curriculum was a reading of the *Parmenides*.[112]

The study of the *de Caelo* came immediately after that of the *Physics* and before that of the *de Generatione et Corruptione* and the *Meteorologica*. There was thus a transition from a discussion of the general principles of nature to a consideration of the universe and of the *first essence* (or 'fifth element') which makes up the heavens,[113] and then to an examination of the general principles of the sublunary world, ending up with the 'meteorological' phenomena that occur in the upper (fiery) part of the sublunary world.[114] But the *de Caelo* was not only read in the framework of this descending order derived from the method of division;[115] it could also be viewed in the context of progressive conversion towards the first principle. So viewed, it allowed its readers to return to knowledge of the Platonic Demiurge (*Timaeus* 28a), whom Proclus numbers with the transcendent gods, identifying the Demiurgic intellect with Zeus in the traditional pantheon. Simplicius, in his interpretation of the *de Caelo*, is fully aware of the close connections it has with the *Timaeus*, although the latter was further up the curricular ladder, coming as it did immediately before the *Parmenides*.

Simplicius offers up his commentary on the *de Caelo* to the Demiurge: the scholarly and encyclopaedic style of the commentary (*hypomnêma*) gives place through a metamorphosis to the more literary style of the hymn, which had an important role to play in Neoplatonist spirituality.[116] This change is apparent in the final prayer.

The offering of the hymn is matched by another religious act: the adoration (*proskunein*) of the god. Here, however, the adoration is entirely internal, and the verb *proskunein* does not denote an external expression, a physical attitude, as in the famous episode in the *Life of Proclus* (ch. 11) where Marinus describes Syrianus and Lachares — soon to be joined by Proclus — preparing to adore the moon on its first appearance.[117] The *in de Caelo* prayer is characteristic of the internal nature of worship that we find in the later

---

notes 19,20) and 486-7. After a first initiation into ethics, the study of Aristotle (or 'the little mysteries' of philosophy) led from the *Categories* to the *Metaphysics*, and then the reading of Plato (the 'great mysteries') was broken down into two cycles, the second of which consisted of the *Timaeus* (physics) and the *Parmenides* (theology).

[111] P. Hadot, 'Les divisions des parties de la philosophie', 220-1; 'Exercices spirituels', in *Exercices spirituels et philosophie antique*, Paris 1981, 44-5.

[112] H.-D. Saffrey, 'Quelques aspects de la spiritualité', 169-71.

[113] The principal aim of the *de Caelo* according to Simplicius: Aristotle's *skopos* has to do with simple bodies, and the first of the simple bodies is the heavenly body, the fifth element, which involves the whole *cosmos* in its perfections (1,2-5,34 Heiberg).

[114] *in Cael* 2,18-3,8.

[115] Cf P. Hadot, 'Les divisions des parties de la philosophie', 201-8. This order of reading was authorised by ch 1 of the *Meteorologica*.

[116] See A.J. Festugière, 'Proclus et la religion traditionnelle', reprinted in *Études de philosophie grecque*, 578-81; H.-D. Saffrey, 'Quelques aspects de la spiritualité', 175-9.

[117] A.J. Festugière, 'Proclus et la religion traditionnelle', 577; H.-D. Saffrey, 'Quelques aspects de la spiritualité', 181.

Neoplatonists. In this connection, H.-D. Saffrey writes:[118] 'as philosophy in Greece was never a purely intellectual activity, but also a life-style, the spiritual life of these philosophers became an unbroken succession of prayer and liturgy', and 'it was philosophical activity itself which, through its very aim, became worship addressed to the gods'. Simplicius' commentary is an exercise that derives from the *religio mentis*, the intellectual celebration of divinity.

At the heart of the prayer lies the Eleusinian image of mystic initiation leading to *epopteia*, the highest grade of initiation, which shows that an exegesis of Aristotle's treatise is in itself a religious act, an exercise that invites the teacher (the author) and his pupils (the readers)[119] to walk the path of intellectual[120] and spiritual progress, of internal change and of revelation. The general drift of the work then becomes apparent. The refutation of Philoponus' treatise *contra Aristotelem* is simultaneously a propaedeutic exercise or 'gymnastic'[121] which allows us to take part in philosophical *philoponia*, an endurance test – Herculean or Odyssean –[122] and a preparatory act of purification before the exegesis of Aristotle's text, which we must approach with a soul free from irrational conceptions. There then follows the exegesis, in its guise as the door to contemplation, whose mysteries are accessible to Simplicius, the honoured guest. Once he has become an initiate (*epoptês*), he reveals (*ekphênai*) what he has seen to those who are worthy of it (*axioi*) – his 'cultivated' readers and 'lovers of knowledge'.

What is the object of initiation and of revelation? 'The greatness of the works of the Demiurge'. The expression sounds like a paean, and the written text which records the revelation can only be a hymn, a worthy counterbalance to Philoponus' blasphemy.

Simplicius' confessed emotion at the greatness of the works of the Demiurge, seen in their totality (*panta*) and in the dependence that links them to the Creator, is certainly not the result of a modern romantic view of nature; it goes back to the ancient category of greatness of soul.[123] Born of an exegetical practice that transformed physics into a spiritual exercise,[124] the greatness of soul of a good interpreter is a worthy partner to Aristotle's intellectual greatness (*megalonoia*), and has nothing but scorn for the feckless flounderings of a Philoponus. But what also overflows, with great dignity, in Simplicius' prayer is the joy of that 'splendid festival' which, according to the aphorism coined by Diogenes the Cynic, the good man can celebrate every day and which is analysed as early as Plutarch of Chaeroneia in a fine

[118] ibid. 169.

[119] The pedagogical framework of the curriculum remains fundamental; the fact that the commentary does not correspond to oral instruction is immaterial.

[120] Thus we can see Simplicius undertaking written exercises in order to clear up his own ideas and those of attentive readers (*in Cael* 102,15-17; 166,14-16). Cf *in Cat* 7,34-8,1 Kalbfleisch.

[121] But Simplicius embarks on it most reluctantly!

[122] *in Cael* 165,7-9: like Ulysses in the storm (*Odyssey* 12.428), Simplicius must continue to refute Philoponus' arguments, that is, to 'pass by Charybdis' again.

[123] P. Hadot, 'Exercices spirituels', 42-3.

[124] ibid.

passage of his *de Tranquillitate Animae*:[125]

> ... the universe is a most holy temple and most worthy of a god; into it man is introduced by his birth as a spectator (*theatês*), to view not hand-made or immovable images (*agalmata*), but those sensible representations (*mimêmata*) of intelligible essences that (according to Plato) a divine intellect has brought to light, representations which have innate within themselves the principle (*archê*) of life and motion: the sun, the moon, the stars, the rivers in which new waters constantly flow, and the earth which sends forth nourishment for plants and animals. A life that is an initiation (*muêsis*) into these mysteries, and a perfect revelation (*teletê teleiotatê*), must be filled with tranquillity (*euthumia*) and joy (*gêthos*) (ch 20, 477C-D).

This joy is a stage on the road that leads to the most perfect happiness that can fall to a man's lot – the final aim of Aristotle's philosophy.

The reading of the *de Caelo* led Simplicius and his readership from physics to theology: from a contemplation of the universe, a visible god, to an ascent above the plane of the senses to the Demiurge himself.[126] This conversion also marks a purification of our ideas about the Demiurge. Human thought is exalted and transformed, losing its human trappings, and thus becomes capable of conceptions about the god which do justice to his transcendency. Similarly, the final prayer of Iamblichus' *de Mysteriis* expresses the desire for a 'participation in the most perfect ideas concerning the gods'.[127] In the final prayer of the *in de Caelo*, two words must give us pause because of their polemic bearing: ' ... so that, conceiving nothing mean (*euteles*) or human (*anthrôpinon*) about you ...'. The adjective *eutelês* (lowly, mean) is applied by Simplicius to precisely those Christian doctrines that inspire Philoponus: 'No one has so wasted his time (*kakoscholôs houtôs*)[128] with the sole aim of seeming to oppose those who prove the eternity of the world – governed by the mean ruling conceptions (*tas kratousas euteleis ennoias*) about him who created the world' (*in Cael* 59,13-15). The expression *hai kratousai ennoiai* refers, in a deliberately archaic style[129] which forbade neologisms, to Christian doctrine, and Simplicius' disapproval is reflected in the word *euteleis*: Christians think that Christ's dead body and the relics of the martyrs, those 'remains worse than dung', are more precious than the heavens, and Philoponus dares to assert that the Demiurge created a world such that the light of the heavens is of the same nature as the light of glow-worms and of fish-scales. Above all, however, Philoponus has an anthropomorphic notion of God (*anthrôpinon*): he 'thinks that God is of the same nature as himself' (*homophuê pros heauton*)

---

[125] ibid. 43; 'Physique et poésie dans le Timée de Platon', 130.
[126] Cf *in Cael* 382,35-383,7.
[127] *de Mysteriis* 10.8 (293,16-294,6); cf I. Hadot, *Le Problème du néoplatonisme alexandrin*, 36.
[128] Another characteristic of Philoponus (cf also 131,21).
[129] Averil Cameron, *Agathias*, Oxford 1970, 75-88 ('Classicism and affectation'); I. Hadot, *Le Problème du néoplatonisme alexandrin*, 31-2.

(90,20). From this fundamental theological error[130] arises the cosmological error, the 'human' idea of the heavens, which are also said to be created and destructible; Philoponus is intent on showing that 'just like him (*paraplêsiôs autôi*), the heavens and the whole universe were born at one precise moment in time and will perish at another' (122,31-33). Human transience is applied both to the Demiurge and to the universe.

The contemplation of the greatness of the works of the Demiurge is a joy; this joy is, however, a step towards a bliss that is even more perfect.

When we have, by reasoning (*logizomenoi*), arrived at a correct idea of the Demiurge, neither mean nor anthropomorphic, we have risen to the level of 'those beings that are beyond nature' (*ta huper phusin*). At this point, our reading of the *de Caelo* becomes part of a general process of conversion.

To understand this process, we must return to Simplicius' description, in the prologue to his commentary on the *Categories* (6,6-18), of the final end (*telos*) of Aristotle's philosophy, and of the means of achieving it. The *telos* can be looked at from an ethical (*kata to êthos*) or a cognitive (*kata tên gnôsin*) standpoint. In the first case it consists in perfection reached by practising the virtues. In the second it becomes identified with the ascent (*anadromê*) towards the unique principle of all things: knowledge is the passage towards the One. These two aspects of the *telos* combine to form a single unit, the perfection of human happiness, which, according to the *Nicomachean Ethics* (10.7, 1177b26ff), allows the title of 'god' to be bestowed on those who attain such happiness. And Simplicius ends by putting all Aristotle's works into perspective: 'All the Philosopher's works *lead to this end*. Some supply the demonstrative method, some put our conduct in harmonious order through the practice of virtue, and *others, through the study of natural beings, make our understanding rise back up to levels above the order of nature.*'

It is plain that Philoponus is very far from this happiness.

III. Simplicius' theory on the nature of the heavens

The description of the physical structure of celestial matter is one of the most rewarding revelations of the initiation experienced by Simplicius and his readers. In the *in de Caelo* we come across the last of the ancient theories of the fifth element.[131]

First of all, it must not be forgotten that Plato's and Aristotle's theories of

---

[130] A complete reversal of the philosophical attitude, which consists in elevating ourselves above individual humanity. Fully to understand Simplicius' reproach, we must remember the central idea of Neoplatonic anthropology: 'Here man is nothing; the private individual is only a corruption of Man with a capital M. ... Man's misfortune is to be an individual, and Philosophy's every effort is aimed at elevating us again to the universal and to the universe' (H.-D. Saffrey, 'Théologie et anthropologie d'après quelques préfaces de Proclus', *Images of Man in Ancient and Medieval Thought* (*Studia Gerardo Verbeke ... dicata*), Louvain 1976, 208). Taking human individuality as a *model* for a theory of god and of the world is a procedure both absurd and disgraceful.

[131] See P. Moraux (1963), 1244-5 and Nachträge 1430-2; *Aristote. Du Ciel*, Paris 1965, xxxiv-lx.

the nature of the heavens are divergent. In the *Timaeus* (39e-40a), Plato asserts:

> The species [of living beings] are four in number: the first is the heavenly species of the gods [= the stars], the second the winged species that flies through the air, the third the aquatic species, the fourth the species which has feet and lives on solid ground. As regards the divine species, the greater part of its substance was composed of fire (*tên pleistên idean ek puros*), so that it should present to our eyes the greatest brilliance and beauty ...

And we know that Aristotle, in book 1 of the *de Caelo*, affirms the existence of a fifth element, which he calls the *first body*, radically different from the four sublunary elements: it moves in a circle, has neither heaviness nor lightness, is unengendered and indestructible, and knows no change either of quantity (*auxêsis, phthisis*) or of quality (*alloiôsis*).

Philoponus, in his denial of the fifth element and of the eternity of the world,[132] contrasts Plato's views with Aristotle's, and has this to say on the passage from the *Timaeus*:

> Plato did not see the heavenly bodies as composed only of fire, but he did think that they partook, principally and for the most part (*pleistou*), of that kind of fire which tempers (*eukraesteran*) the mixture of the other elements (*mixin*); all the substance that is refined and perfectly pure (*pasês ... tês leptomerous kai katharôtatês ... ousias*), and that ranks as form in relation to the rest has, he says, been separated from the elements in order to make up (*sunkrima*) the heavenly bodies, while the more material (*hulikôteras moiras*) and, so to speak, muddy (*trugôdous*) part of these elements has been deposited here below. According to Plato, it is from such fire that both the stars and the sun are made (84,15-22).

It is hard to see, as Simplicius very reasonably remarks, how such an idea could allow us to go on and say that heavenly beings are *of the same nature* as sublunary ones.

In order to understand the doctrine that Simplicius sets against Philoponus' own, we must first of all remind ourselves of the basic principles of his interpretative method.

Plato and Aristotle cannot contradict each other. This is how Simplicius defines the good interpreter: 'it is necessary ..., when Aristotle disagrees with Plato, not merely to look at the letter of the text (*lexis*), and condemn the discord between the philosophers (*diaphônia*), but to consider the spirit (*nous*) and track down the agreement between them on the majority of points (*sumphônia*)' (*in Cat* 7,29-32 Kalbfleisch). This presupposition of a doctrinal agreement between Plato and Aristotle, which originates with Antiochus of

---

[132] We know that Philoponus picks up and elaborates the criticisms indicated by the treatise of Xenarchus of Seleucia *Against the Fifth Essence* (cf P. Moraux, *Aristote. Du Ciel*, lvii and n 7). Philoponus is distinct from all those who attacked the fifth element (the first or fifth essence) before him, because this criticism is a way for him to topple the eternity of the world, something not yet attempted by anyone (*in Cael* 59,6-10).

Ascalon, was almost universally accepted in the Neoplatonist school after Porphyry, and it is summed up thus by I. Hadot: ' ... whenever an Aristotelian theory would be seen by an objective reader to be at odds with Plato's position, it was always, for Simplicius as for the other Neoplatonists, the essentially Neoplatonist interpretation of Aristotle that won the day. This was an interpretation effected through the hypothesis of complete doctrinal agreement between Plato and Aristotle. By assuming that any differences were merely superficial, verbal ones, it was perfectly simple to claim a false harmony between the doctrines of the two thinkers.'[133]

Many texts voice the idea that if disagreements are verbal (and purely verbal), this is due to the different linguistic attitudes of Plato and Aristotle, that is to say, when it comes down to it, to differing philosophical attitudes:

> In my opinion, we ought to consider the intention (*skopos*) and the words (*rhêmata*) together, and to understand that in these matters the opposition of the two philosophers bears not on the things themselves (*peri pragmatôn*), but on the words (*onomatôn*). Because of his love of precision (*to akribes*), Plato rejects the ordinary use of words (*tên tôn onomatôn sunêtheian*), while Aristotle employs it. It is a method which he thinks does no harm to the truth (*in Cael* 69,11-15; cf. also lines 25-29).

> As I constantly say – and it is opportune to say it again now – the disagreement (*diaphônia*) observed between the two philosophers does not relate to fundamentals (*ou pragmatikê*) (*in Cael* 640,27-28).

Plato scorns, and Aristotle uses, the language of the multitude:

> Aristotle relies on the usage accepted by the multitude (*tês tôn pollôn sunêtheias*), which he does not want to stray from .... Plato, on the other hand, despises (*kataphronêsas*) the usage of the multitude, and thinks that, since the world is spherical, he must not speak of above and below, but of periphery and centre (*in Cael* 679,27-31).

This difference of attitude towards language also enshrines a difference of philosophical *method*. Plato's method is based on intellectual intuitions (*apo nou*), Aristotle's on the data of perception (*apo aisthêseôs*):

> Thus the present difference (*diaphora*) between the philosophers bears not on the matter itself (*pragma*), but on the word (*onoma*), and it is the same in most other cases. In my opinion, the reason is that Aristotle often wants to preserve the customary meaning of the words, and sets out, in constructing his argument, from what is manifest to the senses (*apo tôn têi aisthêsei enargôn*), whereas Plato frequently displays contempt for that kind of evidence, and deliberately rises to the level of intellectual contemplation (*tas noêtas theôrias*) (*in Phys* 1249,12-17 Diels).

Convincing evidence can be of two sorts: one is based on intellectual intuition

---

[133] I. Hadot, *Le Problème du néoplatonisme alexandrin*, 195 (see also 68-9, 72-6, 148 n 3).

(*apo nou*), the other on sensation (*apo aisthêseôs*). Because he speaks to beings who live with sensation, Aristotle prefers the evidence that is based on sensation. He is always unwilling to stray from nature, and even objects which are above nature (*ta huper tên phusin*) he studies in their relation to nature. The divine Plato, on the other hand, goes the other way and, in conformity with Pythagorean practice, examines physical objects in so far as they participate in what lies above nature (*in Cat* 6,22-30 Kalbfleisch).

For Simplicius, demonstrative perfection lies in the combination of both methods.[134]

The commentary on the *de Caelo* often describes the inner structure of celestial matter;[135] it is composed (*sunestanai*) of a mixture of the four elements – earth, water, air and fire – in their purest and most perfect form, that is to say their 'superior grades' (*akrotêtes*). The superior grades are the points of perfection (*teleiotêtes*) of the elements, and the superior grade of fire, which is light (*phôs*), is predominant in this mixture. Celestial substance is therefore principally light.

This doctrine is put together from three constituents:

(1) By applying a basic ontological plan[136] to the physical world, Simplicius contrasts imperfect sublunary elements (*atelê*) with their perfect superior grades, from which they emanate: light is the superior grade of fire, resistance to touch is the superior grade of earth.[137]

(2) He uses the notion of a mixture in which something predominates (*epikrateia*), which goes back to Anaxagoras,[138] and which was immensely popular in Stoicism and in Middle and Neo-platonism:[139] when a being is made up of a mixture, it takes its nature and its name from the predominant characteristic.

(3) He reads *Timaeus* 40a in the light of an interpretation 'keyed' to *Timaeus* 58c, a passage in which Plato defines different kinds of fire: ' ... we must think that several kinds of fire (*puros ... genê polla*) have been created: for example (*hoion*) flame (*phlox*), and what comes from flame, but does not burn, giving light (*phôs*) to the eyes, and what, once the flame has gone out, remains of fire in burning bodies (*en tois diapurois*).' Plato's text is carefully constructed: first of all the list is certainly not an exhaustive one, since it is introduced by *hoion*; next, and most important, flame is followed by two other *genê*, given equal weight by the syntactic form, and each described by a positive and a negative characteristic – the presence or absence of burning or of light. Plato does not suggest a hierarchical (ontological) superiority for

---

[134] ibid. 148.

[135] *in Cael* 12,27-13,2; 85,9-15; 360,33-361,2; 379,5-6; 435,32-436,1.

[136] Cf for example Proclus, *Elements of Theology*, propositions 146-148. In the supernatural domain, the superior grade (*akrotês*) is the principle from which the divine orders (*taxeis*) arise and to which they return. In a given divine order, the *akrotês* has a superior unifying power (*henikôtatên dunamin*) which it imparts to the whole order, thus unifying it 'from above' (*anôthen*).

[137] *in Cael* 12,29-30.

[138] Quoted by Aristotle, *Physics* 1.4,187b1-7.

[139] See P. Hadot, 'Être, vie et pensée chez Plotin et avant Plotin', *Les Sources de Plotin* (= *Entretiens sur l'antiquité classique* (Fondation Hardt), vol.V), Vandoeuvres – Genève 21-29 août 1957, Geneva 1960, 124-30; *Porphyre et Victorinus* I, Paris 1968, 239-44.

light, which figures as a derivative entity, making a fleeting appearance within the framework of an optical phenomenon. Indeed, flame is more completely fire than light is. Simplicius, however, deduces from Plato's text a hierarchical classification of the kinds of fire: burning coal (*anthrax*), flame and light,[140] which allows him to see in light the superior grade of fire (*akrotês*) and to describe the heavens as essentially luminous.

Philoponus has therefore failed to grasp Plato's real ideas in the *Timaeus*:

> This individual now seems (*dokei*) to approve of Plato, who asserts that the heavens are made of fire, but he does not know what kind of fire Plato means. He does not realise that it is not the fire of the sublunary world (*touto*), the fire which Aristotle describes as rising upwards (and hence being light) and as floating above all rising bodies (and hence as being the lightest). Plato means rather that this sort of fire (*ekeino*), heavenly fire, is nothing other than light (*phôs*), for light is defined by him as a kind of fire (66,33-67,5).

The bad interpreter has stayed too close to the *lexis*.

Simplicius is the genuine interpreter of Plato's doctrine of the nature of the heavens:

> Plato says that the heavens are made up mostly of fire (*ek pleistou puros*), because since, according to him, there are three kinds of fire – burning coal (*anthrax*), flame (*phlox*) and light (*phôs*) – they are principally (*malista*) made up of the purest (*katharôtatou*) and most luminous (*phôteinotatou*) of these kinds of fire, that is, of light. In fact, just as each of the elements called sublunary is composed of four simple elements which are genuinely elemental, but takes its essence, its character, and its name from the predominance of just one of these elements (*kata tên tou henos epikrateian*), so too the heavens, which are composed of the superior grades of the four elements, take their essence (*ousiôtai*) from that superior grade which predominates (*kata to kreitton tôn akrôn*). They are perfected, very luminous (*phôteinotatos*), and quite brilliant (*hololampês*), and that is why we honour them with the name Olympus (85,7-15).

This physical theory, the fruit of exegesis, is confirmed by a twofold proof, which combines Aristotle's method (*apo aisthêseôs*) with Plato's (*apo nou*).

Sense experience does indeed show clearly that the bodies we call sublunary elements are not simple elements in the true sense (*kuriôs hapla*) but are composite; the predominant element in them brings about a quasi-simplicity:

> The earth maintains its cohesion and does not break up thanks to the water in it; it receives its colour from fire and is regenerated by it; it is always full (*plêrês*), leaving no room for void, even if some of its parts are removed, because then air flows into it; thanks to this it stays upright and remains steady without falling. The same thing can also be observed in the other elements (85,15-21).

And the same analysis can be applied, *mutatis mutandis*, to the substance of the heavens.

---

[140] *in Cael* 12,28; 16,20-21; 130,31-131,1.

The second proof, the 'Platonic' one, starts from an examination of 'elemental nature' (*ek tês stoicheiôdous phuseôs*). Simplicius tries to apply to the description of sublunary elements the doctrine of the three types of plurality that we find, for example, in the treatise *On First Principles* by his master Damascius.[141]

The basic idea put forward in Damascius' dissertation *On Many and Plurality* (*peri pollôn kai plêthous*) is as follows: the triad of Being, Life, and Thought, which coincides with the three orders of reality: the Intelligibles (*noêta*), the Intelligibles and Intellectuals (*noêta kai noera*), and the Intellectuals (*noera*), is matched by three types of plurality, characterised by an increasing separation. To Being (Intelligibles) correspond the elements (*stoicheia*); to Life (Intelligibles and Intellectuals), the parts (*merê*); and to Thought (Intellectuals), the forms (*eidê*). The forms, which are in the Intellectual Intellect, are distinct one from another and each is, on its own (*kath'heauto*), in its proper precinct (*oikeiâi perigraphêi*). The independence of the forms contrasts with the links between the parts, which (the parts) are of necessity at least two in number. They are not and do not wish to be on their own (*kath'heauta*), nor to be parts of themselves (*heautôn*), but they only exist through the close-knit solidarity (*allêlouchia*) that binds them one to another and to the whole (*holon*) of which they are parts; this is what they exist in. They 'incline' one to another and to the whole (*hê pros allêla kai to holon sunneusis*). While the separation of forms in the Intellectual Intellect (*noeros*) is a completed separation (Damascius likes to use the perfect participle *diakekrimena*), the separation of the parts is a separation that has just begun (*arxamena tês diakriseôs*), which is in the course of happening (present participle, middle or passive: *diakrinomena*), and which is not completed: the parts have not got proper precincts (*oikeiai perigraphai*). If we pass finally from the Intelligible and Intellectual order to the Intelligible, the plurality is in the elements. The elements too are at least two in number. But they are not distinct one from another (*ouch hupomenei tên hopôsoun diastasin*), and they have no *perigraphê* (as the forms do), and no separation (*merismos*) (as the parts do); rather, they are a unified mixture (*suncheitai eis krasin kai eis henôsin epeigetai*). A sentence in the treatise *On First Principles* sums up clearly this distinction between three kinds of plurality:

> The extreme terms are, then: (1) What is completely circumscribed in its proper essence (*to perigegrammenon eis oikeian hupostasin*) [the forms]; (2) What is mixed (*sunkekramenon*) and melted in the common unity of all things [the elements]. The middle term is the parts (*merê*) and their separation (*merismos*); they are already (*êdê*), in a certain way, in the process of separation, but are not yet (*oupô*) circumscribed (I.197,15-18 Ruelle).

From this metaphysical analysis, Simplicius retains the idea that, properly speaking, the elements have no individuality and are not separable from the mixture that they form. From this basic fusion emerges, by differentiation, a less tight kind of bond between the terms of plurality which already allows a degree of specific difference.

In the physical world, the elements proper, that is to say not the sublunary

---

[141] Ruelle edition, vol I, Paris 1889, 196ff.

elements which are available for empirical experiment, and which are not absolutely simple, but rather those that are truly simple, never exist on their own (*kath' hauta oudamou esti*), but by mutual interpenetration (*di' allêlon aei kechôrêkota*) form a synthesis (*suntheton*). They are elements *of* the synthesis, just as, in linguistics, one speaks of the twenty-four elements *of* the language or of the eight parts *of* speech. Inside this synthesis, which is a *sunkrasis* of elements which are truly simple but not separable, one becomes predominant, and according to whether this *epikrateia* is that of the true element 'earth', 'water', 'air' or 'fire', the synthesis of quasi-elements is specified – that is to say, defined both in its essence and in its name – as earth, water, air, or fire. This mixture, primordial but governed by a predominating element, forms what Simplicius would call the 'first synthesis' (*prôtê sunthesis*), and this 'first synthesis' brings forth the sublunary elements which await our sensory experience and which we call, according to what predominates (*kat' epikrateian*), 'earth, 'water', 'air' or 'fire'. Their simplicity is not total, it is only predominant, but from their spurious simplicity emerges the 'second synthesis' (*deutera sunthesis*) which composes living beings, plants and their parts, whose biological unity (*hê pros allêla sumphusis*) is caused by the burning desire (*philoneikein*) to safeguard (*diasôzein*) 'the original unity (*hê archegonos henôsis*) that lay in their own causes' – a biological unity that reproduces, in a more clearly differentiated way, the unified mixture of the primordial elements.[142]

A paradox and a difficulty now become apparent. Simplicius intended not only to teach Philoponus something about the interpretation of Plato, but also to re-establish the harmony between Plato's and Aristotle's philosophies and to reaffirm the separation of the heavens and the sublunary world in keeping with the teachings of the *de Caelo*. The doctrine he puts together is a fine example of the philosophical fruitfulness of misinterpretations,[143] and it is interesting to see light defining the heavenly substance, when Proclus saw it, starting in particular from a famous passage in Plato's *Republic*,[144] as defining place (*topos*). But Simplicius gives a very odd version of what we now think were Aristotle's real views; this luminous mixture, however pure it may be, is very far from the simple body described in Aristotle's *de Caelo*, which was radically different from the four sublunary elements. And Simplicius appears to be superimposing his own interpretation of the *Timaeus*, doubtful as it is, on Aristotle's treatise; it is the fire in the *Timaeus* that provides him with

---

[142] *in Cael* 85,21-31.

[143] Cf P. Hadot, 'Philosophie, exégèse et contresens', in *Akten des XIV. internationalen Kongresses für Philosophie*, Vienna 1968, vol I, 333-9.

[144] *Republic* X, 616b. The doctrine of place (*topos*) put forward by Proclus is based on the dual authority of this passage in the *Republic* and on a fragment of one of the Chaldean Oracles (fr. 51, line 3); it arises logically from the problematic of Aristotle *Physics* 4 and from the Stoic division of the 'something' (*ti*) between body (*sôma*) and incorporeal (*asômaton*). See Simplicius, *in Phys* 612,24-613,8 Diels; P. Duhem, *Le Système du monde. Histoire des doctrines cosmologiques de Platon à Copernic*, vol I, Paris 1913, 338-42; Ph. Hoffmann, 'Simplicius, corollarium de loco', in *L'Astronomie dans l'antiquité classique (Actes du Colloque tenu à l'Université de Toulouse-le-Mirail, 21-23 octobre 1977)*, Paris 1979, 149-53.

a model for conceiving Aristotle's first body. We know that Plato's work had more 'authority' than the *de Caelo* because it came before the *Parmenides* at the end of the 'great mysteries' of philosophy.

Yet we must remain convinced that for Simplicius Aristotle's real views, apart from the words he uses in his treatise, do actually tie in with Plato's.[145] Unusually for him, Aristotle is on his guard against the pitfalls of ordinary language, and it is a preventive strategy, aimed at forestalling the blasphemies of the Christians, that leads him to urge the transcendency of the heavens and to fight shy of any phrases that might imply that the heavens and the sublunary world were in any way identical:

> How can Aristotle appear to adopt a position contrary to Plato's where the essence of the heavenly body is concerned, when he admits neither that it is composite nor that it is simple, like fire or any one of what we call the four elements? Doubtless he had foreseen the mad fury (*aponoia*), worthy of the giants (*gigantikê*), of those impious homunculi (*asebôn ... anthrôpiskôn*) who attack celestial beings;[146] and since, for this reason, he wished the heavens to be conceived of as absolutely transcending (*pantelôs ... exêirêmenon*) the sublunary world, and as possessing a divine superiority to it (*theian ... tên huperochên*), he refrained from using words which would tend to pull the heavens downwards (*kathelkein*), that is to say, towards any similarity (*homoiotêta*) with the sublunary elements (85,31-86,7).

Conversely, Plato agreed with Aristotle in ascribing to the heavens a nature of their own, since he matches them up in the *Timaeus* with the fifth of the regular polyhedrons, the dodecahedron (12,16-27).

We know that the theologians at the third council of Constantinople in 680 punned on the surname of John Philoponus, nicknaming him *mataioponos* ('man whose labour is lost'); Philoponus' efforts had in their eyes been fruitless. Simplicius could have made the same pun. In spite of their violence, Philoponus' attacks on Aristotle's fifth element and on the eternal nature of the Universe are, like Christianity itself, only a passing fancy, which will never be a lasting threat to the hellenic tradition. The radical novelty of the Grammarian's ideas was intolerable to one of the great upholders of this tradition, and Simplicius' reactions give us a vivid insight into pagan spirituality at the close of antiquity. His doctrine of celestial matter is the outcome of secular interpretative methods. Yet the strength and clarity that characterise its development, and the solidity of the philosophical and religious convictions that underlie it raise it far above the level of a quirk in the history of thought and elevate it to a consummation of the ancient worship of the heavens.

[145] *in Cael* 91,7-13: Aristotle wanted to hymn (*anumnêsai*) the transcendent nature of the heavens. In the final prayer, as we have seen, Simplicius too offers his commentary to the Demiurge as a hymn (731,25-9).
[146] We know that the gods, with Heracles' help, will finally overcome the Giants. This is another allusion to the inevitable defeat of Christianity.

CHAPTER FOUR

Philoponus and the Rise of Preclassical Dynamics

Michael Wolff

Introduction: Philoponus and the history of impetus theory

Historians of science know John Philoponus as a precursor of impetus theory, and they hold this theory to be the connecting link between Aristotelian physics and classical mechanics. In order to understand this historical progression, we must briefly investigate the term 'impetus theory'.

The expression was introduced by Pierre Duhem, and refers to the doctrine that all motion depends on the transmission of an exhaustible moving force which passes from a moving cause to a movable object and acts on it instantaneously. This doctrine entails (1) that all motion ultimately requires a mover who transmits force, and (2) that the moving force is exhausted as a result of its activity.[1] This second requirement has been interpreted in several ways. The reference to the force's activity can be understood in two ways: first, it can refer to the action of the mover (i.e. of the transmitter of force) on the moved object; secondly, it can refer to the action of the transmitted force on the movable object during its motion. In the former case, exhaustion of force takes place insofar as the mover (at least if he has a *finite* capacity for moving) *loses* some power. In the latter, it takes place insofar as the force in the moved object *weakens*. Early impetus theory seems to have distinguished clearly between these two kinds of exhaustion, but not to have been concerned about the reasons why exhaustion of the second kind takes place. Only later was the question raised as to why the force weakens: is this due *directly* to the force's producing motion, or rather to its overcoming external *resistances* to its producing motion. In the latter case, it

---

* This chapter is an extended version of a lecture given during a conference on Philoponus at London in June 1983. In revising my paper since then, I have become especially indebted to Richard Sorabji, who offered me a set of helpful and stimulating comments.

[1] Duhem introduced the term 'impetus theory' in his *Études sur Léonard de Vinci*, Paris 1906-13; however, he did not offer a general definition. The definition I propose holds for all variants of that theory.

was understood, force would be lost by being transmitted to the resisting bodies.[2]

In the history of dynamics, impetus theory predominated for a long time. It can be traced back from the seventeenth and sixteenth centuries via the Western European Renaissance and the Latin and Arabic Middle Ages to the close of Antiquity. But before we turn to this theory as a historical phenomenon, we must realise that the idea of exhaustion of force is based on a common, repeatable experience: an experience known to anyone who, after strenuous physical activity, has felt the desire for rest. It is an experience which is even expressed by the modern Law of Conservation of Energy, according to which a dynamic quantity (energy) is exhausted and consumed whenever a change of state occurs. Except for impetus theory, however, there is no physical theory which generally *explains motion* by this experience. The inertial motion of classical mechanics does not require any moving force. The ancient atomists also knew of motions caused by nothing. Other ancient theories of motion, notably those of Plato, Aristotle and the Stoics, *did* postulate moving causes, but they did not assume that forces were being exhausted and consumed as a result of their moving activity.[3]

This assumption is peculiar to impetus theory, which is consequently distinguished from modern dynamics in a further respect. Although impetus theory refers to an experience, it is not an 'empirical' theory in the modern sense of the word. It is rather a speculative model like other ancient theories of motion. An exhaustible moving force entails the speculative notion of some dynamic, causal process: the moving force is taken first to proceed from a moving cause, then to be impressed on a moved body and, finally, to be exhausted at the end of the motion.

Because of this speculative feature of impetus theory, it now appears rather

[2] The view that the motive force imparted to a movable body would not be exhausted if the motion took place in a vacuum and, therefore, would not meet any resistance can be traced back to Arabic authors (Avicenna, Avempace etc.). (See S. Pines, 'La dynamique d'Ibn Bajja', *Mélanges Alexandre Koyré* I, Paris 1964, 460, and 'Etudes sur Awhad al-Zaman Abu'l Barakat al-Baghdadi', *REJ* III, 1938, 57-8, repeated by M. Clagett, *Science of Mechanics in the Middle Ages*, Madison 1959, 513.) This view, according to which external resistances, not motion itself, consume force, seems to be a mere consequence of the idea that a transmission of force demands its exhaustion, so that any (finite) cause which transmits a moving force loses it. In the later Middle Ages and the Renaissance, impetus theory was modified by the assumption that, in addition to external force-consuming resistances, there are also internal ones, for instance gravity. Still the young Galileo (in his *de Motu Antiquiora*) held the view that the weakening of impetus is due to its overcoming external *or* internal resistances and, in this sense, is also due directly to motions which overcome gravity (i.e. upward motions).

[3] The Aristotelian theory of projectile motion (a more detailed account of which is given below) *only apparently* implies the idea that moving forces are exhausted as a result of their activity. According to Aristotle, this motion slows and comes to an end because 'the capacity for motion' (*dunamis tou kinein*) 'becomes smaller' (*elattôn*) (*Phys* 8.10, 267a8). But this 'becoming smaller' does not refer to a decrease of a single force, but to a *relation* between the capacities of *different* agents for actively moving a projectile. For Aristotle, projection presupposes not one, but many agents insofar as its continuation is caused by the parts of the medium successively. As agents of motion, these parts are distinguished from each other by their respective capacity for actively moving (*dunamis tou kinein*) the projectile: the further they are away from the original projector, the smaller their capacity.

difficult to regard the theory as a connecting link between Aristotelian physics and classical mechanics. For the words 'connecting link' imply more than the mere temporal order of these theories. They are intended to point to the continuous evolution of one theory into another. But, if we keep in mind the non-empirical character of impetus theory, it is difficult actually to recognise such a continuity. So the words 'connecting link' have to be interpreted differently. Historians of science do not agree which interpretation is to be preferred. Did impetus theory merely lead to a definite break with and to the destruction of certain principles of Aristotelian physics? Was it based on other Aristotelian principles only? Or did it – in spite of its speculative, non-empirical traits – pave the way for certain concepts and structures in classical mechanics? In the former case, one could argue for a discontinuous transition, in the latter for a continuous one.

Both these views present difficulties. If emphasis is placed on the continuity of the transition, it is puzzling that structures which were suited to modern empirical science could be 'prepared for' by a merely speculative theory. If, on the other hand, one puts rather more weight on the discontinuity among these different views of nature, one easily overlooks the obvious affinities between them, at least as regards some of their dynamic assumptions.

These affinities become indirectly apparent through the fact that it was Galileo whose early writings led historians of science to notice impetus theory as a rudimentary form of classical dynamics and to notice John Philoponus as a predecessor of impetus theory. In 1883, Emil Wohlwill realised, on the basis of the incomplete edition of Galileo's works by Eugenio Alberi (published between 1842 and 1856) and Antonio Favaro's preliminary work for the forthcoming Edizione Nazionale (1890-1909), how fundamental the idea of an impressed force (*vis impressa* or *impetus impressus*) was for the young Galileo. The observation that Galileo's early dialogue *de Motu Gravium*, written about 1590, shows the unmistakable influence of Benedetti's theory of *vis impressa*, drew Wohlwill's attention to the long tradition of the doctrine which Pierre Duhem later referred to as 'impetus theory'. In his important essay on the discovery of the Law of Inertia[4] Wohlwill offered a detailed exposition of the material concerning the Renaissance history of impetus theory which led back to Nicholas of Cusa's dialogue *de Ludo Globi*. He conjectured that there must be an older tradition of impetus theory going back through the Latin and Arabic Middle Ages to the Greeks,[5] for 'even in Cusanus the water did not taste of its source. It remained to find the man who would oppose Aristotle with the words "I claim".'[6] Later Wohlwill confessed that 'after a long and fruitless search' he had finally found this man with the aid of hints given in

[4] E. Wohlwill, 'Über die Entdeckung des Beharrungsgesetzes', *Zeitschrift für Völkerpsychologie und Sprachwissenschaft* 14, 1883, 365-410; 15, 1884, 70-135, 337-87. In 1888, a complementary essay by Wohlwill on the same issue was published: 'Hat Leonardo da Vinci das Beharrungsgesetz gekannt?', *Bibliotheca Mathematica* (Neue Folge II) 19-26. This and other essays by Wohlwill, which are little known today, had a lasting influence on Duhem's investigations into the history of impetus theory. See P. Duhem, 'De l'accélération produite par une force constante', *Congrès d'histoire des sciences*, Genève 1904, 859-915, and his *Études sur Léonard de Vinci*.

[5] E. Wohlwill, op. cit.

[6] E. Wohlwill 1905, 24.

the literature of the sixteenth century. He had followed up a remark by Francesco Buonamico, made in his *de Motu* of 1561, that 'Philoponus and other Latin scholars have, in their doctrines of projectile motion, attacked Aristotle most vehemently, so that one can say: "They have abandoned the flag of their teacher." '[7] Wohlwill's article on Philoponus, in which he drew attention for the first time to Hieronymus Vitelli's edition of Philoponus' commentary on Aristotle's *Physics* as the earliest source of impetus theory, followed in 1905.

That the analysis and reconstruction of Galileo's achievements is what drew attention to impetus theory and its ancient origins, is further revealed by the following circumstance. Vitelli's edition of Philoponus' *Physics* commentary was already published in 1887, but Wohlwill did not encounter Philoponus until 1905. Yet in studying *Scholia in Aristotelem* (edited by C.A. Brandis in 1836), Wohlwill had noticed as early as 1883 that some other Greek commentators – Alexander of Aphrodisias, Themistius and Simplicius – used a Greek term (*dunamis endotheisa*) which apparently corresponds to the notion of impressed force. Wohlwill, however, realised that what lay behind the terminology of these commentators was not impetus theory, but the Aristotelian idea that, in projectile motion, the capacity to move a movable object (*dunamis tou kinein*) is passed on to the medium through which the object moves.[8] It is strange that Wohlwill, in this context, did not mention Philoponus, who also uses the term *dunamis endotheisa*, although in a sense which obviously deviates from Aristotle. This is peculiar because Wohlwill was well acquainted with Alexander von Humboldt's *Kosmos*, which contains many interesting notes on the early history of physics, many of which Wohlwill quoted.[9] In the second volume of the *Kosmos* (1847), Humboldt had drawn attention to Philoponus' *de Opificio Mundi*, book I, chapter 12, where the movements of celestial bodies are explained by a transmission of force (*kinêtikên entheinei dunamin*), a force directly imparted to the celestial bodies. Here Humboldt had praised Philoponus' theory as the only ancient theory which made a contribution to the transformation of physical astronomy into a celestial mechanics.[10] It is strange that Wohlwill, in his essay of 1883, failed to mention this passage from Humboldt's *Kosmos* even though it was of particular interest for the history of impetus theory. In his article of 1905 he *does* mention the passage, but he gives an incorrect translation and does not realise that Philoponus' 'contribution to a celestial mechanics' has something to do with impetus theory.[11] Philoponus' impetus theory seems to have interested him only insofar as it contained literary points directly relevant to Galileo.[12] A single historical observation like Humboldt's, although important in itself, was not enough to reveal impetus

[7] F. Buonamico, *de Motu*, lib V, cap XXXVI, 504; cf A. Koyré, *Galileo Studies*, The Harvester Press 1978, 12 and 44.

[8] E. Wohlwill, 'Über die Entdeckung des Beharrungsgesetzes', loc. cit. 14, 379-80. As for Alexander of Aphrodisias, Themistius and Simplicius, see below, pp 103-4.

[9] ibid. 371.

[10] A. von Humboldt, *Kosmos II*, Stuttgart & Tübingen 1847, 348-9.

[11] E. Wohlwill (1906) 24.

[12] Before 1630, there existed neither a Greek nor a Latin edition of *de Opificio Mundi*; and Galileo's own contributions to celestial mechanics do not actually relate to the explanation of translatory motions of celestial bodies.

theory as a unique explanation of motion. The observation did not suffice to uncover Philoponus as a precursor of impetus theory; that had to wait until a connection was made between impetus theory and sixteenth-century mechanics.

However, the later theses advocated by Wohlwill in his article on Philoponus have proved significant for all subsequent discussion of Philoponus and the history of impetus theory, and they still merit attention. Wohlwill contended that:

(1) Philoponus is, in a certain sense, a predecessor of Galileo. Where he departs from Aristotle, this is due to scientific criticism and 'in the direction in which we see the progress of science'.[13]

(2) Philoponus is the real initiator of impetus theory. Earlier traces of this theory cannot be found except, perhaps, in Hipparchus of Nicaea.[14]

(3) Philoponus arrived at impetus theory on the basis of experience and experiment. Apart from that, his theory expresses a 'natural feel for mechanical facts'.[15]

These three theses contain a direct comment on the problems mentioned above. According to Wohlwill, Philoponus' criticism of Aristotle reveals the continuity of impetus theory with classical mechanics. In his view, Philoponus' criticism shows how the puzzle about the origin and success of impetus theory is to be solved. Philoponus' writings demonstrate that impetus theory, like classical mechanics itself, emerged from experience and experimental method, from natural observation and scientific argument.

Since Wohlwill, our knowledge about Philoponus and the prehistory of classical mechanics has increased. How many of his views are still tenable?

1. Philoponus – 'a predecessor of Galileo'?

Wohlwill saw the modern traits of Philoponus' dynamics primarily in three theoretical achievements: (a) a non-Aristotelian theory of projectile motion, (b) the formulation of a law of falling bodies, and (c) a new interpretation of natural motion. What reasons did Wohlwill have for these judgments? And what should we think of his reasons?

(a) Philoponus' theory of projectile motion

Philoponus' most detailed treatment of the theory of projectile motion is to be found in his commentary on Aristotle's *Physics* 4.8 (639,5-642,26 (Vitelli)). Unfortunately, Philoponus' commentary on book 8 has been lost. There he had apparently discussed the problem of forced motion in greater detail.[16] Concerning the commentary on book 4, S. Sambursky pertinently points out

[13] ibid. 31.
[14] ibid. 26-7.
[15] ibid. 26.
[16] This follows from Philoponus *in Phys* 639,7ff.

that Philoponus 'rejected the main contention of the Peripatetics that in every forced motion there must always be an immediate contact between the mover and the body forced to move in a direction other than that of its natural motion. ... He argued that if string and arrow, or hand and stone, are in direct contact, there is no air behind the missile to be moved, and that the air which is moved along the sides of the missile can contribute nothing, or very little, to its motion. Philoponus concludes that "some incorporeal kinetic power is imparted by the thrower to the object thrown" [cf *in Phys* 642,4-5], and that, "if an arrow or a stone is projected by force in a void, the same thing will happen much more easily, nothing being necessary except the thrower" [*in Phys* 642,6-9].'[17]

Physics 4.8, to which Philoponus refers, concerns the existence of the void. Aristotle had argued (214b31-215a22) that the existence of the void entails an antinomy. According to his argument, motion in the void is not possible because there is no moving cause in the void; but, on the other hand, there is no cause for resting at a particular place in the void either – why should a body in the void be here rather than there, or be at rest rather than in endless motion? Philoponus, however, is not at all interested in the structure of this antinomy. He is only interested in Aristotle's theory of projectile motion, which this antinomy seems implicitly to presuppose. According to Aristotle projectile motion implies a medium for two reasons; first, the medium *causes* the continuation of motion and, secondly, it *terminates* it. As media, water or air can fulfil this twofold function because they are light in one respect and heavy in another. Insofar as they are light, they facilitate motion; insofar as they are heavy, they hinder it (*de Cael* 3.2, 301b16-31).

Philoponus' *endotheisa dunamis* is intended to take over both functions. 'I claim', Philoponus declares in a self-confident, rhetorical attack against Aristotle, 'in the same way as you say that the thrust of the air is responsible for unnatural motion, and that such motion continues until the motive power, which flows from the agent into the air, diminishes – in just the same way it is apparent that, if something is unnaturally moved in a vacuum, then it will also move until the locomotive power (*kinêtikê dunamis*) initially imparted to it (*endotheisa*) diminishes (*exasthenêsêi*)' (*in Phys* 644,17-22).

Wohlwill's interpretation of this passage shows greater caution than is customary. While recent scholars are inclined to say that Philoponus here paves the way for 'the modern vectorial term "momentum"' or 'the scalar term "kinetic energy"',[18] Wohlwill compares Philoponus' diminishing locomotive power with the young Galileo's notion of impressed force as a *vis naturaliter deficiens*.[19] And this comparison is indeed appropriate, because Philoponus refers to a decreasing, *not* a conserved, quantity.

The real novelty of Philoponus' idea of the transmission of force will not be understood if we take it to be more modern than it actually is. Nor will we understand its novelty if we read into it that text against which Philoponus argues – as if it were Aristotle himself who discovered the idea of the

[17] S. Sambursky (1970) 136.
[18] ibid.
[19] Cf Galileo Galilei *Opere* (Ed. Naz.) I. 315ff.

transmission of force. The first of these two misinterpretations by no means excludes the second, and there are modern interpreters who make both mistakes. They simply agree with Philoponus that there is only one respect in which his own theory departs from Aristotle and that for Aristotle too there is an impressed force, although not imparted to the stone thrown but to the air which moves it (*in Phys* 644,19). In a similar way, Samburksy writes that Aristotle 'proposes an hypothesis which contains the first germ of the celebrated idea of "impetus".'[20] So it seems to follow that Aristotle arrived at some notion of 'impulse'.[21] But if this rendering were correct, Aristotle's argument would be contradictory. Aristotle intended to use his theory of projectile motion to maintain a principle which was, in fact, incompatible with the transmission of force, namely the principle that 'anything moved is moved *by something*' (*Phys* 8.10). With regard to projectile motion, he applies this principle by saying that the moved object ceases to be moved *at the same time as* the motion of the moving cause comes to an end (*Phys* 8.10, 266b28-267a1). Therefore it is not possible that a stone or the air will still be in motion (passively) after the motion of that which moves it has come to an end, unless it is moved by another cause. However, it is *not* impossible (267a5-7 is explicit) that the air (actively) moves something after it has itself ceased to be moved. The capacity which Aristotle thus attributes to the air (as a medium able to move something) is obviously comparable to the capacity of an elastic body, e.g. a spring, actively to move another object, for a spring is able to cause motion *after* it has itself been moved, i.e. stretched by another (moved) body. Its passive motion ceases at the same time as the active motion by which it was stretched. But then it will itself be capable of active motion. It seems that, at this point, Aristotle tacitly derives from the possibility of projectile motion and from the principle that anything which is moved presupposes something that moves it, the conclusion that the medium of motion (air or water) does not consist of inelastic parts which partly resist and partly follow the projectile passively. Furthermore, Aristotle suggests that, by every thrust that moves the projectile forward, the parts of the thrusting medium are one after another put into gradually decreasing tension[22] which thrusts the body further. This seems to be the core of Aristotle's theory of projectile motion as outlined in *Physics* 8.10 and *de Caelo* 3.2 (301b16-31).[23] There is no room in this theory for the transmission of force. Motive force is understood by Aristotle to be nothing other than the capacity of one object (A) to move another object (B), with which, during its motion, A is permanently in contact.[24] He postulates that, by touching B, A generates not only the motion of B but also (if B consists of air or water) B's capacity for moving a third

[20] S. Samburksy (1962) 70-1. ,

[21] G.A. Seeck, 'Die Theorie des Wurfs', in G.A. Seeck, ed, *Die Naturphilosophie des Aristoteles*, Darmstadt 1975, 386.

[22] Simplicius (*in Phys* 1349, 11-1350,9) uses the term '*epimenousa kinêsis endotheisa*' in order to denote this tension. Air and water are susceptible to a '*monimôtera kinêsis*' (to a more persistent motion) than earth.

[23] Cf also Aristotle *Phys* 4.8, 215a14 and *Insomn* 2, 459a29-459b1.

[24] Cf Aristotle *Phys* 8.1, 251b1ff and 4, 255a34; *GC* 1.6, 322b21 and 9, 327a1; *GA* 2.1, 734a3; *Metaph* 9.5.

object C (which B, again, both touches and moves). This postulate can be called the Aristotelian idea of contact causality. It must be clearly distinguished from the idea of the transmission of force.

According to modern physics, the capacity of A to change B's state of motion depends on the relation of the masses of A and B. However, the modern concept of mass, as it was formed in the seventeenth century, presupposes the invalidity of Aristotelian physics, especially the distinction between absolutely heavy and absolutely light substances. But it is this very distinction with which Aristotle operates in his theory of projectile motion: the air, which thrusts the stone and overcomes its heaviness, can do this, not despite, but precisely because of its own absolute lightness (*Cael* 3.2, 301b23f). Galileo's juvenile writings show how difficult it still was to challenge the presuppositions of this assumption by argument.[25] Not surprisingly, the criticism of the Aristotelian principle of contact causality put forward by Philoponus (who accepts Aristotle's distinction between absolute gravity and absolute lightness) is not really cogent. In place of an effective refutation, he reads his own concept of impressed force into Aristotle's theory of motion and concludes that the resulting theory is inconsistent and in need of modification.

At this point we are confronted with an odd situation. On the one hand, Philoponus seems to anticipate Galileo's notion of *vis impressa*. On the other hand, his introduction of that notion lacks legitimacy. For this reason, it seems necessary to qualify Wohlwill's claim that Philoponus is a 'predecessor of Galileo'. Wohlwill wanted not merely to emphasise that Philoponus was *using* the Galilean notion of *vis impressa*, but still more to show that Galileo's *arguments* against the Aristotelian theory of projectile motion had been anticipated by Philoponus.[26] But this view seems scarcely tenable.

(b) Philoponus' 'law of falling bodies'

A similar qualification seems to be called for concerning Philoponus' so-called 'law of falling bodies', which can be found in the 'Corollarium de inani' of his commentary on the *Physics* (*in Phys* 682,30-684,4).[27] What Wohlwill rightly considered modern about this passage is its observation regarding the speed of fall (*in Phys* 683,16-25) and the use of this observation to criticise Aristotle. Once again, it is surely no coincidence that Wohlwill's attention was drawn to this passage by reflection on a contemporary of Galileo's, the Florentine Peripatetic Vincenzio Di Grazia. In his *Considerazioni* of 1613 (reprinted in Favaro's edition of Galileo's works),[28] Di Grazia had

[25] Cf Galileo *Opere* I. 285-6. Here, Galileo is directly discussing Aristotle *Cael* 3.2, 301b16-31. In his arguments, he makes use of some principles of hydrostatics, which, in a comparable context, had already been used by Benedetti. The application of these principles to cosmology in fact arose from Copernican assumptions. Cf M. Wolff, 'The cosmological application of impetus mechanics and the physical argument for Copernicanism: Copernicus, Benedetti, Galileo', *The Interrelations between Physics, Cosmology, and Astronomy: their tension and its resolution, 1300-1700*, ed Y. Elkana & A. Funkenstein, forthcoming.

[26] E. Wohlwill (1906) 24 and 27.

[27] For an analysis of this passage see M. Wolff (1971) 23-37.

[28] Vincenzio Di Grazia, *Considerazioni*, republished in Galileo *Opere* IV. 379-439.

discussed Galileo's work on floating bodies and had defended Aristotle against Galileo's rejection of the assumption that the density of the medium is proportional to the time of the fall of a heavy body (*Physics* 4.8,215b3-12). In this context, Di Grazia blamed Galileo for uncritically adopting the false argument of Philoponus.[29]

In his *de Motu* of *c.*1590,[30] Galileo himself mentions Philoponus as one of those who, forced by the 'power of truth', realised the falsity of Aristotle's views regarding the relation of quantities which play a role in the motion of falling bodies. But he notes that Philoponus arrived at his views 'by belief (*fides*) rather than real proof or by refuting Aristotle'. 'And, indeed,' he writes, 'if one were to accept Aristotle's assumption about the ratio of the speeds of the same body moving in different media, one could scarcely hope to be able to refute Aristotle and upset his proof. For Aristotle assumes that the speed in one medium is to the speed in the other, as the rareness of the first medium is to the rareness of the second.'[31] Galileo proudly adds: 'and no one has so far ventured to deny this relation.'[32] Wohlwill, however, became aware that this remark of Galileo's was not entirely true: Philoponus *does* deny the relation when it is expressed in the form 'time required in one medium is to time required in the other medium as density of the first medium is to density of the second'.[33]

This is a peculiar situation: Di Grazia blamed Galileo for having believed Philoponus uncritically; Galileo blamed Philoponus for merely believing the truth without having real insight into it; Philoponus himself seems not merely to believe what is, in Galileo's view, true, but also to have what Galileo would take to be a real understanding of the truth. Could there be any better validation of the thesis that Philoponus *was* a 'predecessor of Galileo'?

On closer examination, the situation proves more complicated. It is true that both Philoponus and Galileo deny with arguments that the density of the medium and the speed of fall are proportional, but they do so for different reasons. Like Benedetti, Galileo calculates the density of a body by means of its specific weight. One can observe that bodies having equal specific weights but unequal absolute weights float equally well or equally badly. From such observations Galileo concludes, like Benedetti, that unequal heavy bodies must fall with the same speed in a vacuum.[34] Furthermore, he concludes that the fall of a body *A* as well as its extrusion or buoyancy in a given medium *B* is a result of the difference between the specific weights of *A* and *B*. This leads

---

[29] ibid. 432.

[30] Galileo *Opere* I. 284.

[31] ibid. 284. Cf the English translation by I.E. Drabkin in *Galileo Galilei, On Motion and on Mechanics*, ed. I.E. Drabkin and S. Drake, Madison 1960, 49.

[32] ibid.

[33] Cf also Drabkin, loc. cit., who is in agreement with Wohlwill.

[34] Galileo is aware of the novelty of this conclusion when, in a footnote on *de Motu*, he says: 'And, I pray, let not Themistius laugh, who says (cf Themistius *in Phys* 132,21-26): "If, for example, the man who drops anchor should be asked why, in the same depth of sea, the ten-pound anchor sinks more swiftly than the three-pound one, will he not answer with a laugh: Surely here is a question important enough to be referred to Apollo – why the ten-pound anchor is heavier than the three-pound one!?" ' (*Opere* I. 265; cf *On Motion* ... 29).

to the disproportionality between the time of fall of the given body *A* and the density of a given medium *B* (the time of fall being proportional not to the density of *B* but to the difference between the specific weights of *A* and *B*).

The calculation of the density of the medium by means of specific weights indicates a real break with Aristotelian principles. It did not leave these principles intact but was rather a consequence of Benedetti's and Galileo's Copernicanism. The acceptance of the Copernican system implies a rejection of the Aristotelian distinction between absolute heavy and absolute light bodies. This distinction was not really compatible with Copernicus' geokineticism. Therefore Copernicus and his adherents assumed that all elemental bodies are (more or less) heavy and that all celestial bodies, like the earth, consist (exclusively) of heavy matter.[35]

This assumption was incompatible with Aristotle's cosmology which acknowledged the existence of absolutely light bodies,[36] and which could not therefore accommodate the concept of specific weight. It is true that, for Aristotle, light elements such as fire and air are less dense than heavy elements such as water and earth, but since they are absolutely light and have no weight at all, one cannot say that their specific weight is smaller. For this reason, Aristotle's notion of density is, in a strict sense, incommensurable with that of the Copernicans. For the same reason the Aristotelian notion of the void is, in a strict sense, incommensurable with that employed by Galileo. For Galileo the void is as it were a body whose specific or absolute weight is zero.

For Philoponus, on the other hand, there seems to be no reason to deny Aristotle's doctrine of the absolutely heavy and light. So it is not surprising that Philoponus does not put forward the idea of specific weight.[37] What is surprising, however, is that Philoponus none the less has a reason for attacking Aristotle's assumption about the ratio of time and density. Philoponus himself points out that it is very difficult to argue against this assumption, since it is (on the basis of Aristotle's doctrine of absolutely light bodies) not possible 'to estimate (*lambanein*) the ratio of the densities of air and water' (*in Phys* 683,1-3). This remark on the impossibility of estimating the ratio of densities is not yet intended to be a criticism of Aristotle, although Philoponus intends to argue against the proportionality of time and density. Indeed, Aristotle's assumption that these quantities are proportional (215b6ff) would be misunderstood if interpreted as an empirically verifiable hypothesis about measurable quantities. Aristotle merely intended to refute the atomists and their fictions regarding the motion of atoms in the void. And this refutation is implicitly based on the impossibility of estimating the ratio of densities. In Aristotle's view, the atomists had illegitimately presupposed a fixed quantitative relation between the velocity (i.e. the time of fall) of an atom and the density of its surroundings, presuming that the velocity of

[35] Cf my article 'The cosmological application of impetus mechanics …'.

[36] Cf e.g. Aristotle *de Cael* 4.4, 311b27.

[37] E.A. Moody (in his 'Galileo and Avempace. The dynamics of the leaning tower experiment', *Journal of the History of Ideas* 12, 1951, 172) insisted that Aristotle, Philoponus and Avempace treated both the falling body and the resisting medium in terms of densities of specific weights. But E. Grant (1965) 81-8 has corrected this error.

motion in the void is finite. Regarding this presumption, one may rightly ask what is meant by 'motion in the void', when a standard or measure of density is completely lacking. If there is no such standard (and without the modern concept of mass it was in fact not possible to define such a standard), then there will not be any fixed ratio between the densities of two different media. One may, then, *arbitrarily* equate this ratio with the ratio of the times required for a body to fall through these different media. But this equation (which entails the proportionality of time and density) leads to a contradiction with atomistic assumptions, because velocity in the void will then be infinite. This seems to be Aristotle's argument.[38]

Philoponus is not turning against Aristotle in order to save the actual existence of the void. He is not defending the atomists. He merely wants to reject the assumption that it is legitimate to equate the ratio of densities (arbitrarily) with the ratio of times. But before we turn to his arguments, we must seek the motive behind his arguments.

The theory of projectile motion already seems to indicate that Philoponus wants to exclude the medium as an efficient cause of motion: motion is to originate exclusively from forces within the moved body. The proportionality of time and density, on the other hand, suggests that the medium plays an active role. For if we assume that the medium is an efficient cause of differences not in motion, but only in resistance, then time and density cannot be simply proportional. It will not be the time of motion, but only the *resistance* that increases proportionally to an increase in the density of the medium. In that case, the velocity will decrease, but not proportionally: only a *part* of the time can be proportional to density.

It is just this assumption that Philoponus tries to justify by the observation 'that two unequal weights dropped from a given height strike the ground at almost the same time' (*in Phys* 683,16-25). From this observation Philoponus does not derive what Galileo assumed to be true, namely that bodies of different absolute weight fall equally fast in the same medium (or in the void). Instead Philoponus, like Aristotle, thinks that velocity increases with heaviness. 'The more heavy bodies are combined, the greater will be the speed at which they move' (*in Phys* 420,13f). And 'the same space will consequently be traversed by the heavier body in a shorter time and by the lighter body in a longer time, even though the space be void' (*in Phys* 679,20-21). Philoponus even denies the additivity of weights, saying that the weight of two bodies combined is greater than the sum of the weights of the individual bodies (*in Phys* 420,8ff). For this reason, his observation explicitly excludes consideration of weights which differ considerably from each other (*in Phys* 683,21f). If the difference is small, the (merely resistant) medium will almost entirely compensate for the difference of velocities resulting from the difference of weights. Philoponus' observation, as he interprets it, confirms precisely this compensation.

In Philoponus' view, observation demonstrates what was to be proved: '*If*

[38] His assumption that time and density are proportional is, therefore, not a (false) 'law of falling bodies' (as has often been claimed), but rather an elenctic argument. A more detailed reconstruction of this Aristotelian argument is given in M. Wolff (1971) 11-22.

the density of the medium acts (only) as a source of resistance (*empodistikon aition*), *then* the natural heaviness (alone) will act as an efficient cause (*poiêtikê aitia*)' (*in Phys* 681,10-12). Since he takes observation to prove the antecedent of this conditional, the efficient cause of fall is nothing other than the heaviness of the falling bodies.

The aim of Philoponus' argument obviously differs from that of Galileo. The conviction that heaviness is an efficient cause, and the medium merely a source of resistance to motion, seems to have at best been presupposed by Galileo. For him, it never seems to have been an object of proof. One can even say that Galileo's buoyancy theory actually denies the assumption that, in upward motion, the medium acts merely as a source of resistance. So it seems to be an amazing coincidence that, although Philoponus and Galileo were arguing with entirely different aims, they made use of a partly identical observation and drew a partly identical conclusion from it.

This coincidence becomes even more striking if we note with Edward Grant 'that the observation is introduced by Philoponus in a completely *ad hoc* manner, for the sole purpose of discrediting the Aristotelian position'.[39] This *ad hoc* manner becomes apparent if we consider a contradiction in Philoponus' considerations. Philoponus would be correct in resting the argument on his observation if he also assumed that the medium offers, if not a stronger, at least an equally strong resistance to the larger weight as it does to the smaller. But in the 'Corollarium de inani' (*in Phys* 679,5-11) he maintains the contrary: 'Clearly, then, it is the natural weights of bodies, one having a greater and another a lesser downward tendency, that cause differences in motion. For that which has a greater downward tendency divides a medium better. Now air is more effectively divided by a heavier body. To what other cause shall we ascribe this fact than that that which has greater weight has, by its own nature, a greater downward tendency, even if the motion is not through a plenum?'[40] Again, in the same passage Philoponus tries to show that heaviness alone constitutes the efficient cause of motion. He therefore argues that a difference of velocities results from the difference of weights because the medium offers less resistance to a heavier weight (velocity and weight are proportional because weight is inversely proportional to resistance). This consideration implicitly leads to a contradiction with his interpretation of the phenomenon observed by him. But he does not seem to realise that he is depriving his argument of all cogency. Thus the non-empirical, speculative character of his reasoning becomes evident. His *ad hoc* reliance on observation has no real experimental character, as Wohlwill thought, but turns out to be in all probability rhetorical.

Not only the Corollary on Void but also the Corollary on Place (*in Phys* 557,8-585,4) exhibits rhetorical traits. In this context too, Philoponus avoids discussion of Aristotle's dialectical arguments against the existence of an actual void. Paradoxically, he uses the idea of a non-actual empty space to argue that motion does not require a medium. Both corollaries argue for this

---

[39] E. Grant (1965) 86.
[40] English translation in M. Cohen & I.E. Drabkin (1958) 217-18.

thesis by using thought experiments which take place in that (hypothetically actualised) empty space. The Corollary on Void introduces experiments with archers and runners who shoot or run through a vacuum. It shows motion to be possible there and ascribes to it velocities that depend on and are proportional to the internal forces in the moving objects.[41] While these experiments use projectile and animal motion (i.e. 'violent' and 'arbitrary' motion), the 'Corollarium de loco' appeals for the same purpose to 'natural' motions, that is, to the motions of elemental bodies towards their natural places.[42] The paradox of a natural motion and a natural place *in the void* prompts Philoponus to the *ad hoc* conception of three-dimensional, 'incorporeal' empty space as a kind of 'order' (*taxis*) in which all species of elemental bodies have a certain (natural) place towards which their inner forces make them strive.[43]

(c) Philoponus' interpretation of natural motion

This takes us to Philoponus' third achievement which, according to Wohlwill, points in 'the direction in which we see the progress of science'. It is true that Philoponus still adheres to the Aristotelian doctrine that 'natural' and 'unnatural' (or 'forced') motions must be distinguished, but he is, in fact, levelling out this distinction. He argues that forced and natural motions alike are caused by internal causes found in the moved objects themselves. Thus he decides that heaviness is a *phusikê dunamis* (*in Phys* 499,12) and a 'non-relational quality', *autê kath hautên poiotês* (*in Phys* 678,23f), which does not depend on a *schesis allou*, a relation to something else (*in Phys* 678,22f; cf *in Phys* 679,27ff). Because of this decision, which coincides with Philoponus' interpretation of heaviness as a *poiêtikê aitia*, an efficient cause of motion, Philoponus denies that natural place is a concurring cause of natural motion. The understanding of heaviness as a *poiêtikê aitia* contradicts traditional Aristotelian assumptions. Simplicius, relying on *Physics* 8.4, 255b30ff, explicitly rejects this definition (Simp. *in Phys* 287,29f and 288,6ff). In his view, heaviness is not an efficient cause of motion, nor a capacity for moving a body, but a mere fitness for (passively) being moved (*epitêdeiotes pro to kineisthai*). According to the Aristotelian principle that everything which is moved is moved by something else, naturally moved bodies are moved by external causes. Their motions are not simply released, i.e. 'accidentally' caused, by something else, but also (substantially) caused by an external mover, i.e. by some final cause.[44] According to Philoponus, on the other

---

[41] Philoponus *in Phys* 692,27-693,6 and 691,9-26.

[42] ibid. 581,19-31.

[43] ibid. 581,19-21.

[44] Apart from his doctrine of the 'unmoved mover' of celestial bodies Aristotle (*Phys* 8.4, 255a1-6; 255b29) explicitly *denies* the perfect spontaneity of natural motion. Light and heavy bodies have a principle of motion (*archê tês kinêseos*) in themselves, not in the sense that they can spontaneously move themselves but only in the sense that they are, in one way or another, capable of being (passively) moved (255b30-31). Thus an external cause of motion, which Aristotle denotes as its natural 'where to' (*poi*), is required (255b15). Aristotle explains the finality of natural motion by conceiving the natural place of an elemental body as being its

hand, there are only internal causes of natural motion.

Some analogies between Philoponus' theory of natural motion and his theory of projectile motion are obvious. Gravity and lightness, for example, or the 'efficient cause' of natural motion, is in Philoponus' theory an exact replica of *dunamis endotheisa*. The 'force' of gravity resembles the force with which an arrow shot by an archer through empty space moves to a target, or the strength with which a runner runs through the void. The greater the moving force, the faster and farther the arrow will fly or the runner run (*in Phys* 691-2). Analogously, the heavier a body, i.e. the greater its downward tendency (*mallon katôphoros*), the faster and more efficiently it overcomes the resistance of the medium and the more it moves downward (*in Phys* 679,5-18). Furthermore, as Simplicius reports, Philoponus subscribed in his *contra Aristotelem* to Themistius' view that bodies which have reached their natural places cease to be heavy or light (Simp. *in Cael* 70,2ff). In a similar way he holds that the motive force of a projectile has been consumed at the end of its journey. The analogy between Philoponus' explanations of natural and forced motion reveals itself most impressively in *de Opificio Mundi* 1.12 (already mentioned by Humboldt) where the causes of celestial motion and even heaviness and lightness are interpreted as forces impressed on bodies by the Demiurge. *De Opificio Mundi* was written between 557 and 560. But rudiments of Philoponus' quasi-mechanistic interpretation of celestial motions can already be found in his commentary on the *Physics* of 517, in which the rotation of the sublunary sphere of fire is said to have its origin in the force imparted to it.[45] In *de Aeternitate Mundi contra Proclum* of 529 (13.5, 492,20-493,5 (Rabe)), we even find an explicit comparison of celestial and sublunary rotations with rotations created by 'machines' (*mêchanêmata*), e.g. centrifuges. In Philoponus' view there is, corresponding to the hierarchy of supra- and sub-lunary spheres, a transmission of force imparted by the superior to the lower spheres.[46] He calls their movements 'supernatural': they are neither natural, nor contrary to nature, and they have their source in a moving cause which is superior to the nature of the moved sphere.

The principle of the transmission of force obviously undermines the Aristotelian dichotomy between natural and unnatural motion. Philoponus no longer seems to accept it easily. This is perhaps already shown in a passage of his *Physics* commentary of 517. There he addresses the following question to the advocates of the Aristotelian theory of projectile motion: 'When one

---

energeia (255a28-b24). In *de Cael* 4.3 3,310b24-311a9, Aristotle understands the natural place of light and heavy bodies even more clearly as their *entelecheia*. The thesis that the *topos*, as natural place, has the capacity or the force to move bodies, was brought up for discussion in *Phys* 4.1, 208b10ff. It was explicitly rejected by Philoponus (*in Phys* 581,19-31). According to him, the natural tendency for motion (*rhopê*) does not fall under the category of relation (*pros ti*) (*in Phys* 680,22; cf. 678,22-3 and 679,27ff). For an analysis of Philoponus' theory of natural places see M. Wolff (1971) 81-5.

[45] Philoponus *in Phys* 384,18-385,4.

[46] In his *Divina Comedia* (Paradiso, Canto Secondo II.127-32), Dante Alighieri modified the mechanistic comparison used by Philoponus by comparing the force transmitted from the heavenly spheres to the sublunary regions with the force transmitted from the gearing of a mill to the blacksmith's hammer.

projects a stone forcibly by pushing the air behind it, does one compel the stone to move in a way contrary to its nature (*para phusin*)? Or does the thrower impart a motive force to the stone too?' (*in Phys* 641,13-16).[47] Philoponus apparently demands that, before deciding the issue of how it is that the forced motion of projectiles can take place, supporters of the Aristotelian theory should first say whether forced motion is natural or unnatural. He does not seem to accept the traditional classification of forced motions as unnatural.

Even if Philoponus is not perfectly conscious of invalidating the Aristotelian meaning of 'natural' and 'unnatural' motions, his interpretation of these motions does in fact invalidate it and prepares for an integrated dynamic theory of motion. In this sense, Wohlwill is perfectly correct when he says that Philoponus' interpretation actually points 'in the direction in which we see the progress of science'. Using words similar to Wohlwill's, Shapere says with regard to impetus theory in general: 'Impetus is not yet inertia, nor is it momentum, but it is, in the ways described above, a visible move away from fundamental Aristotelian conceptions (e.g. it allows a sustaining force in projectile motion to be internal and incorporeal) and in the definite direction of classical mechanics.'[48]

2. Philoponus and Hipparchus

If we now try to discover the reasons and causes of this 'move away from ... in the direction of ...' and consider all the apparently modern characteristics that Wohlwill rightly discerned in Philoponus' dynamics, we must first examine the two-fold answer offered by Wohlwill: (a) Philoponus arrived at impetus theory on the basis of experience, experiment, and natural sense experience (see above, p 88, thesis (3)); and (b) he has taken over certain elements of his theory from Hipparchus of Nicaea (see above, p 88, thesis (2)).

[47] M. Cohen & I.E. Drabkin (1958) 222 translate this passage differently: 'When one projects a stone by force, is it by pushing the air behind the stone that one compels the latter to move in a direction contrary to its natural direction? Or does the thrower impart a motive force to the stone too?' In a letter to me, Richard Sorabji has pointed out that 'it makes a world of difference which interpretation we follow'. Cohen and Drabkin imply that the motion is on *either* hypothesis *para phusin*, whereas my translation allows that on the second hypothesis (the one Philoponus goes on to accept) the motion is *not para phusin*. Only this second interpretation makes Philoponus consciously violate the distinction between motion in accordance with, and contrary to, nature. I think that the syntactical structure of the first question (641,13-5), and the content of the second one goes against Cohen's and Drabkin's translation: The phrase 'by pushing the air behind the stone' (*tôi ôthein ton katopin tou lithou aera*) is part of the subordinate clause 'when one projects a stone by force' (*hotan tis rhipthêi biai lithon*); in the second question, Philoponus raises the problem of whether a motive force is imparted *to the stone too* (*kai tôi lithôi*), hence he is not – as Cohen and Drabkin assume – calling into question that a motive force is imparted to the air.

[48] D. Shapere, 'Meaning and scientific change', *Mind and Cosmos*, ed R.G. Colodny, Pittsburgh 1966, 79.

(a) Experience, experiment and natural sense experience

The first answer is not very convincing. Philoponus refers to a single empirical observation, but he does it, as we have seen, in a purely *ad hoc* manner which lacks real cogency.[49] His reference to observation or, as he calls it, to (empirical) 'evidence' or *enargeia*,[50] is in this context a rhetorical remark.

His mention of 'natural sense experience' is equally unconvincing. Why should the principle of the transmission of force come closer to natural sense experience than Aristotle's principle of contact causality and his view that anything which is moved is moved by something else? With respect to Galileo, Wohlwill himself criticised certain attempts to explain scientific discoveries in psychological terms rather than examine them 'historically': 'A purely psychological explanation is, in such a case, all the more in danger of being misguided, as anything that presents itself to the scholar as a simple, evident or even necessary connection of ideas and facts, has come to be what it is only under the influence of the new knowledge and, for the discoverer, existed at least in a different combination.'[51] It was just this dissatisfaction with nineteenth-century attempts to explain the scientific achievements of Galileo by certain psychological considerations, that led Wohlwill to recognise impetus theory. But what makes it now legitimate to apply the same psychological explanations to John Philoponus?[52]

The difficulty of explaining the origin of impetus theory might be pushed back a step, although not overcome, by an attempt to extend the chain of its precursors further into the past.[53] The search for ancient precursors particularly suggests itself if one tries to find the origin in the field of scientific observation. For it seems more appropriate to associate the prehistory of classical mechanics with the tradition of ancient science than with that of Aristotelian and Neoplatonic speculation. Further, Galileo preferred to regard himself as continuing the tradition of Archimedean method and ancient observational science. It is, therefore, not surprising to find, in the context of a passage in *de Motu*, where he applied impetus theory to the problem of retardation and acceleration, that Galileo claimed this application 'had also been the view of the very able philosopher Hipparchus, who is cited by the learned Ptolemy'.[54] According to Galileo, Hipparchus, the great ancient astronomer, had already taught that the gradually decreasing impressed force was the cause of retarding forced upward motion and of accelerating natural downward motion. Galileo did not forget to add that he himself

---

[49] See above, pp 94-5.   [50] Philoponus *in Phys*, 683,16-17.

[51] E. Wohlwill, 'Über die Entdeckung des Beharrungsgesetzes', loc. cit. 14, 366-7. The idea that 'immediate sensation' and 'instinct' paved the way for the law of inertia has been criticised also by Emile Meyerson. See his *Identity and Reality*, New York 1930, 137-40.

[52] Nowadays the psychological manner of explaining progress in science is widely held to be old-fashioned. Perhaps it would be more fashionable to speak of an 'immanent logic of facts' (*immanente Sachlogik*), instead of 'natural sense experience of facts'. Cf H.G. Gadamer's review (*Ph Rundschau* 27, 1980, 152) of M. Wolff (1978).

[53] Indeed, this solution has a questionable advantage: the assumption that the precursors' chain stretches indefinitely into the past can never be strictly falsified.

[54] Galileo *Opere* I. 319-20; *On Motion* ... 89-90.

rediscovered his theory independently of Hipparchus and that it was not until two months after this discovery that he came to know the views of that greatest of ancient astronomers, who 'is, in fact, greatly esteemed and is extolled with the highest praises by Ptolemy throughout the whole text of his *Almagest*'.[55]

(b) Hipparchus of Nicaea (c. 190-c. 125)

If Galileo's interpretation of Hipparchus' theory of acceleration were correct, then the central tenets of impetus theory would have been advocated as early as 700 years before Philoponus' time. Careful analysis of the source from which Galileo received his views on Hipparchus seems to indicate, however, that Galileo merely read his own theory into the text of his source.[56] This is not surprising, for, first, it was a Renaissance convention to pretend to continue the glories of ancient celebrities and, secondly, Galileo's source, Simplicius' commentary on Aristotle's *de Caelo* 1.8 (*in Cael* 264,25-265,6), suited this purpose very well. Time and again, historians of science read Simplicius with Galileo's eyes and consequently regard Hipparchus as a precursor of impetus theory. After Wohlwill, in 1884 (before he became aware of Philoponus), had drawn attention to Galileo's source material and thus to Hipparchus,[57] this material was repeatedly used to show that impetus theory was originally and substantially an achievement of that great ancient astronomer. In his article on Philoponus, Wohlwill himself held to this view, although he now emphasised the originality of Philoponus' anti-Aristotelian arguments. Against this concession Arthur Erich Haas directed a polemical essay which denied Philoponus' originality and argued that Hipparchus had completely anticipated him.[58] In his *Études sur Léonard de Vinci*, Pierre Duhem replied to Haas' essay with a cautious counter-criticism.[59] Only E.J. Dijksterhuis' *Val en Worp* of 1924 advanced good arguments against the view that Hipparchus was a precursor of impetus theory.[60]

In his commentary on *de Caelo*, Simplicius discusses various attempts to explain the acceleration of falling bodies. He writes here that 'Hipparchus in his work entitled *On Bodies Carried Down by their Weight (barutês)* declares that

[55] ibid.

[56] In a footnote on the passage in *de Motu*, Drabkin notes: 'As can be seen from that paragraph, it is questionable that Galileo knew the material at first hand, for he seems to assume that Alexander criticised Hipparchus for not treating the case of free fall from rest.' (*On Motion* 90 n 7).

[57] E. Wohlwill, 'Über die Entdeckung des Beharrungsgesetzes', loc. cit. 14, 383-4. Heiberg's edition of Simplicius' commentary on *de Caelo* (Berlin 1894) was not yet accessible to Wohlwill at that time: Wohlwill's attention was drawn to Hipparchus not by the Greek text of Simplicius, but by Galileo, whose source was the Latin translation of Simplicius' commentary by H. Scotus.

[58] A.E. Haas (1905) 337-42.

[59] P. Duhem, *Etudes sur Léonard de Vinci* 3, 61-2.

[60] E.J. Dijksterhuis, *Val en Worp* Groningen 1924. In spite of Dijksterhuis' arguments, many authors have upheld the opinion that Hipparchus implicitly used the concept of *vis impressa*; most notably Willy Hartner and Matthias Schramm, see their article 'La notion de l' "inertie" chez Hipparque et Galilee', *Collection des traveaux de l'académie internationale d'histoire des sciences* 11, Florence 1958, 126-32.

in the case of earth thrown upward it is the projecting force (*anarripsasa ischus*) that is the cause of the upward motion, so long as the projecting force overpowers the downward tendency of the projectile (*tou rhiptoumenou dunamis*), and that to the extent that this projecting force predominates, the object moves more swiftly upwards; then, as this force is diminished (1) the upward motion proceeds but no longer at the same rate, (2) the body moves downward under the influence of its own internal impulse (*oikeia rhopê*), even though the original projecting force (*anapempsasa dunamis*) lingers in some measure, and (3) as this force continues to diminish the object moves downward more swiftly, and most swiftly when this force is entirely lost.'[61] If one compares this text with the remarks Aristotle makes on fall and projection in the *Physics* and *de Caelo*, one will notice a striking difference. Hipparchus interprets the change of velocity of these motions as the effect of a composition of 'forces' or, to be more precise, of a certain 'capacity' of the projected body (*tou rhiptoumenou dunamis*) and the projecting force (*anarripsasa ischus*). Indeed a similar theory of composition of forces seems to reappear in medieval impetus theory after Buridan.[62] However, Hipparchus is not led to this view by regarding the projecting power as an impressed force but, as Simplicius reports, because of his peculiar notion of heaviness (*barutês*), which seems to have been the main subject of Hipparchus' work, as its title certainly indicates: 'According to Hipparchus, bodies are heavier the further they are removed from their natural places.'[63] This view is the exact opposite of Aristotle's, as Simplicius explicitly points out. According to Aristotle, bodies become heavier the nearer they come to their natural places.[64] Simplicius reports, on the other hand, that Hipparchus agrees with Aristotle that the change of heaviness has an influence on the velocity of motion.[65] This makes it necessary for him to interpret the changing velocity of fall and projectile motion differently from Aristotle. Like Aristotle, Hipparchus seems to assume that, when an earthy body is thrown upward, the external moving force becomes smaller, so that a deceleration takes place. But unlike Aristotle, he holds that the heaviness of the projected body increases with the increase of distance from its natural place, so that this increase also contributes to the deceleration. Hence, when the body has reached its highest point, there will still remain a little of the projecting force although it no longer predominates over the downward tendency.

Hipparchus conceives the composition of heaviness (or torpor) and projecting power in a new way. But his explanation of changing velocities in projectile motion and fall does not, in principle, depart from the Peripatetic doctrine. He simply decides in favour of one of the alternatives considered in the Peripatetic *Quaestiones mechanicae*. There we read in Quaestio 32, for

[61] Simplicius *in Cael* 264,25-265,3 (the English translation from M. Cohen and I.E. Drabkin (1958) 209).

[62] See A. Funkenstein, 'Some remarks on the concept of impetus and the determination of simple motion', *Viator* 2, 1971, 329-48.

[63] Simplicius *in Cael* 265,10-1.

[64] ibid. 265,9-10. Cf Aristotle *de Cael* 1.8, 277a12-33.

[65] Simplicius *in Cael* 266,27-9.

example: 'Why is it that an object which is thrown eventually comes to a standstill? Does it stop when the force which started it (*apheisa ischus*) fails, or because the object is drawn in a contrary direction, or is it due to its downward tendency, which is stronger than the force which threw it (*rhipsasa ischus*)?'[66] Further, there is no reason to assume that Hipparchus rejects the Aristotelian doctrine that the medium has a projecting power transmitted to it. On the contrary, Simplicius reports that Alexander of Aphrodisias accepted Hipparchus' explanation of acceleration, although he did not accept his notion of heaviness.[67] One may infer from this that Alexander, whom we know to have advocated Aristotle's theory of projectile motion,[68] did not object to Hipparchus' assumption that, even after the projectile has reached the highest point of its motion, some of the force counteracting its downward motion remains and contributes to a retardation of the speed of fall.

Hipparchus takes more into account than the natural downward motion which is immediately preceded by forced motion. In his view, any acceleration in fall of bodies presupposes a cause which acts contrary to downward motion. 'For', he says, 'the force which held them back (*tou kataschontos dunamis*) remains with them up to a certain point, and this is the restraining factor (*enantioumenê aitia*) which accounts for a slower movement at the start of the fall.'[69] Again, Simplicius reports that Alexander agreed with Hipparchus' view, albeit with the reservation that his agreement did not extend to those cases in which the falling body had come into being through a light body being transformed into a heavy one. For in such cases acceleration is caused, in his view, by the transformation itself, or by the fact that during the (gradual) transformation the body is gradually becoming heavier.[70]

Alexander's criticism gives a reason for interpreting Hipparchus' theory as a departure from the Peripatetic doctrine only in the points explicitly mentioned, and Simplicius' text does not offer an obstacle to such an interpretation. On the contrary, Dijksterhuis has pointed out that, for the Galilean interpretation, the text presents difficulties. The aorist-forms *anarripsasa ischus, anapempsasa dunamis* and *tou kataschontos dunamis* indicate that Hipparchus regarded the projecting and the restraining power as *external* forces. They are forces which *have* thrown or restrained the object.[71] Internal moving forces are not mentioned in this context. So it seems plausible that Hipparchus' theory is in full accord with Aristotle's principle of contact causality. It is certainly no coincidence that interpreters who were not acquainted with Galileo's reading did not attribute to Hipparchus a principle of transmission of force. In this respect, T. Henri Martin's detailed and subtle interpretation of Simplicius' text, given in his *Études sur le Timée de Platon* of

---

[66] Aristotle *Mech* 32, 858a13-6; cf ibid. 33, 34 and 35.

[67] Simplicius *in Cael* 265,6-8. Cf Duhem, *Études sur Léonard de Vinci* 3,62-3.

[68] Cf Simplicius *in Phys* 1346,37-1347,16.

[69] Simplicius *in Cael* 265,4-6.

[70] ibid. 265,6-29.

[71] E.J. Dijksterhuis, *Val en Worp*, 34-5. In a comparable context (Aristotle *Mech* 32, 858a14-5), similar aorist forms can be found. F. Krafft (see his *Dynamische und statische Betrachtungsweise in der antiken Mechanik*, Wiesbaden 1970, 78) takes these aorist forms into consideration, but his conclusions differ from those of Dijksterhuis.

1841,[72] is instructive. And it is noteworthy that a contemporary of Galileo's, Francesco Buonamico, still seems to have understood Hipparchus in a Peripatetic sense: 'In fact Hipparchus (according to Simplicius in a small work devoted to this problem) thought that natural motion is faster at the end, because at the beginning of its motion the body is hindered by an external force: as a result of this it cannot exercise its native force and consequently it moves idly; later, as this extrinsic and external force fades away, the natural force builds up again and, freed, as it were, from obstacles, acts more effectively.'[73]

Indeed, Simplicius' words suggest that the 'capacity of the cause which has restrained [the falling body]' (*tou kataschontos dunamis*) is the capacity of an external cause. This capacity will not stop acting on the falling body once its motion is released, rather a certain (constantly diminishing) amount of this capacity will 'remain' (*paramenein*), i.e. continue to act on the falling body, up to a certain point on the trajectory. This continuation need not necessarily be interpreted as a result of a transmission of force. Hipparchus could simply have meant that the cause which prevents the heavy body from falling is also the cause of its position in a medium and that this medium impedes the natural downward motion of the heavy body with a constantly diminishing *dunamis*. Therefore the tendency of the heavy body to approach its natural place is not fully developed at the beginning (*kat' archas*, 265,6) of its downward motion.[74]

Now, concerning the ancient precursors of impetus theory, I should like to suggest a criterion for deciding whether or not an ancient author may be said to have been such a precursor. The author must explicitly *argue* for the principles of impetus theory. We must presuppose that ancient authors addressed themselves to people who, rather than being acquainted with impetus theory, were familiar with other theories of motion. Therefore impetus theory had to be introduced by argument. Except for Philoponus, there seems to be no ancient author who argues for it. In this sense, we cannot follow those[75] who take Alexander of Aphrodisias, Themistius or

---

[72] T.H. Martin, *Études sur le Timée de Platon* 2, Paris 1841, 272-80.

[73] F. Buonamico, *de Motu* lib IV, cap XXXVII, 410-1; cf A. Koyré, *Galileo Studies*, 16 and 46.

[74] It cannot be ruled out that Hipparchus' notion of heaviness (the *dunamis* of a heavy body for downward motion) developed out of hydrostatic considerations. In any case, the idea that the heaviness of heavy bodies at their natural place is zero is reminiscent of the 'Archimedean principle', according to which that *dunamis* is equal to the difference between the respective weights (or downward tendencies) of the displacing and displaced body. According to this principle, (1) the (displaced) medium must be regarded as an efficient cause of the upward motion or rest of the displacing body, and (2) heaviness must be distinguished from the 'true weight', i.e. the real downward tendency. It cannot be ruled out that, in Hipparchus' theory too, heaviness is distinguished from *rhopê* and, being not the 'true' but only the 'apparent weight', is not simply the downward tendency, but also a capacity to resist motion, a certain kind of inertia or torpor. Stoic influences on Hipparchus have also been assumed. The Stoics were interested in a more consistent application of the Aristotelian principle of contact causality. Cf S. Sambursky (1962) 38 and 76; G.E.L. Owen, *Scientific Change*, ed A.C. Crombie, New York 1963, 100-2.

[75] S. Sambursky (1962) 71-3. Already Girolamo Cardano regarded Alexander of Aphrodisias and Simplicius as his predecessors as they, in his view, subscribed to the idea of imparted force. Cf E. Wohlwill, 'Über die Entdeckung des Beharrungsgesetzes', loc. cit. 14, 387.

Simplicius to be predecessors of impetus theory, even though they make use of the term *endotheisa dunamis* or related expressions in their commentaries on Aristotle. It is true that Philoponus may have taken up this terminology. But the other commentators employed it exclusively for characterising that *dunamis* which, according to Aristotle's doctrine, is passed on to the medium through which the projectile moves. Like Aristotle, they took this *dunamis* to be the capacity of something actively to move something else.[76] How far removed they were from impetus theory has already been shown by the fact that they did not consider the *possibility* of the transmission of force. Simplicius was not an exception, even though, in a passage from his commentary on Aristotle's *Physics* (*in Phys* 1349,16), he raised the question: 'But if we say that the man throwing the missile transfers to the air a steady motion, why don't we say that this motion is given to the missile without having recourse to the air and therefore without our being forced to assume that it is not only moved but also moving?' The transmission of force appears neither in the question nor in its exposition. Although, or just because Simplicius explicitly questions why the medium as a moving cause is not dispensable, this passage reveals most clearly how far that idea was even from the mind of a contemporary of Philoponus.[77] If we did not know anything about Philoponus' impetus theory, we could not infer the existence of that theory from Simplicius' arguments. Apart from Philoponus, there seems to be no ancient author who argued for the principles of impetus theory, nor one who argued *against* them. Therefore we cannot even indirectly infer the existence of ancient precursors of Philoponus. Hence there is no reason for denying the central role in the history of dynamics which can be attributed to Philoponus in the light of the ancient sources available to us. Even in Philoponus, there is no hint that he was aware of any precursors. Instead, there is the passage quoted by Wohlwill in which Philoponus implies impetus theory to be his own original view when he attacks Aristotle with a self-confident undertone in his words: 'But I claim ...'[78]

---

[76] Wohlwill clearly saw that, like Alexander of Aphrodisias (whose opinion regarding this matter we know through the commentaries of Themistius and Simplicius), Themistius upheld the Aristotelian theory of projectile motion. ('Über die Entdeckung des Beharrungsgesetzes', loc. cit. 14, 380). However, Wohlwill took Themistius' comparison of the motive force of air with the capacity of heating, which a body acquires when heated itself, to be a departure from Aristotle in the direction of impetus theory. In fact, Themistius just emphasises that a body, being heated, not only becomes warm but also acquires the capacity (which, then, is its *oikeia dunamis*) to warm up other things (Themistius *in Phys* 234,29-235,12). This capacity is an exact analogue of Aristotle's 'capacity of something to move something else'. It must be noted that Aristotle used just the same comparison (*Insomn* 2, 459a29ff). And it is remarkable that Aristotle used there the phenomenon of heat transfer *explicitly* as an analogue of *motion by contact*, not of motion caused by the transmission of force.

[77] Another interpretation of this passage is offered by S. Samburksy (1962) 72 and H. Carteron, 'Does Aristotle have a mechanics?', *Articles on Aristotle, 1. Science*, ed J. Barnes *et al.*, London 1975, 170 n 39. Carteron further refers to Simplicius *in Phys* 440,8, a passage which is, however, even less able to show that Simplicius discussed the idea of transmission of force.

[78] See above, p 86.

3. Philoponus and the origins of impetus theory

The real originality of Philoponus' dynamics does not, however, become evident unless we take into consideration the historical context of and the motives for Philoponus' criticism of Aristotle's theory of motion. Above all we should not fail to notice the Neoplatonic background of his criticism, and in particular observe that Philoponus' concept of transmission and exhaustion of force seems to be a transformed Neoplatonic idea.

(a) Simplicius' controversy with Philoponus

In order to throw light upon this Neoplatonic context, we must look more closely at the relationship between Philoponus and the Neoplatonic philosopher Simplicius. Simplicius' commentaries on the *Physics* and on *de Caelo* were written after 532 (after his return from the court of King Chosroes), that is some years later than Philoponus' commentary on the *Physics* of 517 and the *de Aeternitate Mundi contra Proclum* of 529.[79] We have pointed out how apparently unfamiliar Simplicius was with impetus theory. Now, if we take into account the date of his commentaries and if we consider how extensively Simplicius discusses Philoponus' views there, it appears strange that Simplicius does not actually pay attention to Philoponus' impetus theory. It may be no exaggeration to say that Simplicius ignores Philoponus' views concerning dynamics. When he was writing his commentary on the *Physics*, some of Philoponus' writings were obviously known to him, including the *de Aeternitate Mundi contra Proclum*, which Simplicius quotes. In this work, Philoponus had explicitly advanced his new idea of impressed force. The fact that Simplicius passes over this in silence is scarcely less strange than the unusually edgy and piqued tone with which he polemicises against Philoponus. Even though this polemic can partly be explained by reference to certain external incidents (Simplicius might have regarded Philoponus' *contra Proclum* of 529 as a political defamation of the Academy),[80] this does not explain why Simplicius leaves Philoponus' impetus theory wholly undiscussed. Could he, by accident, not have noticed it? Or does Simplicius think of these ideas as aberrations beyond rational discussion which are to be dismissed as manifestations of that abnormal, uneducated, ignorant and quarrelsome mind which he likes to attribute to Philoponus?

Simplicius avoids confrontation with Philoponus' dynamic principles even when attacking arguments which directly presuppose those principles. The principles play a role in at least one of Philoponus' proofs of the non-eternity of the world. This proof is to be found in his *contra Proclum* as well as in his *contra Aristotelem*, and Philoponus devoted a special, albeit shorter treatise[81] to

---

[79] Cf A. Cameron, 'The last days of the Academy of Athens', *Proceedings of the Cambridge Philological Society* 195, 1969, 23f.

[80] Cf Simplicius *in Phys* 1117,15 and H.-D. Saffrey, 'Le chrétien Jean Philopon et la survivance de l'école d'Alexandrie', *REG* 67, 1954, 396-410.

[81] Cf H.A. Davidson, 'John Philoponus as a source of medieval Islamic and Jewish proofs of Creation', *JAOS* 89, 1969, 358-60.

that proof. Simplicius deals with it in his *Physics* commentary. Philoponus
had put forward the argument that the heaven, being a corporeal entity, can
only contain a moving force of finite quantity and as a result can only be
moved for a finite time.[82] As far as the formal structure of this argument is
concerned, Philoponus was able to refer to Aristotle who, in *Physics* 8.10, had
established the principle that a finite body cannot have infinite force
(*dunamis*).[83] However, for Aristotle this principle entailed that an infinite
motion could not be caused by a corporeal force, and, therefore, the unmoved
mover of the heavens must be incorporeal in nature. For him, this is, in fact,
the only possible conclusion, since in his view anything which is moved is
moved by something else, and this moving cause cannot be a force located
and actively at work in the moved body itself. In contrast to Aristotle,
Philoponus implicitly presupposes that the active force by which the celestial
bodies are moved is located inside them. Hence it follows that if this force
cannot be of an infinite quantity (as Aristotle had implicitly postulated) it
must have been transmitted to the celestial bodies at a given moment in the
past. Moreover, it will be exhausted at a given moment in the future.

Simplicius does not attempt to criticise the dynamic presuppositions of
Philoponus. He is merely indignant that Philoponus fails to consider the aim
of Aristotle's argument and especially that he does not understand the
difference between the actually infinite power of the unmoved mover to move
and the potentially infinite power of the celestial bodies to be moved.[84]
According to the Peripatetic view, the capacity for *being moved* for an infinite
time is, unlike the infinite capacity for *actively moving, not actually* (but only
potentially) infinite. The objection advanced by Simplicius implicitly draws
attention to the fact that the presuppositions of impetus theory obstruct an
adequate understanding of Aristotle. Philoponus' argument against the
eternity of the world is based on assumptions which have no foundation in
Aristotle's philosophy. In this respect Simplicius' criticism is completely
legitimate. It is legitimate insofar as Philoponus distorts the meaning of an
important Aristotelian principle, blatantly disregards certain Aristotelian
distinctions, and consequently arrives at conclusions which contradict
Aristotle's doctrine of the eternity of the world. Philoponus' criticism of
Aristotle is not based on Aristotle's own presuppositions.[85] His manner of
arguing even makes it understandable why Simplicius does not have a high
opinion of him as an exponent of Aristotle, but rather takes his interpretations
to be an expression of 'Egyptian mythology'. And if what Simplicius tells us is
true, namely that Philoponus' criticism of Aristotle was exclusively – and
with great success – addressed to the philosophically uneducated layman,[86]

---

[82] Simplicius *in Phys* 1326,38ff. Here the idea of an exhaustion of force is applied to the theory
of celestial motion.

[83] Aristotle *Phys* 8.10, 266a24ff.   [84] Simplicius *in Phys* 1327,29ff.

[85] G. Verbeke is not right in saying that Philoponus' criticism is always based on Aristotle's
own presuppositions. See G. Verbeke (1982) 52.

[86] See the quotations from Simplicius in W. Wieland (1960) 300f; cf G. Verbeke (1982) 50.
The influence of Philoponus, who never had a chair of philosophy in Alexandria (see L.W.
Westerink (1962) xiii), was unimportant (see A. Cameron op.cit. p 105). It is at least noteworthy

then we can understand the indignation with which Simplicius reacts to the success of Philoponus' rhetoric.

On the other hand, Simplicius' criticism is rather naive, insofar as the arguments advanced by Philoponus are due neither to mere ignorance and lack of philosophical education, nor, as Simplicius suggests, to simple stupidity or foolishness. Philoponus is fully aware that his own presuppositions diverge from Aristotle's. At least his commentary on the *Physics* shows that in his opinion these presuppositions can be defended with arguments against Aristotle, whereas Simplicius does not seem even to notice them, much less attack them.

(b) Philoponus and Christianity

The fact that Philoponus' dynamics have a function within the context of his proofs of the non-eternity of the world, i.e. of its *creation* in time, could lead one to assume that Philoponus' impetus theory has its origins in Christian theology. This functional explanation of the origins of impetus theory is widely accepted. It has recently been modified by Fritz Krafft. Although Krafft, like Haas, traces Philoponus' idea of impressed force back to Hipparchus' theory of projection and fall, he wants to explain Philoponus' *application* of this theory to the interpretation of moving forces *in general* 'solely'[87] by reference to the fact that Philoponus was a Christian and that 'according to Christian belief, the universe was created by God'. In Krafft's view, it was this belief that prompted Philoponus to interpret all moving forces, including the 'natural tendencies' of heavy, light and celestial bodies to move in a certain 'natural' way, as forces imparted to these bodies by the creator of the universe.

Now, as Philoponus' arguments against the eternity of the world show, it cannot be denied that the transmission of force serves a particular function with respect to his conviction that the world was created in time. But there are two reasons for denying that Philoponus' dynamics has its origin ultimately in the Christian dogma of creation.

First, Philoponus supplies several proofs of the world's non-eternity, which are completely independent of the principle of transmission and exhaustion of force.[88] These proofs are even better and more acute than the one with which we have dealt, i.e. the proof from the finitude of the force imparted to heavenly bodies. Unlike this proof, they are – as *ad hominem* arguments against Aristotle – partly cogent, as Richard Sorabji has shown.[89] In this respect, Philoponus' principle of transmission and exhaustion of force is redundant.

Secondly, this principle of Philoponus is not even necessary in order to regard God as creator, or as the 'efficient cause' of the universe (and of the

---

that, of the Christian successors of Ammonius Hermeiou and Olympiodorus, neither Elias nor David nor Stephanus of Alexandria took over Philoponus' criticism of the doctrine of the world's eternity (see Westerink op. cit. xxiii-iv).

[87] F. Krafft (1982) 60.

[88] Cf W. Wieland (1960) and R. Sorabji in Chapter 9 below.

[89] R. Sorabji, Chapter 9 below.

motions in it). From Simplicius[90] we know that Ammonius, the teacher of both Philoponus and Simplicius, wrote a treatise in which he advanced many arguments to show that Aristotle had already considered God to be the 'efficient cause' (*poiêtikon aition*) of the universe, and not just its final cause. Simplicius himself finds this interpretation quite acceptable. Obviously the idea of God as creator or 'efficient cause' did not include the idea of God as a mover transmitting force. Whether or not Philoponus was a Christian is therefore probably irrelevant to his having developed impetus theory. (Perhaps it is not even relevant to his arguments for the world's creation in time.)

Philoponus' arguments often reflect Christian convictions, but one must not forget that Philoponus was not simply a Christian but a Christian with heterodox (monophysitic) tendencies. He was involved in controversies, not just with pagans but also with Christian contemporaries such as Cosmas Indicopleustes, who accused him of latent paganism.[91] The usual, but in fact somewhat unclear and unreflective distinction between 'Christian' and 'Greek' thought seems to be inadequate to characterise all these controversies. This applies also to Simplicius' dispute with Philoponus. Wolfgang Wieland has pointed out[92] that, in the irreconcilable antagonism between Simplicius and Philoponus, we see not a mere 'transition from pagan to Christian belief', but a much more significant 'metamorphosis in the way of thinking'. In this metamorphosis 'the fact that Philoponus argues as Christian is, in substance, of little importance'.[93] We have to make clear what distinguishes Philoponus' new way of thinking. Wieland has observed that, in their arguments for and against the eternity of the world, there is an interesting difference between Simplicius and Philoponus. Whereas Simplicius considers all that happens in the world to be a cycle of coming into being and of passing away and regards its causes as simultaneous with what happens, Philoponus, in contrast, considers the causation of events to be a linear process. He regards the succession of time and the succession of

---

[90] 'The view that the world was created in time was not a function of Christian readings of Plato, but can be traced back to Plato's immediate pupils – if not to Plato himself – and reappears at intervals thereafter. ... Christian conceptions are compatible with the view offered and may have influenced its choice but ... the internal history of the Platonic tradition offers sufficient explanation of the facts.' H. Blumenthal (1982) 59; see also Ph. Merlan, *Greek, Roman and Byzantine Studies* IX.2, 1968, 197. As for the history of the problem, see Proclus *in Tim* I, 276, 10ff; different views are held by F.M. Cornford, *Plato's Cosmology*, London 1937, 34ff and G. Vlastos, 'The disorderly notion in the *Timaeus*', 1939, reprinted with a postscript as 'Creation in the Timaeus: is it a fiction?', in *Studies in Plato's Metaphysics*, ed R.E. Allen, London 1965, 379-99 and 401-19.

[91] Cosmas Indicopleustes, *Topographia Christiana* 7, 1 (340A), *Sources chrétiennes* 197, Paris 1973, 56; cf W. Wolska-Conus, *La Topographie chrétienne de Cosmas Indicopleustes*, Paris 1962, ch 5, esp pp 183ff. With respect to one of Philoponus' commentaries on Aristotle, Henry Blumenthal has shown that, 'whatever one's assessment of Philoponus' Christianity might be, it is clear that it was not otherwise sufficiently pervasive to prevent him from producing Neoplatonic material that is not strictly compatible with Christianity' (op. cit. 59). 'So it is', writes Sambursky (1962, 174), 'no wonder that he provoked indignation among the Christians, no less than he hurt pagan feelings.'

[92] W. Wieland (1960) 309-14.

[93] ibid. 309.

motions as a linear sequence of conditions and causes.[94] Wieland has drawn attention to the fact that it is just this analysis of the linear process of time and motion, and consequently of the world, which Kant, in his 'Antinomies of Pure Reason', exploits in order to present evidence for the thesis that there is a beginning of the world in time.[95]

(c) Causality and spontaneity

Philoponus' interpretation of motion as a successive causal relation is expressed nowhere so clearly as in certain elucidations which he gives of the principle of the transmission of force. A section in his commentary on the third book of Aristotle's *Physics* is of particular interest. There Aristotle had declared that the movement of the mover was identical with the movement of the moved object, because it is one and the same event when *A* moves *B* and when *B* is moved by *A*.[96] Philoponus gives a dynamic interpretation to this identity of movements. According to him there is a capacity for motion in the mover and a capacity for motion in the moved object, and these capacities are identical, because the capacity in the moved object is a force which the mover passes on to it: it is an 'impressed force'. 'This force being one and the same starts the whole process (*proodos*) in the motive subject and finds its end (*telos*), goal (*apoteleutêsis*) and, as it were, abode (*hoion monê*), in that which is moved. For when it enters the movable object, it will not leave it, but, remaining there, it perfects it – where this perfection (*teleiôsis*) consists in the production of force (*proagôgê tês dunameôs*), i.e. motion.'[97]

Here Philoponus gives a detailed account of how a movement is caused in one thing by another. As the context shows,[98] he considers it important that the cause of motion (or force) is *neither* 'violent', *nor* 'according to the nature' of the moved object. The 'nature' of the object is entirely irrelevant to the process of motion which is caused by an 'impressed force'. What its movement depends on is the 'process' through which the moving principle (imagined as a real capacity or force) passes. At first, the force is located in a corporeal mover which is different from the object to be moved; on account of the mover's activity, the force leaves it and passes into the object, where it remains. The moved object is not disturbed in its natural behaviour by this transfer of force; on the contrary, it is 'perfected' by the impressed force and this perfection is shown in its motion (it is not shown by the attainment of some natural or artificial external goal).

---

[94] ibid. 311-12. The difference between the views of Simplicius and Philoponus is reflected in the words used by Simplicius when he contrasts the view of *anakukloumena tôi eidei ginomena* with the view of *ginomena ep' eutheias proionta* (Simplicius *in Phys* 1181,13).

[95] W. Wieland, (1960) 315-16.

[96] Aristotle, *Phys* 3.3.

[97] Philoponus *in Phys* 384,35-385,4.

[98] See ibid. 378,23-8: 'As regards natural motion, it is impossible for that which is moved in a straight line to move in a circle. For even though the sphere of fire and, connected with it, the sphere of air, are moved with the universe, this motion is not natural but supernatural, as is the case with animal bodies which, being naturally heavy, do not move sideways by nature, but on

The peculiarity of analysing a causal relation in this way is that it depicts this relation as a temporal process. This view is vividly expressed in the (above mentioned) parallel passage of the *de Aeternitate Mundi contra Proclum*, which compares the causal relation between the celestial spheres as they rotate to the 'mechanism' (*mêchanêma*) by which a centrifuge creates motion. As water is set in motion by it and then continues to be moved, so is the sublunary world moved by an impressed force which starts at the periphery of the universe and ends at its centre.[99]

The novelty of Philoponus' understanding of causality is shown by the fact that he creates his own conceptual tools for articulating his understanding. He does so by reinterpreting a traditional terminology, for he adopts the Neoplatonic terminology of the Athenian school, but at the same time destroys its original meaning. (Here we are, once again, witnessing Philoponus' propensity for 'misunderstanding'.) *Proodos, monê, teleiôsis* – 'process', 'abode', 'perfection' are a typical constellation of Neoplatonic expressions. Similar combinations are to be found, for instance, in Proclus' *Elements of Theology*. Proclus too spoke of forces which involve a process, abode and perfection. He even said that such forces can be 'impressed' by a cause (for instance a moving cause) on something else (for instance on a movable body). In this respect, Proclus' technical term '*dunamis endidomenê*' belongs to a constellation of expressions which is similar to Philoponus' own.

But these similarities are merely superficial, for Philoponus changes the connotations of the Neoplatonic concepts. He explicitly draws attention to this change by using the phrase '*hoion monê*' ('as it were, its abode') instead of the simple word '*monê*'. Apparently he does not want to attribute '*monê*' – in the usual Neoplatonic sense of the word – to the impressed moving force. According to Proclus a moving cause impresses a *dunamis* on a moved object, but in his opinion this impression of a *dunamis* does not mean that the moving cause *transmits* an impressed force. It rather means that the *dunamis* although impressed on the moved object 'abides' in the moving cause.[100] This view of Proclus agrees with the traditional Peripatetic interpretation of motion. It claims that all things which are in motion are moved by something else and that every internal capacity for motion presupposes an external cause, which retains in itself the capacity to move something else, and acts instantaneously as either an efficient or a final cause. The idea that the force always '*abides*' in the cause implies that the force can never be separated from it and, in *this* sense, can never be transmitted. Proclus' use of the technical term '*dunamis endidomenê*' is in accordance with the use of Alexander of Aphrodisias and Themistius,[101] who had applied this term to the capacity of a body to move

---

account of a soul.' In 384,18-385,4 Philoponus explains the 'supernatural' motion by transmission of force.

[99] *aet* XIII 5, 492,20-493,5; see above, p 97.

[100] Cf Proclus, *The Elements of Theology*, ed E.R. Dodds, Oxford 1963, prop. 26,27,30 and 35. Jean Christensen de Groot, (1983) 177-96, has examined the similarity between Proclus' and Philoponus' theories of causality more closely, but I think she did not bring out the contrast between these theories clearly enough.

[101] See above, pp 103-4.

something else, a capacity which the mover imparts – not to the projectile – but to the medium in which forced motion occurs. In accordance with Aristotle they considered that an object which is moved always receives with its motion the capacity for moving, but this capacity never means the capacity to continue its own motion actively and spontaneously. Rather the capacity for active motion always '*abides*' in an external cause which moves by acting instantaneously. It is from this view that Philoponus deviates. By saying that the impressed force 'abides' not in the mover, but rather in the object which is moved, he emphasises that the 'process of force' means not only that force is transmitted, but also that, as a result of this transmission, the moved object receives a *capacity for spontaneous motion* (i.e. motion without an external cause which acts instantaneously). In the Neoplatonic context the metaphor 'process' designated a *momentary* causal relation. In Philoponus' usage it means a succession of events in time. In this respect Philoponus' 'process of force' seems to anticipate the modern idea of a temporally successive causality, as it is found, *mutatis mutandis*, in the philosophies of Hume and Kant.

In the following I intend to show how Philoponus' idea of successive causality is connected with his attempt to promote a new idea of spontaneity and of spontaneous motion, which does not require the momentary presence of an external (efficient or final) cause.

It seems obvious to me that one of the common features of Philoponus' criticism of various doctrines of Aristotle's philosophy is that he tries to abandon the requirement that all motion depends on the presence of external causes. According to Aristotle (and the Neoplatonists) there are basically two kinds of causal relations between that which moves something and that which is moved. I called the first kind 'contact-causality'. Contact-causality refers only to forced motions: something which moves another thing by force must be in contact with it. A mover which acts by contact-causality is an efficient cause of motion. The second kind could be named 'final causality' and refers to natural (i.e. elemental or animal) motions. All natural movements are ultimately caused by external objects, which determine the direction and intensity of these movements by being present at every instant and acting not as efficient, but as final causes.[102]

Philoponus in his various writings conspicuously gives up both these kinds of causality. In his commentary on the *Physics* he gives up contact-causality, as we have seen, by abandoning the medium as an efficient cause of forced motion. Moreover, in the same commentary as well as in *de Aeternitate Mundi contra Proclum* and *de Opificio Mundi*, Philoponus abandons final causality with respect (1) to elemental movements towards a natural place and (2) to celestial motions.

(1) His theory of empty space allows him to do away with those external final causes that Aristotle had called 'natural places'. It enables him to regard the natural upward and downward tendencies of elemental bodies no longer as relations to external causes (*scheseis allou*) but as exclusively internal forces

---

[102] See above, n 44.

which strive to establish a certain 'order' (*taxis*) within empty space. He reinterprets the Aristotelian 'natural places' as mere positions within that order.

(2) Philoponus' mechanistic theory of celestial motions leads, in effect, to a result comparable with that of his theory of empty space. It too enables him to do away with an external final cause of motion, because it makes it possible for him to replace the unmoved mover of the celestial bodies with an internal moving force.

All these internal forces are, according to Philoponus, *impressed* forces and originate ultimately from an external cause, i.e. from God who is an efficient cause and who, having acted *qua* efficient cause, has set the goals. But this cause acts neither *as* a final cause nor by contact. God as the mover of things is a 'prime' mover only in the temporal sense of the word, he is not a *causa finalis*. This idea of God is one which Simplicius finds inconceivable. 'How could anybody with a normal mind', he writes, 'possibly conceive of such a strange God who first does not act at all, then in a moment becomes the creator of the elements alone, and then again ceases from acting and hands over to nature the generation of the elements one out of another and the generation of all the rest out of the elements?'[103]

Now, it is interesting to observe that animal motions, too, are interpreted by Philoponus as exclusively caused by internal forces. The Peripatetic tradition had regarded animal motion as a special case of natural motion. According to Aristotle, bodies have a 'natural' motion insofar as it belongs to their 'nature', or insofar as they have the principle of their motion in themselves. The 'soul' is simply a special case of such a principle. It does not move the body by a truly spontaneous action, rather it is the (internal) capacity of the body to be moved by certain external final causes.[104] Philoponus now tries to interpret animal motion as perfectly spontaneous. The main motive of his criticism of Aristotle in his commentary on *de Anima* seems to be the defence of the perfect spontaneity of the soul, its *autenergeia*, as opposed to Aristotle's doctrine that animal motions are teleologically determined.[105] Aristotle had taught in *de Anima* 3.10 that the rational soul, as a motive cause, merely mediates between the moved body and a moving cause, which is ultimately an external unmoved object. This first mover is

---

[103] Simplicius *in Phys* 1151,28.

[104] According to Aristotle, not even animal motion can be said to be perfectly spontaneous. Rather, it requires final causes which are external. On the account of *Phys* 8.2, 253a7-20 and 6, 259b1-20, living beings are able spontaneously to generate locomotion, but no other kinds of motion. Other kinds are rather determined ultimately by external causes. But even the spontaneity of animal locomotion is not a perfect spontaneity (259b7); for it is generated in the animal only as a result of other kinds of motion in the organism, among them the motions of the sense organs. These motions make it possible for the organism to be moved by certain external final causes. In *de An* 3.9, 432b15-6, Aristotle says that (loco)motion of a living being always has a final cause and that it is bound up with *phantasia* and *orexis*. He describes this final cause as the *prakton agathon* (433a29), as that which is good for the living being. The final cause itself is unmoved (434a16), whereas the *orektikon* is in motion and has an efficient cause of its motion (434a16-17; *Phys* 8.2, 253a17). Aristotle obviously means that locomotion of living beings has always *both* an efficient cause *and* a final cause.

[105] As regards further details, see M. Wolff (1971) 72-9.

'the good which is to be practically realised' (*to prakton agathon*).[106] Philoponus' view is opposed to this doctrine. Already in the prooemium of his commentary on *de Anima* he declares that it must be possible voluntarily to depart from the good (*hekousa tou agathou ekpiptein*), and then also to purge one's soul (*kathairein*). According to Philoponus, that would not be possible if there were no spontaneous motion of the soul.[107] A similar view is adopted by him in *de Opificio Mundi*.[108]

However, it is noteworthy that Philoponus agrees with Aristotle's criticism of the Platonic doctrine of the soul's self-movement. Philoponus accepts that the soul, being a principle of motion, can never itself be moved. In this respect he also keeps aloof from Proclus, who had said that the soul is substantially (*kat' ousian*) moved by itself or 'self-moved'.[109] But against Aristotle Philoponus insists that the soul is not just the 'form', or formal cause of an animal, but really the first cause of its motion.[110] For this purpose, he reinterprets Plato's (and the Neoplatonists') concept of the 'self-moved' (*autokinêtos*) soul in such a way that it turns into the doctrine of the soul's acting spontaneously or being 'self-acting' (*autenergêtos*).[111] He characterises the *energeia* of the soul by the same words which, in his commentary on the *Physics*, he applies to impressed force. It is an incorporeal moving cause, and does not need any corporeal substratum in order to move the animal body. By moving the animal, it also perfects (*teleioi*) it.[112]

Again, the Neoplatonic terminology does not conceal the novelty of Philoponus' concept of animal motion. For Proclus and the Neoplatonists, 'self-motion' did not imply perfect spontaneity. The self-motion of animals, like the natural motion of elemental bodies, was not caused by contact, rather it was based on an internal capacity which the Neoplatonists took to be 'incorporeal' (whereas the capacity of elemental bodies to move according to their corporeal nature was corporeal). But, for the Neoplatonists, too, that incorporeal capacity of animals for self-motion was a capacity for being moved by certain final causes and, to this extent, it still required the presence of an external 'mover' as the goal, the practical purpose, the *telos* of motion.

---

[106] Aristotle *DA* 3.10, 433b16.

[107] Philoponus *in DA* 18,18-9.

[108] Philoponus *Opif* 300,20-301,10. In Philoponus' view, a bad (or evil) action is not just something voluntary, (*hekousion*), but also lacks a final cause, i.e. it pursues an evil not by seeing it as an *apparent* good (*phainomenon agathon*, as Aristotle would say, *EN* 3.4, 1113a16ff.). This view is most clearly expressed in *Opif* 301,11-303,19. Here Philoponus argues that the *proairêsis* (deliberate choice) is the *sole* (*monê*) cause (*aitia*) of the existence (*parupostasis*) of good and evil (302,14-16 and 19-20; cf 301,15-18). He further says that the *proairêsis*, being the cause of good and evil, is *not* itself bad (301,13-4 and 302,19). Aristotle, on the other hand, argues that any *proairêsis* already presupposes a goal and therefore cannot be the cause of this goal; it merely decides on the means of attaining it (*EN* 3.4, 111b26-7). He also holds that the *proairêsis* is partly good and partly bad (1111b34). The bad *proairêsis* is that which springs from an ignorance of what is good (1110b31-1111a2). The idea of a free will being the *cause* of good and evil is unfamiliar to Aristotle, and seems to anticipate certain principles of modern ethics.

[109] Philoponus *in DA* 114,23-4. Proclus *Elements of Theology*, loc. cit. 22,11.

[110] Philoponus *in DA* 109,24ff.

[111] ibid. 114,24ff.

[112] ibid. 206,26ff.

The peculiarity of Philoponus' view becomes clearer when one tries to understand the problems it raises. His conception of the soul as an incorporeal *energeia* or force, which spontaneously moves the animal body, raises problems which in certain respects resemble the Cartesian mind-body problems. If we notice them, we shall realise that Philoponus' doctrine of forced and natural motion, and especially his idea of successive causality, probably results from his attempts to solve these problems.

(d) Mechanicism and the idea of spontaneity

Aristotle's belief that animal motion is teleologically determined was in accord with three general principles of his theory of motion:

(1) All things, which are in motion, are moved by something else.
(2) The motions of the moved object and the mover (or the motive cause) are simultaneous.
(3) All motions demand a first motive cause (the chain of motive causes does not involve an infinite regress).

According to Aristotle, the prime mover in animal motion is not the soul itself.[113] The soul is rather the form or structure of the animal organism which enables it to be moved by certain perceptible or intelligible external objects, for which the organism is striving. In this sense animal organisms too are moved by something else (according to principle (1)), and ultimately by an external final cause. Ultimately it must be an (unmoved) final cause which, according to principle (3) is the prime mover in animal motion. According to principle (2) it acts simultaneously with the motion it causes.

Philoponus denies the general dependence of animal motion on final causes, for, as we have seen, the soul is free to *depart* from striving for a real or apparent good. However, Philoponus too assumes that there is a first cause of animal motion (according to principle (3)), but in his opinion this first cause must be the soul itself, which he regards with Aristotle as unmoved. This assumption does not conflict with principle (1), but it collides with principle (2) (the principle of the simultaneity of active and passive motion). How is it possible that something unmoved, although it is not the final, but the efficient cause of motion, acts instantaneously, or simultaneously with the motion it causes? If the first cause of animal motion does not act as final cause, what does it mean to say that its activity is simultaneous with the motion of the animal body?

Philoponus' commentary on Aristotle's *de Anima* suggests an answer to just this problem. He recognises that the unmoved soul cannot act on the animal body unless it does so *directly*. The soul does not need a movable corporeal substratum in order to move the animal. Indeed, if it did an infinite regress would arise.[114] Philoponus considers the soul to be a dynamic entity which

---

[113] See the final chapters of *DA* 3, also *Phys* 8 and *de Motu Animalium*.
[114] Cf Philoponus *in DA* 12,22-31.

is able to effect motion not as a result of being moved itself, but by its mere presence in the animal body. He occasionally compares the effect of the soul on the body to that of an immediate push, without taking this effect to be a result of locomotion.[115] This comparison indicates a difficulty, which Philoponus tries to resolve by offering a kind of vitalistic theory of living organisms. He develops this theory in the context of his criticism of Aristotle's theory of light (*de Anima* 2.7). It may be summarised as follows. The soul does not act by a real push or pressure, but by propagating its incorporeal energy over the organs of the animal body. This propagation is comparable with the emission of light, which emanating from the sun successively (but with infinite speed) penetrates the diaphanous bodies.[116] Unlike Aristotle, Philoponus considers light and the energy of colour to be an incorporeal, transmissible *energeia*. He is therefore able to compare the action of the soul's *energeia* with the emission of light. In his commentary on the *Physics* he normally compares the action of impressed moving forces on bodies with the emission of light, or the *energeia* of colour.[117] The impetus theory which Philoponus develops in this commentary seems merely to draw conclusions from the analogy between the energies of light and soul. Philoponus illustrates this analogy in his commentary on the *de Anima* in the following way: 'The energy of the carpenter, which passes through the axe and into the wood, in some way or other affects its shape, yet nothing similar takes place in the axe. We can say, analogously to this example, that light transmits a certain incorporeal energy from the luminous body to the transparent medium, whose nature is such that it absorbs this energy.'[118] Like the transparent medium, which receives and transmits the *energeia* of light, the organs of an organism receive and transmit the incorporeal moving force of the soul. And just like these organs, artificial organs or tools (the carpenter's axe) are able to receive and transmit this moving force. Here, impetus theory is already anticipated *in nuce*.

Philoponus' vitalistic explanation of animal motion uses some additional analogies between light and soul. In his *de Motu animalium* Aristotle had explained the physiological process by which a certain desire of an animal leads to its motion. According to Aristotle, desire leads to a heating of blood around the heart and heating leads to expansion which causes spatial motion. Aristotle did not explain the transition from desire to heating and from heating to expansion. Philoponus, now, attempts to explain this physiological process by an analogy. Just as light by its mere presence heats the air and, as a result of heating, generates the motion of the air, in the same way the

[115] ibid. 108,7-15 and 24-9: 'Those explaining motion by some sort of mechanism alone do not attribute to the body any fitness or natural capacity for motion. We, however, assume that through the presence of the soul some vital force is implanted in the body, and accordingly its absence leads to a collapse of the body. It is therefore probable that the soul cannot re-enter the body after the removal of that fitness which had been implanted in the body at the beginning ... . Similarly a stick pushed against a door cannot move the door when the door has not the fitness necessary for being moved ... .'

[116] ibid. 329,3-330,27.

[117] *in Phys* 642,9-20.

[118] *DA* 329,30-37.

energy of the soul heats and moves the animal body by its mere presence.[119] Just as sunlight increases the natural heat of the air and thereby sets it in motion, the energy of the (rational or irrational) soul heats and moves the animal's body.[120] In this context Philoponus draws attention to burning mirrors, which in his opinion show that heat can be generated without friction, and consequently without spatial motion.[121] For him, it is a characteristic feature of incorporeal energy that it can be concentrated (like light by burning mirrors) and its effects thus increased. Philoponus intends to show that the soul, as an incorporeal efficient cause of motion, is able to generate spontaneously, by its mere presence in the body, the body's locomotion.

In the same way, in modern times, especially in seventeenth- and eighteenth-century physics, actions resulting from the mere presence of forces come to be called 'sollicitations'. The technical term 'sollicitation' derives from ecclesiastical penal law and originally means the sin of adultery committed within the confessional. It means an action caused by the mere presence of a manifestly wicked will. It is perhaps no mere accident that the language of later physics points to a connection which is already in Philoponus' commentary on the *de Anima*. His forces, which by their mere presence act on bodies which have a specific fitness (*epitêdeiotês*) to receive them, anticipate the concept of sollicitation. Whereas sollicitation has its origin in a certain spontaneous psychology, Philoponus seems to develop impressed force in order to make the perfect spontaneity of the soul possible.

This connection of ideas seems to me to be very interesting. Usually impetus theory is said to be an early type of mechanicism. The view which I believe to be legitimate is ultimately based on the fact that impetus theory has promoted the idea of successive causality. But it seems to be no less legitimate to regard Philoponus' impetus theory as an expression of the idea of spontaneity. That sounds paradoxical, but it is not. We need only remember Kant's doctrine of the 'Antinomies of Pure Reason'. In its structure, Kant's third Antinomy[122] resembles Philoponus' argument. To prove that there are in the world spontaneously acting causes (i.e. free individuals) Kant assumes that all that happens in the world is a result of a mechanical process. Kant's 'Kausalität aus Freiheit' resembles Philoponus' idea of '*autenergeia*', i.e. the soul's self-activity. For Kant's 'Kausalität aus Freiheit' is also based on a spontaneous action which is the first link in the chain of a successive physical process. Kant, therefore, defines 'Freiheit im kosmologischen Verstand' (freedom in the cosmological sense) as a dynamic entity, namely as 'das Vermögen, einen Zustand von selbst anzufangen' (a capacity to initiate a state spontaneously).[123] In his proof, Kant is actually drawing a conclusion which is just one of the latent and fundamental convictions of the modern mechanistic outlook. This outlook is defined by the conviction that, in

[119] ibid. 332,4-21.
[120] ibid. 332,12-14.
[121] ibid. 332,17ff.
[122] I. Kant, *Critique of Pure Reason*, B472-76 = A444-49.
[123] ibid. B561 = A533.

principle, one has to explain all natural phenomena by reference to efficient causes (mechanical forces). But we find that even the strongest advocates of the mechanistic outlook hold that, ultimately, there are mechanical causes which cannot be understood as the effect of a mechanical cause. This conviction has never been given up without giving up the mechanistic outlook itself.

The close connection between the mechanistic outlook and the idea of spontaneity appears to be of great importance not only for the problem of how the continuity between impetus theory and modern mechanics (or mechanical explanation) is to be understood, but also for the problem of explaining the original motives for advancing impetus theory.

It does not seem likely that Philoponus in his commentary on the *Physics* (and elsewhere) propounded his impetus theory solely to solve certain problems within the problematic of Aristotle's physics. As we have seen, Philoponus' arguments against Aristotle's assumptions concerning projectile motion and concerning the 'law of falling bodies' have a merely *ad hoc* character and sound rather rhetorical. On the other hand, in all his arguments against Aristotle's theory of motion, Philoponus endeavours to show that motion of all kinds (forced, natural elemental and natural animal motion) can dispense with external causes, and is to this extent perfectly spontaneous. There are good reasons to assume that this endeavour results from Philoponus' conviction that the soul is absolutely free and acts spontaneously, i.e. not determined by external final causes. As we have seen, this conviction entails certain assumptions concerning forced motion insofar as the forced motion of inanimate objects (like tools) is, according to Philoponus' commentary on *de Anima*, not essentially different from animal motion.

If we now take into account the fact that in all probability Philoponus wrote his commentary on the *de Anima* earlier than his commentary on the *Physics* (and therefore earlier than all the other writings which concern impetus theory),[124] it looks as if his theory of forced motion propounded in detail in the *Physics* commentary merely draws conclusions from assumptions already laid down in the earlier *de Anima* commentary. Unlike the commentary on the *Physics* (written in 517), the *de Aeternitate Mundi* (529), the commentary on the *Meteorology* (after 529), and the late *de Opificio Mundi* (between 557 and 560), the *de Anima* commentary still accepts the Neoplatonic view that the celestial motions and the rotation of the spheres of fire and air are caused by souls.[125] In his later writings Philoponus interprets these motions analogously to artificial or forced motions. As a result, he deprives the heavenly bodies, as well as the spheres of fire and air, of their Neoplatonic souls, and instead gives them forces, which act spontaneously. He must then reinterpret the traditional (Aristotelian as well as Neoplatonic) 'prime mover' as the first member in a (temporal) series of efficient causes. Ammonius' doctrine of God as an efficient cause did not yet draw these conclusions.

---

[124] Regarding the date of the commentary on *DA*, see M. Wolff (1971) ch 4, and R.B. Todd, 'Some concepts in physical theory in John Philoponus' Aristotelian commentaries', *Archiv für Begriffsgeschichte* 24, 1980, 152.

[125] See M. Wolff (1971) ch 4.

Perhaps only Philoponus' examination of Aristotle's arguments for the existence of the prime mover, carried on in his lost commentary on *Physics* 8.10, could have led him to these conclusions.[126]

Now, if Philoponus' impetus theory originates from a certain idea of spontaneity, i.e. from convictions concerning the freedom of action and will, it seems no less reasonable to consider the basic principles of his impetus theory to be derived from moral philosophy or ethics rather than from natural philosophy or philosophical psychology. For impetus theory opposes some basic assumptions not only of traditional ancient physics and psychology, but also of traditional ancient ethics. According to Philoponus the spontaneity or freedom of the will is not, as the Stoics held, a mere 'inner' freedom, the freedom of consent to inescapable and indisposable fate. Nor is it, as Alexander of Aphrodisias (the critic of the Stoics) held, a mere freedom of choice, or, as Plotinus and other Neoplatonists held, the *energeia* of the soul by which it orders itself and which is not able to act on external objects. According to Philoponus, the freedom of the will is rather the spontaneous activity of the soul by which it is able to act immediately on corporeal nature. God (or 'Providence') has provided the soul with the ability not only to order itself, but also to order physical things 'here below'.[127]

Obviously, Philoponus' concept of spontaneous activity (*autenergeia*), like Kant's cosmological concept of freedom, has certain ethical connotations. Like Kant, Philoponus connects the idea of spontaneity with the idea of a (morally) free will. And like Kant, he takes the morally free will to be a *force* which is a prime mover. It is a capacity to begin a process spontaneously, and it influences the external, corporeal world directly. This not only makes Philoponus an opponent of traditional physics and psychology but also an opponent of traditional ethical views. In contrast to Aristotle's ethics, Philoponus does not believe that free will is determined by an external final cause and conditioned to pursue the (real or apparent) good.[128] The soul is free 'deliberately to depart from the good' and voluntarily to pursue evil, because 'evil rests on a free will'.[129] The soul which does evil, thereby bears all guilt and is responsible for ridding itself of this burden by 'purging itself'.[130] Like Kant, Philoponus advances a free will which is the origin of good and evil. This idea is hardly compatible with a Manichaean view of evil. It does not accept that something in the world is *by nature* evil, for doing evil is a matter of free will. Philoponus attacks the Manichaean view by pointing out that evil exists neither as a substantial property nor as a substance; it consists solely in a 'misuse of physical forces', and such misuse is based on our free choice.[131] In his arguments for this view, Philoponus refers to current moral and legal concepts, such as guilt and merit and the demand that an

---

[126] Philoponus' whole theory of natural motion (especially the analogy between natural and projectile motion) may be a consequence of this examination.

[127] Philoponus *in DA* 6,20.

[128] See above, n 108.

[129] Philoponus *in DA* 18,18 and *Opif* 300,19.

[130] *In DA* 18,19.

[131] *Opif* 301,16-17.

individual's rights be respected.[132]

Philoponus' idea of free will also has certain political implications. In his *de Opificio Mundi* he derives sovereign power and political government from the free will of human beings and denies that a particular individual is determined by nature or by being in God's image to be sovereign and governor.[133] It happens, however, among animals, he says, that authority exists by nature (*phusei*) and not as a result of choice (*hairesei*). The queen bee, the bellwether and the leader of a flock of cranes are truly images of God insofar as their sovereignty is by nature. But human sovereignty, the authority of a human king or judge, is nothing natural, but is conventional (*thesei*) and rests on 'the will of human beings, which is, however, often neither right nor reasonable, for which reason many rulers fall'.[134] God did not subject men to natural authority, 'but it depends rather on us to make somebody God's image just as we like, by vesting him with authority and kingdom.'[135] Philoponus' view of sovereignty and political authority in resting on the free will of governed people, seems to be unique in the political literature of late Antiquity.[136]

It is generally true for ancient philosophy that its particular interpretations of nature and natural phenomena had a moral significance and were not irrelevant for ethics and practical life.[137] Now, if one subjects Philoponus' writings to close analysis, one gets the impression that the motives for his peculiar interpretations of natural phenomena overlap with his motives for devising a new concept of freedom and spontaneity. Even in his commentary on the *Physics*, there are a number of passages which reveal the modern traits of his specific theory of spontaneity and of his 'moral philosophy', to which he did not devote a special inquiry. Here are three examples:

First, in his *Corollary on Void*, Philoponus attempts to show that in animal motions caused by a free will (*proairetikai kinêseis*) the ratio of the quantities of motion is equal to the ratio of the forces consumed in these motions.[138] That is, the difference in the quantities of physical activity of a human being is due not to external conditions, but only to the exhaustion of the force which one has spent.

Secondly, we have seen that, according to Philoponus, forced motions are not simply contrary to nature; they are not mere disturbances of nature. Motions which are caused 'by violence', and thus by impressed forces, are rather *perfections*.[139]

Thirdly, Philoponus' assumption that animal motions are not determined by given external causes, entails that artificial creation (*technê*) is not

---

[132] ibid. 300,19-303,19.

[133] ibid. 261-264.

[134] ibid. 263,17ff.

[135] ibid. 263,27ff.

[136] Cf F. Dvornik, *Early Christian and Byzantine Political Philosophy* (Dumberton Oaks Studies 9) vol 2, Washington 1966, 710f; F.C.H. Frend, *The Rise of the Monophysite Movement*, Cambridge 1972, 52.

[137] Cf J. Ben-David, *The Scientist's Role in Society*, New Jersey 1971, 33-44.

[138] Philoponus *in Phys* 691,9-26 and 692,27-693,6.

[139] ibid. 384,33-385,4.

determined by final causes given by nature. Philoponus does not accept Aristotle's assumption that every *technê* either imitates nature or assists it. He explicitly denies that there is always a natural *telos* corresponding to an artificial creation.[140]

Conclusion

If we are prepared to assume that the basic presuppositions of impetus theory can be traced back not to observational experience which Aristotle missed, but rather to a certain concept of man and to certain ethical principles, we need not attempt to explain the emergence of the theory solely by reference to new observations of falling bodies and the like. Is it not more appropriate to ask about the origin and kind of *ethical problem* to which impetus theory originally helped to provide an answer? The experience that forces are exhausted in all physical activities of human beings could have been just such a problem. Earlier society, which had left this experience chiefly to slaves, could not really have had such a problem. But, by the close of Antiquity, times were changing.

[140] ibid. 310,16-29.

CHAPTER FIVE

Philoponus' Impetus Theory in the Arabic Tradition

Fritz Zimmermann

Shlomo Pines has been arguing over the years that (1) Avicenna's theory of inanimate motion resembles the so-called impetus theory of Philoponus; (2) that some of the arguments advanced by Philoponus in his commentary on Aristotle's *Physics* were known to Arab philosophers before and after Avicenna; and (3) that the impetus theory of the later Latins is likely to have been inspired by the Arabic tradition of Philoponus.[1] Having looked into the matter at Richard Sorabji's request, I feel that Pines is bound to be right on all three counts. The whole range of the available evidence has yet to be examined in detail, but here is a preliminary report.

1

As Pines remarked, Avicenna's theory of inanimate motion turns on the notion of *mayl*, 'inclination'; *mayl* translates Greek *rhopê*; and *rhopê* is often used by Philoponus for impetus. For according to Philoponus, it is the natural *rhopê* of the four elements that causes, say, fire to rise and stones to fall.[2] The idea that weight (heaviness or lightness) is a body's tendency towards its natural place is Aristotle's, as is the use of the word *rhopê* for that tendency. But Aristotle had balked at making weight a 'mover', i.e. a cause or source of motion. He had also thought that a body could not continue in motion unless kept going by a mover. As a result, he had been at a loss for a satisfactory answer to the following questions. What makes a stone thrown up into the air continue on its upward journey once it has left the throwing hand? And what makes it return to the ground thereafter? The answer to the first question, he thought, must be that the stone was propelled by a kind of shockwave sent by the throwing hand through the surrounding air.[3] His answer to the second question seems to have been that since the returning stone was merely moving towards a level at which it would naturally be at rest, its fall was really caused by the mover that had first lifted it above that level.[4]

[1] See, in particular, S. Pines (1938a),(1938b), (1953), (1961).
[2] Philoponus *in Phys CAG* XVI 195,29.
[3] *Phys* 4.8; 8.10; *de Caelo* 3.2.
[4] *Phys* 8.4,255a30ff. Why, Aristotle asks (255b13ff), do light and heavy bodies rise and fall? Because it is part of their *nature* to do so (*pephuke poi*). Steam has an unactualised potential for rising above the level of water, either when it has come down as water for lack of heat, or when it is held down by a lid. When it does rise, one cause of its motion is the application of heat, another

Philoponus preferred to say that the stone's rise was caused by an upward tendency imparted to it by the throwing hand, and that its fall was caused by a downward tendency engendered by its elevation from its natural place. In other words, he dismissed Aristotle's misgivings about admitting internal movers in inanimate bodies. Weight, no longer denoting the mere fact that if you let go of a stone it would fall,[5] became a positive cause of motion. Avicenna (d. 1037) takes a similar view, counting *mayl ṭabī ʿī* 'natural inclination' (lightness and heaviness) and *mayl qasrī* 'forced inclination' among the causes of motion. I cannot attempt to do justice to the ramifications of Avicenna's theory of *mayl*, or to Pines's discussion of it. Instead, I shall confine myself to the short account produced by Ghazali (d. 1111) in his *Maqāṣid al-Falāsifa* (*Intentions of the Philosophers*), which is an abstract of Avicenna's philosophy.

Every motion, says Ghazali, must have a cause. When a stone rises or water is heated, it is called a compulsion. When a stone falls or water cools down, it is called a nature. In the case of voluntary motion, it is called an act of will.[6] What it means for a body to be moved by its 'nature' is illustrated as follows. When you hold a bag full of air under water, you feel its *inclination* in the pressure it exerts against your hand. When you let it go, it *moves* to the surface. When it floats on the surface, with no inclination or motion, it is still possessed of the *nature* on account of which it will have inclination and, if unimpeded, motion whenever it is removed from its natural place.[7] We see that the 'nature' in question is the (Aristotelian) property of inanimate bodies

---

the removal of the lid. Aristotle winds up by saying (256a1f) that when light and heavy bodies rise and fall, the mover is either (a) that which causes them to become light or heavy (*hupo tou gennêsantos kai poiêsantos kouphon ê baru*) or (b) that which removes an obstacle. I take it that (b) is meant to comprise both the application of heat and the removal of the lid in the example of steam; and that (a) refers to the creation of an unactualised potential of lightness or heaviness, viz. by removing a body from its natural place (here, the bringing down of steam in the shape of water). But of course, it is also possible to relate (a) to the application of heat, or to the fashioning that caused bodies to be as they are (viz. light or heavy) in the first place. (For medieval readers that would have been the act of the creator. Unfortunately, the one MS we have of the Baghdad *Physics* (see p 128 n 25 below) includes no comment on that passage.) On no account does Aristotle here supply a mover corresponding to the air in his explanation of projectile motion. If a stone, when simply dropped, falls to the ground in a straight line *because it is made that way*, then so, when flung in another direction, does it fall in a curved line *because it is made that way*. In that sense, both types of motion are equally 'natural'. In another sense, it is 'natural' for the stone to fall vertically rather than in a curved line. In either case, if it is the air that propels it along its curved trajectory, what is it that propels it down its vertical path? If again the air (as Aristotle indeed suggests at *de Caelo* 3.2, 301b22-30), then what provides the initial impulse corresponding to the push initiating projectile motion? If the stone's 'nature', then inanimate bodies will have a source of motion within them after all. And if that internal mover is active at the *beginning* of the fall, why not throughout? Readers of Aristotle dissatisfied with the asymmetry of his accounts of natural and unnatural motion in inanimate bodies will thus be led to question his aerial-propulsion theory, and to wonder whether the mover responsible for either type of motion might not be an impetus implanted, now by nature, now by force.

[5] *Phys* 8.4, 255b15-17: *tout' esti to kouphôi kai barei einai to men tôi anô to de tôi katô diôrismenon* 'what it is for a body to be light or heavy is to be distinguished by an upward or downward tendency'.

[6] Maqāṣid al-Falāsifa, ed S. Dunyā, Cairo 1960, 309f. = *Algazel's Metaphysics: A Medieval Translation*, ed J.T. Muckle, Toronto 1933, p 134.16ff.

[7] ibid 263f. Dunyā = 99.18ff Muckle.

of being naturally at rest at a particular level between centre and periphery of the sublunary world. Remove them from that level, and their 'nature' will take the form of an 'inclination' to return. Being a cause of motion residing in the elements themselves, Avicenna's natural *mayl* is clearly the same as Philoponus' natural *rhopê*.

According to Ghazali, no body can move without inclination.[8] His 'proof' reproduces Aristotle's argument at *de Caelo* 3.2,301a23ff, which is based on the erroneous notion that a body's speed is proportional to its weight. For example, other things being equal, a body half as heavy as another will fall half as fast. If, then, speed is a product of factors including weight, zero weight will always produce zero speed. In other words, a weightless body will not move at all, even if pushed or pulled.

Ghazali also reproduces a parallel argument – against the possibility of a void – from *Phys* 4.8,215b19ff.[9] Aristotle's error there is to assume that a body's speed is inversely proportional to the density of the medium through which it travels. For example, other things being equal, if air were half as dense as water, the same body would fall twice as fast through air as it would through water. If, then, speed is a product of factors including the density of the medium below the line, zero density will produce infinite speed. In other words, motion through a void would have to be instantaneous, which is impossible. Hence there can be no void.

As Pines remarked, Aristotle's error at *Phys* 4.8 had been spotted by Philoponus, and was spotted again in Muslim Spain by Avempace (d. 1139). In his *Notes on the Physics*, Avempace points out that the proportion of the density of air to that of water equals the proportion, not of the *speed* of body X in air to its speed in water, but of the *retardation* of X in air to its retardation in water.[10] Hence zero density will produce zero retardation, not zero speed. Motion in a void is possible. For in Avempace's view, moving bodies are possessed of a 'motive force'.[11] He, too, apparently subscribed to an impetus theory akin to that of Philoponus.

If Avicenna fell for Aristotle's faulty arguments from zero weight and zero density, he rejected the claim – made in the same chapters *de Caelo* 3.2 and *Phys* 4.8 – that the rising stone, once separated from the throwing hand, was propelled by the surrounding air. The objections of Avicenna and his follower Abū l-Barakāt of Baghdad (d. *c.* 1165), which recall Philoponus' discussion in his commentary on the *Physics*, have been noted by Pines. Ghazali does not mention the discarded thesis and has little to say about *motus separatus* other than that exemplified by the stone's fall to the ground or the bag's rise to the surface of the water. To the impetus theory of Avicenna, as of Philoponus, projectile motion presents no special problem. If a stone travelling downwards by itself is moved by a 'natural inclination', one travelling by itself in any other direction will be moved by an 'unnatural (forced) inclination' imparted

[8] ibid 264f. Dunyā = 100.6ff Muckle.

[9] ibid 316. Dunyā = 139.15ff Muckle.

[10] *Shurūḥāt al-Samāʿ al-Ṭabīʿī libn Bājja al-Andalusī*, ed M. Ziyāda, Beirut 1978, 144.

[11] ibid 142 penult.

by the throwing hand or instrument.[12] The issue between Aristotle and Philoponus is whether inanimate bodies can be said to be moved by themselves (i.e. have their mover within them) at all. Ghazali briskly settles the point in a few lines. Why, he asks, can we not say that the falling stone is pushed or pulled by an external mover, such as the air through which it moves or the ground towards which it moves? Because in that case, small bodies would have to move faster, large bodies more slowly. But the reverse is the case, which shows that their motion comes from themselves (*ex ipso est motus*).[13]

Thus what we find is that Avicenna in the East and, independently, Avempace in the West both share Philoponus' view that *motus separatus* in inanimate bodies is caused by an internal mover. They also both produce objections, reminiscent of the arguments of Philoponus, to certain claims connected with Aristotle's *denial* of internal movers in inanimate bodies. It is therefore natural to assume that some of Philoponus' criticism of Aristotle was available in Arabic.

2

Curiously, though, Arab impetus theorists do not say that Aristotle is *against* internal movers of inanimate bodies, Philoponus *for* them; that Aristotle is *wrong*, Philoponus *right*. Neither Avicenna nor Avempace acknowledge a debt to Philoponus. Both play down their disagreement with Aristotle, introducing their internal movers almost as though they were a piece of Aristotelian orthodoxy. Apparently, Arab Aristotelians were less than completely aware of Philoponus' influence on their reading of Aristotle, and reluctant to say that they were following Philoponus even when they knew they were. The result is a medieval tradition of Philoponan impetus theory with little credit given to Philoponus.

To take the second point first, Philoponus is best known in the Arabic tradition as a critic of the philosophical belief in the eternity of the world. As such he is referred to by both Avempace and Avicenna. Avempace mentions Philoponus' attack on Aristotle's argument for the eternity of motion.[14] At one point, he speaks of 'Yaḥyā ibn 'Adī al-Naḥwī',[15] conflating the names of two philosophers: Yaḥyā ibn 'Adī (d. 974), a Christian Aristotelian of Baghdad (not well known, apparently, in Muslim Spain), and Yaḥyā al-Naḥwī 'John the Grammarian', i.e. Philoponus. Pines, noting the muddle, discovered a plausible explanation in the unique Leiden manuscript of what I

---

[12] For Ghazali that goes almost without saying. He comes close to actually saying it at 264.18f Dunyā = 100.12-15 Muckle, where he speaks of the resistance offered by the natural downward inclination of the rising stone to the 'inclination of the motion forced upon it' (*mayl al-taḥrīk al-qahrī: corpus habens inclinacionem verbi gracia deorsum, cum prohicitur sursum est illi inclinacio naturalis resistens inclinacioni mocionis violente.* The idea that projectile motion is caused by an *inspired* force is strongly implied.

[13] *Maqāṣid* 310 Dunyā = 135.2ff Muckle.

[14] *Shurūḥāt* 180ff.

[15] ibid 194 ult.

shall call the Baghdad *Physics*.[16] The Arabic version of Aristotle's text by Isḥāq ibn Ḥunayn (d. 910) is there accompanied by a large amount of annotation. *Inter alia*, there are some references to 'Yaḥyā ibn 'Adī', some to 'Yaḥyā al-Naḥwī', and very many to plain 'Yaḥyā'. It would appear that Avempace, using a similarly annotated version of the *Physics*, assumed that the 'Yaḥyā' of the notes always referred to Philoponus.

Pines also discovered that in connection with *Phys* 4.8, certain notes marked 'Yaḥyā' reproduce, though not *verbatim*, objections raised by Philoponus in the surviving portion of his commentary. It seems to me that much if not most of the material marked 'Yaḥyā' derives from that commentary in a fairly straight line.[17] The truth, I hope, will be revealed by an Oxford thesis now in hand on the relation of 'Yaḥyā' to Philoponus. For the most part, it will be just a matter of tracing the passages underlying the Arabic digest in what survives of the Greek original. It will be less easy to determine to what extent, if any, our digest contains extracts from portions of the commentary now lost in the original. The fact that the language of the digest differs in many points from that of the tenth-century Aristotelians of Baghdad will help to tell Yaḥyā al-Naḥwī from Yaḥyā ibn 'Adī. Unfortunately, that help looks like deserting us towards the end. An antecedent of the Leiden manuscript (1130) for some reason switched exemplars in the middle of book 6. The generous annotation of earlier parts gives way to the much flimsier commentary of the Baghdad Nestorian Abū l-Faraj ibn al-Ṭayyib (d. 1043). He too quotes 'Yaḥyā'. But that Yaḥyā does not seem to speak the distinctive language of the digest quoted in the earlier parts.

At all events, the Leiden manuscript includes, at *Phys* 4.8, a summary of Philoponus' defence of impetus as an internal mover complete with the so-called corollary on void. Here is a crucial sentence on the *motus separatus* of an arrow in flight:

Yaḥyā. We must therefore say that its force <is one which> arises in the thrown <body> itself as a result of <the action of> the thrower – <an> incorporeal <force>[18] which, having taken it so far, expires.[19]

---

[16] The Leiden MS has since been published by 'Abdarraḥmān Badawī, *Arisṭūṭālīs: Al-Ṭabīʿa*, 2 vols, Cairo 1964-65.

[17] As was pointed out by G. Endress, *The Works of Yaḥyā ibn 'Adī*, Wiesbaden 1977, pp 36-8. Professor Pines tells me that he now inclines to a similar view, as he will explain in the forthcoming second volume of his *Collected Works*.

[18] Supply *wa-hiya quwwatun*.

[19] Vol 1, p 371,21f Badawi. The translator has yet to be identified. A possible candidate is Qusṭā ibn Lūqā (ninth century). It is hard to tell whether the Arabic of 'Yaḥyā' is drawn immediately from the Greek of Philoponus, or from a fuller Arabic version, or whether it simply translates an antecedent Greek epitome. Usually it abbreviates (not without occasional gains in clarity); here for once we have an amplification, and an intelligent one at that. The relative clause added at the end no doubt accords with Philoponus' view. Philoponus is no less innocent of our concept of inertia than is Aristotle, whom a little earlier he has represented as holding that a projectile in flight is propelled by the surrounding air 'until the motive force imparted to it (*sc.* the air) is spent' (*CAG* XVII 641,10f: *mechris an ekluthêi hê endotheisa autôi kinêtikê dunamis*). Like Aristotle, Philoponus thinks that a body in motion has to be pushed or pulled every inch of the

In the Greek of Philoponus that sentence reads as follows:

> ... rather, some incorporeal motive force must be imparted (*endidosthai*) to the thrown <body> by the thrower.[20]

The Arabic digest does nothing to conceal the fact that Philoponus contradicts Aristotle. Occasional glosses by Baghdad Aristotelians tend to side with Aristotle. To the passage just quoted the Leiden manuscript adds a rather fatuous comment in the name of Yaḥyā ibn ʿAdī's pupil Ibn al-Samḥ:

> An arrow met by the slightest obstacle in its path drops dead. If there were a force in it which carried it forward, it would not stop so easily. Thus we know that the force is in the air.

It would appear that for the Aristotelians of Baghdad Philoponus was there to be read but not to be agreed with. Whether some of them none the less permitted themselves to be swayed by his defence of impetus as an internal mover remains to be seen.[21] Avempace, at any rate, did – if we are right to see, with Pines, the source of his impetus theory in the 'Yaḥyā' passages of the Baghdad *Physics*. The language of the two sets of notes (Avempace's and Yaḥyā's) has yet to be compared. For the time being, the most likely theory is that Avempace knew that he was siding with Philoponus but did not like to say so.

Avicenna too was in two minds about Philoponus. In a reply to his compatriot the astronomer and polyhistor Biruni (d. *c.* 1050), he claimed that Philoponus had feigned dissent from Aristotle over the question of the eternity of the world in order to deceive the Christians.[22] In reality, as was clear from his commentary, for example, on the final section of the *de Generatione et Corruptione*, he quite agreed with Aristotle. Biruni, no one's fool, retorted that Philoponus had expressed his disagreement with Aristotle, not only in his *contra Proclum* and his *contra Aristotelem*, but also in his commentaries.[23] Presumably, Biruni was thinking above all of the commentary on the *Physics* (excerpted, as we saw, in the margins of at least one major edition of Aristotle's text). Presumably, Avicenna was no less familiar with that commentary than was Biruni. And presumably, he ignored its open disagreements with Aristotle so as not to upset his neat distinction

---

way; finite forces produce finite motions. Unlike Aristotle, he thinks that the impetus responsible for projectile motion resides in the projectile, not in the surrounding air. In either case, it cannot be infinite and therefore will spend itself in due course.

[20] *CAG* XVII 642,4f.

[21] In what is ostensibly a mere summary of Aristotle's argument at *Phys* 8.4, Ibn al-Ṭayyib (in the Leiden MS) flatly contradicts Aristotle, no doubt under the influence of Philoponus, as follows: 'Although the matter (*sc.* of the natural motion of the elements) is so obscure, it is not beyond us to understand that things do have a mover. For they are possessed of forces by which they are moved as animate things are moved by their souls' (ii.842.21-23 Badawi).

[22] The implication is not that Philoponus himself was not a Christian, but that in a Christian world he could not afford to be honest.

[23] S.H. Nasr and M. Maghegh (eds), *Al-Biruni and Ibn Sina: al-Asʾilah waʾl-Ajwiba*, Tehran 1973, 13,51f.

between the Philoponus of the commentaries, loyal follower of Aristotle, and the Philoponus of the public refutations, Christian detractor of the philosophical tradition. Now, on the issue of the eternity of the world, Avicenna found himself persuaded by Aristotle; on the issue of impetus, by Philoponus. Unlike (I suppose) Biruni, Avicenna thought that the second issue did not affect the first. He also thought the second rather less important than the first. The core of Aristotle's cosmology was sound and had to be defended. If Avicenna agreed with Philoponus that some revisions were required at the periphery, he was not going to advertise the fact.

The lesson is that the name of Philoponus did not in general inspire trust and admiration. It has often been assumed that his concern for the truth of scripture was bound to endear him to the Middle Ages. In reality his reputation was flawed. For in the eyes of posterity he had doubly disgraced himself by embracing the short-lived Tritheist faction within the Monophysite party and by attacking, so to speak, his own school (the philosophical tradition from Aristotle to Proclus) from behind. It is remarkable, and has been remarked, that the Christian representatives of the philosophical school of Alexandria in its final phase followed in the footsteps, not of the Christian Philoponus, but of the pagan Olympiodorus. Elias and Stephanus are likely to have been Chalcedonians at war with all brands of Monophysitism. No doubt they disapproved of Philoponus. His writings, then and later, enjoyed notoriety rather than authority. The influence they undoubtedly had was not one to be acknowledged gladly. His impetus theory seems to be a case in point: it was adopted without due credit given to its author.

The other point is that, under the influence of Philoponus, it was possible to mistake his impetus theory for Aristotle's own. Does not Aristotle quite regularly speak of the 'impetus (*rhopê*) of heaviness and lightness'? The passages in which he denies the existence of internal movers in inanimate bodies are few, scattered, and confused. One natural reaction would have been to shrug them off. It is true that in the face of Yaḥyā's outspoken criticism of *Phys* 4.8 it was impossible to overlook the folly of Aristotle's aerial-propulsion theory. But could that not be seen as just a momentary lapse from his own better judgment?

Avicenna certainly had reason to believe that *mayl* was an authentically Aristotelian term. The word translates *rhopê* at *Metaph* 10.1,1052b29 in the Arabic standard version from the early ninth century. More important, he had reason to believe that his *concept* of *mayl*, too, was an authentically Aristotelian one if he knew a treatise on the causes of motion extant in Arabic under the name of Alexander of Aphrodisias. In it the following sentence was found by Pines:[24]

---

[24] The text has since been edited and translated by N. Rescher and M.E. Marmura, *The Refutation by Alexander of Aphrodisias of Galen's Treatise on the Theory of Motion*, Islamabad 1965. The quotation is at p 17. Whether the Arabic is likely to reflect the underlying Greek accurately is hard to say. Some Arab translators of Alexander took liberties. The identity of the translator of the present piece has yet to be established.

He (Aristotle) likewise made clear (in *Phys* 8) in the case of bodies that move
naturally through the inclination (*mayl*) within them that their source of motion
is from the inclination existing in them by virtue of which they move naturally.

That sentence would have suggested to Avicenna that his own theory of *mayl*
had indeed been shared by Aristotle. But would it have suggested that theory
to him if he had not already known it? If, as one would expect, he had a more
substantial source, it has so far eluded us.[25] The most suggestive instances of
rhopê occur in the *de Caelo*. But the Arabic standard version lacks the word
mayl.[26] And so does Isḥāq's version of the *Physics*. A hint worth following up
may be contained in Avicenna's reference, mentioned earlier, to Philoponus'
commentary on the *de Generatione et Corruptione*. Aristotle's text does not speak
of *rhopê*, but Philoponus' commentary, occasionally and uncontroversially,
does.[27] The Arabic version of that commentary (as of the text itself) has not,
apparently, survived. Perhaps it was something in the Arabic of that
uncontroversial commentary which gave Avicenna the impression that
Aristotle believed the *motus separati* of inanimate bodies to be caused by
'inclinations' arising within them?

At all events, in the absence of the word *mayl* from the Arabic *Physics* and *de
Caelo*, it seems unlikely that Avicenna would have conceived his theory of
inclination merely by reading Aristotle. We must seek the source of his
inspiration among the commentators. Alexander may have helped. But as
Avicenna was clearly familiar with Philoponus, the similarity of their views
on inanimate motion can hardly be an accident. Presumably, Avicenna's
mayl is a straight descendant of Philoponus' *rhopê*.

3

It would be odd, as Pines pointed out, if the emergence in thirteenth/
fourteenth-century Europe of Latin impetus theories owed nothing to
the discussions stimulated in antecedent Arabic and Hebrew philosophy by
Avicenna's version of the impetus theory of Philoponus. (Avempace's
argument against the impossibility of motion in the void was also much
discussed in Spain; but it is not clear to me whether it was recognised as an
argument for impetus.) Anneliese Maier objected that none of the Arabic
sources mentioned by Pines were known to the Latin impetus theorists

---

[25] *Mayl* occurs, occasionally and not very suggestively, in the 'Yaḥyā' notes of the Baghdad
Physics. Avempace, who seems to rely on those notes, does not speak of *mayl*. I therefore think it
more likely that Avicenna derived his terminology from a source not shared by Avempace. For
the same reason, I am reluctant to see that source in some lost work by Farabi (d. 950), an
author valued no less by Avempace than by Avicenna (indeed, at *Shurūḥāt* 195.1f Avempace refers
his reader to Farabi's (lost) *Book of Changeable Entities* for a refutation of Philoponus' attack on the
eternity of motion). Pines lists one use of *mayl ṭabīʿī/qasrī* in Farabi. But that occurs in a
collection of treatises including pieces of doubtful authenticity.

[26] At *de Caelo* 3.2, 301a22 the Arabic (ed Badawi, Cairo 1961) leaves the word *rhopê*
untranslated. Two lines later it has *thaql* 'heaviness'. At 4.1, 307b33 it has *al-wazn* 'weight'
(Badawi reads *alwān* 'colours'; I am indebted to Professor G. Endress of Bochum for the correct
reading).

[27] See the *index verborum* to *CAG* XIV (2).

studied by herself.[28] But that is no reason to despair of the possibility of Arab influence on the Latins. It was possible to know Avicenna's concept of 'inclination' from writings other than his own. For example, Ghazali's *Maqāṣid al-Falāsifa* was translated into Latin (as *Logica et Philosophica Algazelis*)[29] as early as the second half of the twelfth century. It was widely read. And although it treats explicitly only of *natural* inclination (*inclinacio* in the Latin), it does make, as we saw, the crucial point that inclination is a cause of motion residing in the moving body itself.

Ghazali's book proved particularly popular in Hebrew. It was translated and expounded more than once, and often referred to. About 1400, the Spanish Jew Crescas, writing in Hebrew, adopted the view that inanimate bodies moved towards their natural places by means of a force inherent in their form. That view was associated by his predecessors with Avicenna and Ghazali.[30]

Ghazali's *Maqāṣid* is not the only link between Avicenna and medieval Europe. A great deal of Arabic philosophy lived on in the Hebrew literature of Spain, France, and Italy. Crescas, as Wolfson's study of his critique of Aristotle's natural philosophy impressively shows, had many well-read predecessors. Their erudite discussions will not have been without effect upon their Latin contemporaries. If Ghazali's intimations went unheeded, perhaps some further links are waiting to be discovered in the writings of those learned Jews.

[28] A. Maier, *Die Impetustheorie der Scholastik*, Wien 1940.
[29] Modern edition, without the section on logic, by Muckle (n 6 above).
[30] H.A. Wolfson (1929) 299, 672f.

CHAPTER SIX

Summary of Philoponus' Corollaries on Place and Void

David Furley

Corollary on Place

CAG
Vol 16
I. Critique of Aristotle's argument that place cannot be a
three-dimensional extension

557,8   (a) Aristotle's first objection (211b19-23): place will then go through the whole of the contained body and divide it into an actual infinity of parts.

    Even if place were a *body*, this would not follow, although it might seem plausible; but since it is incorporeal it does not follow *either* that the void goes through the contained body (because the body *fills* it), *or* that it divides it. Cf whiteness, heat, etc., which go through bodies without dividing them.

557,28   Does void divide by being an *extension*? No: lacking the qualities of body (hardness, etc.) it does not divide, just as the application of a surface or line to a surface or line does not divide them, nor the application of thousands of them – because they are incorporeal.

558,10   If a three-dimensional incorporeal extension divides a body when applied to it, then the same must be true of two-dimensional extensions. So Aristotle's own theory of place, which holds that the surfaces of the container and the contained coincide, is vulnerable to his own objection. This is not because the surfaces are in place (they are not), but just because they are applied to each other. There cannot be a special dispensation that says a length may be applied to a length without dividing it, or a breadth to a breadth, but not a depth to a depth. Hence extension as such is not a cause of division.

559,9   It is not extension, but extension with matter, that causes division. All acting and being acted on is caused by matter.

559,19   But further, even if place, being three-dimensional, were a body, it would not follow that the contained body is infinitely divided, although it would follow that body passes through body, which is impossible, or that one of the bodies is void. Aristotle himself objects to body passing through body, not that it divides it to infinity, but that the largest can then be contained in the smallest. Recall his good objection to the theory of growth through empty pores: he says it would follow either that the whole body is empty, or that there is no growth, but only filling of the pores – not that the void passes through the food and divides it to infinity.

That, then, is how those who say place is an extension would respond: 560,3
body, passing through the void, *fills* it, and the void neither divides nor is
divided by the body. The void does not come to be in the body, but the body
in it, because the void, being incorporeal, is immovable. If you insist that the
void passes through the body that comes to be in it, so be it: this does not
entail that the void moves.

(b) Aristotle's second objection (211b24-25): there will be several places in 560,16
one place.

This gets some plausibility from its compression, but it is not altogether
true. Why should it be considered absurd that two or more extensions
coincide? There is no difficulty in understanding how one line or plane can be
applied to another, or how many qualities can coexist in the same object:
hence, there is nothing absurd in supposing that incorporeal three-dimensional extensions coincide.

Three-dimensional extension does not entail being a body; a body is a 561,3
substance, composed of matter and form, that has the property of
three-dimensional extension. Aristotle makes this clear in his criticism of
Melissus' claim that what *is* is infinite (*Phys* 1.2), that infinity may be a
property of a quantity but not of a substance.

(We shall postpone consideration of the problem whether we sometimes 561,25
separate quantity from substance.)

There is nothing impossible, then, in the coincidence of two extensions. But 561,27
those who suppose place is an extension do not have to say there are more than
two extensions together, nor that place is in place, nor that place moves. When
a jar of water is moved, it does not take the internal extension, where the
water is, with it, but *changes* place, as a whole; the void is immovable. A solid
ball fills a place the same size as itself; when moved, it fills another extended
region; it does not shift any part of the void. In this respect there is no
difference between a solid ball and a hollow jar whose contents are in contact
with it.

And there is no need to suppose place is in place, or a plurality of extensions 562,29
coincide. A body, when moved, leaves its previous region of place and occupies
another. But only corporeal extension can occupy place-extension so that they
coincide; there cannot be a third, because two bodies cannot occupy the same
place. Just as a plurality of planes can coincide, but not more than two *actual*
plane surfaces can coincide, since three bodies cannot touch each other in the
same (place), so it is possible for a plurality of extensions to coincide, but this
never happens in actuality, since two bodies cannot be in the same place – nor
void in void, because the void is immovable. Even if it be conceded, falsely,
that void may be in void, it will be only in the way that lines or planes
coincide so as to make one *single* line or plane.

II. Philoponus' argument against Aristotle's proposition that place
is the limit of the surrounding (body)

(a) If it is *qua* body that a thing is in a place, but not *qua* body that a thing is 563,26 in the surface of the surrounding (body), place is not a surface. It is the limits of the body, not its inner depth, that are in the surface.

564,3   (b) A place must be equal to the body that occupies it, but a surface cannot be equal to a body.

564,14   (c) If place must be immovable, but the surface is moved along with the body whose surface it is, the surface cannot be place. To say that it is immovable *qua* place, although movable *qua* surface, will not do, because it is a place by virtue of being in contact with the contained body – and while I remain stationary the surface of the air in contact with me moves (i.e. changes) when the air around me moves. Not even the outer heaven provides an immovable surrounding surface: nothing is immovable except the earth.

565,1   (d) *Kinêsis* is either alteration, or growth, or in place. The last is either circular or straight motion. Since the outer heaven has circular motion, it must move in place. But what can be its place, since there is no container outside it? Aristotle's obscurity on this allowed his interpreters to twist his words as they wish.

565,21   Some say the convex surface of Saturn's sphere is the place of the fixed-star sphere, although this ignores Aristotle's insistence that place is the exterior container and is equal to the contained. They also confuse the roles of place and what is in the place; but this relationship cannot be reversed, any more than that between father and son.

566,7   Others say that the continuous parts of the sphere are places for each other. But Aristotle denies that the parts of a continuum are in place *per se*. Furthermore, if the parts can be place for each other, there is no necessity that place not be a part of what is in place. For if the whole is precisely a whole *of parts*, and each of the parts contribute to the whole, the parts are related to each other – for they contribute to the being of each other, since each when taken apart from the whole apparently loses its being.

567,8   (e) When a thing moves through a corporeal medium, we say the parts of the medium yield their place in turn to the moving body. If place is a surface and not a three-dimensional extension, then when I move from Athens to Thebes, the parts of the air through which I move yield nothing but surfaces to me. But even infinitely numerous surfaces, applied to each other, do not make a three-dimensional whole. So how does the moving body move forward?

III. Philoponus' defence of the claim that place is a three-dimensional extension

567,29   If we eliminate matter and form, as well as the limit of the surrounding (body), the only remaining option will be that place is a three-dimensional extension, distinct from the bodies that occupy it, incorporeal in its own definition – an extension empty of body.

568,1   This can also be shown directly, (a) In an exchange of place by bodies, as described in II (c), when a body moves through a corporeal medium, e.g. air, an amount of air equal to the body is displaced. Since the measure is of the same size as what is measured, if the displaced air is 10 cubic metres, then the extension occupied by it is also 10 cubic metres, and that is what it yields to the moving body. But that was its *place*. Hence, place is cubic, i.e. three-dimensional. Place is a measure of the occupants of place; hence, it is equal to them. A jar contains a certain amount of water: when we measure the amount, we measure not the perimeter, but the volume, not the shape but the contents. Moreover, when we measure the space in a jar, it is not the air it contained previously that we measure, because the water did not pass through the air and was not measured by it: the air got out of its way.

Hence there *is* an extension distinct from the contained bodies – not a space that either is or can be empty of body, but one that is distinct from the contained bodies and empty by its own definition, just as matter is distinct from form but can never be without form. And this extension while receiving a succession of different bodies remains unmoved both as a whole and in part – as a whole because the cosmic extension occupied by the whole cosmic body could never move, and in part because an incorporeal extension, void by definition, could not move.   569,5

(b) The force of the void (vacuum) proves both that this extension exists   569,18 and that it is never without body. Consider the clepsydra (vulgarly known here as the 'snatcher' – *harpagion*). The water in it does not flow out through the filter if the top vent is stopped, but rests on the air at the perforations, and stays up, contrary to nature, until the vent is opened. The cause is the force of the vacuum. Since the holes in the filter are tiny, air and water cannot change places through them: the water tends to fall, the air resists it, and a position of rest occurs because of the natural force of the vacuum in the whole.   570,14

How is it that water, which naturally falls, can be raised by sucking on a tube dipped into it? Sucking the air out would create a vacuum, but the famous force of the vacuum lifts the water against its nature. In this way by making a siphon with a bent tube one can empty a whole vessel. Once the water has started to flow in the tube, water is drawn up the tube out of the vessel continuously by the force of the vacuum.

But if there were no extension, distinct from the contained body and empty   571,9 by its own definition, in the clepsydra and the tube, how does nature prevent the water leaving the clepsydra or raise it in the tube? 'So that a vacuum may not occur'. But *what* vacuum, my dear fellow, if there is no extension between the walls of the vessel other than the extension of the water? When the water flowed out, nothing was left but the hollow surface of the vessel, and a surface cannot be either empty, or full. How much air is needed to fill the *surface* of the earth? What fills the exterior surface of the heavens, if it is agreed there can be nothing outside, as well as no void?

(c) A possible Aristotelian reply. We might say that a void occurs when the   572,7 surface does not touch a body – since it is a necessity of nature that every body touches something. But that would imply that every bounded thing is bounded *against* something, and Aristotle has adequately shown the falsity of that. Touching is relative to something else, but being bounded is not. There is nothing outside the cosmos, and so the outside of the heaven is touching nothing. Even if bodies inside the cosmos always touch something, Aristotle says, this is an accident, not an essential property, and they could lose it without infringing on their nature.

However, the following shows that the void is not just the absence of contact with a body. If this were the case, there would be a void outside the cosmos, since the outer heaven touches nothing; and we agree there is not. And Aristotle could not have understood void in this sense in his arguments against void.

573,22   (d) If bodies do not need a three-dimensional extension equal to themselves to move into, if the Creator made only bodies that need no extension to occupy, why do skins and jars burst when grape-juice turns to pneuma?

574,13   (e) A thought experiment. If we imagine all the earth, water, air, and fire in the cosmos to be non-existent, what would be left in the middle except a void extension? It would be possible to extend a radius from the centre to the periphery in all directions: what else would the radius pass through but a void three-dimensional extension? Let it not be said the hypothesis is impossible. For when a hypothesis has an impossible consequence, from the impossibility of the consequence we refute the hypothesis; we often posit impossibilities in order to examine the nature of things in abstraction. Aristotle imagines the heavens to be at rest in order to refute those who say the earth is supported by the swift rotation of the sky; he also strips bodies of every quality and form, to get an idea of matter all by itself. Plato asks what would the cosmos be like without God. My procedure is the same. If there were nothing between the exterior limits of the cosmos, then when the contained bodies were imagined away the limits would collapse upon themselves.

575,21   (f) This argument about the whole cosmos can be applied to a part also. Let there be a bronze sphere, not solid but containing air. If we suppose the air changed to earth or water (not an impossibility), then since that will occupy a smaller place than the air, there will be a remainder in which there was air but now there is nothing. Themistius says the sphere would collapse. He is right, but unwittingly: the reason is not that it was held in shape by the body inside it, like a balloon, but that the extension inside does not remain void.

576,12   Themistius objected wrongly to Galen, too. Galen imagined that no other body flows in when water is taken out of the vessel, in order to study the consequence. Themistius accused him of begging the question, and creating an extension to suit himself. But actually he was not saying 'let there be an extension empty of body', but rather 'let no other body enter and let us see if there is anything between or not'. If you think, Themistius, that if we suppose no other body enters when the water leaves there must be a separate extension there, you are caught on your own wings. You are saying that there *is* another extension inside, but one that is never empty of body – and that is just what I say.

577,10   (g) Defence of Philoponus' theory against the claim of Themistius and others, that local extension is just corporeal extension without qualities. It will follow on their hypothesis that body passes through body, whereas we do not say that extension is a body, but that it is the extension occupied by bodies and extension void of all substance and matter.

This extension has all the commonly agreed properties of place: it is equal to its occupant, immovable, no part of the occupant, separable, and the container of the occupant. It does not entail that body passes through body, or that parts of a body are in place in their own right.

(h) Dismissal of a possible objection: if place is extension without substance   578,5
or matter, consisting wholly in its dimensions, and if dimensions are in the category of quantity, then we are committed to the existence of a quantity without substance, which is impossible. I reply that our theories must conform to the facts, not vice versa. Just because we see quantity is always accompanied by substance and declare that it cannot exist in its own right, it does not follow that nature is in fact like that. We would have as much right to say that substance cannot exist in its own right, because we see it always accompanied by quantity. All the categories need each other. Matter, and three-dimensional, qualityless, corporeal extension, however much they can exist in their own right, nevertheless do not exist without qualities. So also local extension, however much it exists in its own right, never remains empty of body: one body always replaces another. Indeed, the force of the void may be just this – the inseparability of this kind of quantity from substance.

(i) How does extension, on our theory, accommodate the differences of up   579,19
and down, light and heavy? How can the limits of up and down be defined? How can this space have the power (*dunamis*) by which heavy and light seek their own place by virtue of their natural impulse? In Aristotle's theory, this is explained: the better and higher is related to the more deficient as form is to matter; so fire seeks to be contained by the surface of the moon's sphere, so as to be ordered and informed and perfected by it, just as the female desires the male or the bad desires the good.

This is easily answered. The same problems arise on Aristotle's theory.   580,18
How do you define the limit of the 'down' place? Imagine a mountain reaching to the moon: is the top of it up, or down? Imagine it first floating, detached from earth, then continuous with the whole earth. What is the limit of the 'down' place? Similarly with 'up'. If only what is in contact with the moon's sphere can be said to be up, then air is not up, unless there are two kinds of 'up'.

The truth is that there is no up and down by nature in the universe, but   581,7
only (as he says elsewhere) the circular and the circumference. Down is just the place taken by heavy bodies, up, that taken by light ones. To say place has power is ridiculous: things do not move through desire of a surface, but through desire of the order assigned them by the Creator. When they reach it, they most attain being and perfection.

I say that *up* is the region of space occupied by light bodies, and *down* the   581,32
region occupied by heavy bodies – and Aristotle cannot define up and down absolutely. If up is the concave surface of the moon's sphere, then only fire is up, not air, or there are two kinds of up; similarly with down. If the heavens below the fixed stars are to be in place, some must be up, some down, some light, some heavy. If a thing is contained by the surfaces adjacent to it, the same thing must be container and contained, place and occupant. My theory avoids these difficulties.

582,18   (j) It is objected that if there is such a void extension, it must be infinite. The limit of three-dimensional things is a surface, and only bodies have surfaces. Hence extension must extend to infinity beyond the body of the cosmos, and that is in itself irrational (*alogon*) and is refuted by Aristotle here and in *de Caelo*.

582,28   But I do not understand how it is that this objection appears plausible. Why is it not possible to conceive of a surface of the same kind as the extension in question? Even if this were impossible, it would not follow that the void is extended to infinity: being a place for bodies, it is as big as the possible extension of the cosmic bodies, and is limited by their limits. Similarly in a jug filled with non-continuous corporeal pieces each part of the extension within is limited by the limits of the occupying piece, and so with the whole.

583,13   (k) 'The philosopher' defended Aristotle thus: being an inquirer into nature (*phusikos*) he considers only what has a principle of motion and rest in it, and therefore excludes from consideration as the place of natural bodies such an extension as we propose.

583,30   My reply is, first, that Aristotle does not fail to consider such an extension, but considers it and explicitly denies its existence; secondly, that place according to Aristotle's conception also is immovable (212a16); and, thirdly, that there are many other immovable things in nature, such as the poles of the

585,4   spheres, and the souls of living things, both rational and irrational.

Corollary on Void

I. Philoponus' criticism of Aristotle's arguments against the existence of void extension

675,12   Introduction: I will first criticise those of Aristotle's arguments about void extension which start from the premise that bodies move with unequal speed. The goal is not to establish the existence of an extension that is actually void, but rather of a void that is actually always filled with bodies.

676,4   (a) Summary of Aristotle's arguments.
     (A) *Phys* 4.8, 215a24-b22. Assuming the same moving body, the same distance moved, but different media, the body will take a proportionately shorter time to move through a thinner medium. If there is motion through a void, it must take time; that time will have some proportional relation to the times for corporeal media; and hence the void will have a proportional relation with those media. But that is impossible.
     (B) *Phys* 4.8,215b22-216a4. (Alternatively), still assuming the same moving body but granting that to move through a void *does* take time: there must be a very thin medium such that a body will take the same time to move through it as through a void, which is impossible.
     (C) *Phys* 4.8, 216a11-21. Assuming the same medium but different bodies: in a corporeal medium bodies of different weight or different shape move at different speeds, because they part the medium differently. But in a void, where there is no medium to part, they must move at the same speed, which is impossible.

(b) Criticism of these arguments.   677,9

Against argument (C): It is inconsistent with his own statements. He concedes that differences between heavier and lighter bodies, as well as differences in the medium, cause differences in motion; but in that case this difference in the bodies themselves should make a difference even in a void. If not, then they should also move with the same speed through air, since the air makes no difference to their internal impulse, and this is plainly (*enargôs*) false.

He himself explains (215a29) that the medium is a cause through *resistance*:   678,13
the cause internal to the moving body must then be the *active* cause; and this remains when the resistance of the medium is removed.

The heavier body moves through air faster. It cuts the air faster just   678,29
because it has greater impulse: its greater impulse is not caused by its cutting through air faster. Hence it will also move faster through void. Qualities like heaviness are not in the category of relatives (*tôn pros ti*).

Against argument (B): speed of motion varies according to differences of   681,1
impulse and differences in the medium. If there is no resistance from the medium, a given body will move a certain distance in a certain time according to the magnitude of its impulse. If it moves through a medium, some time is *added* to this time because of the resistance of the medium. If the medium is thinned to infinity, the added time is reduced to infinity, but never to zero.

Hence, a given body does not take the same time to move the same   681,29
distance through a thin medium and through void. Aristotle was led into this mistake by supposing that the time of motion varies in direct proportion to the thickness of the medium. This is not easily seen to be false, because thickness is hard to measure. But if it were true, it would be reasonable to think (*eulogon*: 683,9) that in the *same* medium the time would vary in direct proportion to the impulse of the moving body, and this at least can be seen to be false: a body twice as heavy as another may fall faster but certainly not twice as fast.

Against argument (A): Aristotle mistakenly thinks that the media through   684,10
which motion takes place vary in thickness in direct proportion to the times taken. But this is not true. What varies in direct proportion to the medium is the time that is *added* because of the medium to the time taken to move through a certain distance in the void. No time is taken to get through the void as such; hence there is no proportional relation between the time taken to get through the void as such and the time taken to get through a medium as such. Hence, the void has no proportional relation to the full. The mistake is to suppose that the proportional relation of the times (not properly analysed) applies to what the times are predicated of – e.g. to suppose that because I take an hour to think of a theorem, and a stone takes two hours to drop through water, thinking is twice as fast as sinking, and water twice as thick as thought.

(c) Criticism of the argument in *Phys* 4.8, 216a26-21.

This aims to prove that the void is an unnecessary assumption, in that it is   686,30-
not needed as the place of bodies. A body occupies a place because it has a   687,13
certain extension (*diastêma*). If you abstract from it everything else, the

extension that is left will be indistinguishable from the postulated void place. So what need is there for this extra extension? And if *one* extra is needed, why not more?

687,14-   Criticism of this argument. I have already shown in the Corollary on Place
689,25   that there is such a thing as void extension, and so the argument that it is unnecessary shows no more than that the explanation still needs to be found. But the extension of a body is not the same as the void. If you perform the operation of abstraction postulated by Aristotle's argument, it is *matter* that is left, and this is not the same as void. And matter cannot be abstracted without causing the thing to cease to be a body at all.

688,9   The nature of properties depends on their being *in* a substrate body: e.g. in a substrate body, opposites cannot co-exist, although they can perfectly well co-exist in your mind, when you theorise about them. So this corporeal extension, if it is not *in* a body, will lack the properties it has when it is: it will not be in place, indeed it will not be extended at all.

688,25   Hence the body that is in place is the physical body, and since its own extension does not qualify, it needs a void space to function as its place – but only one, because whereas a corporeal extension needs an incorporeal one to be in, an incorporeal one does not need another.

II. *Philoponus' defence of the existence of void extension*

689,26   We shall now show that motion through void extension is not an impossibility, but motion is an impossibility without void extension. We have already shown the first in criticising Aristotle's arguments, but will now demonstrate it independently.

690,3   (a) The heavenly spheres, although they do not move by parting any medium, nevertheless take time to move, and move at different speeds. Hence the time taken by a motion depends on the impulse of the moving body, not only on the medium, and things may take different times in moving through a void. We have shown in the Corollary on Place that all bodies occupy an empty space (*chôra*). Hence the heavenly bodies do so, and space is what they are actually moving in with their different motions.

691,9   (b) An argument from willed motions. Imagine that the extension between Athens and Italy were void. Could I walk from one to the other, and all points in between, in *no* time and be everywhere simultaneously? These questions cannot be answered by saying that in walking or running one foot is always at rest, and it is the resting that takes time: in fact, the movement of the whole body is continuous. But a walk through a void cannot take *no* time, since one cannot be everywhere at the same instant, and yet it cannot be impossible to walk through a void, since there is nothing to impede such a walk. But if willed motions can occur in a void and take time, so can natural motions.

692,27   (c) An argument from forced motions. If two archers, one strong and one weak, shoot arrows in a void, it is impossible that both arrows reach the mark simultaneously. Forced motion in a void must take time: so must natural motions. Imagine the region between fixed stars and earth to be void, and a stone to be dropped from the stars: the stone must (a) take *no* time to fall to earth, and so be up and down simultaneously; or (b) stay up, unnaturally, supported by the void; or (c) take some time to fall to earth. Since (a) and (b) are impossible, (c) must be true.

(d) An argument to show that motion is impossible without void. When a body moves through air, it displaces a volume of air equal to its own volume. But it would not need to displace anything, if it did not need a space equal to itself in volume to occupy, and the air would not need to be displaced from this extension if it did not need another space to move into. Hence the locomotion of a body requires a succession of void spaces. These extensions are not, of course, ever actually void: the volatile air fills them too quickly for a void to be left.   693,28

A concluding note. When two differently shaped bodies move through a corporeal medium, the difference in shape may cause them to move at different speeds, because of a difference in the resistance of the medium. This is not true of motion in a void: in this case, only a difference in natural impulse will cause a difference in speed.   694,28

CHAPTER SEVEN

Philoponus' Conception of Space

David Sedley

1. The problem

Philoponus' conception of space takes much of its impetus from Aristotle. In *Physics* 4 Aristotle argues against the identification of place with form, with matter, and with a three-dimensional extension (*diastêma*) coextensive with the occupying body, and defends instead his own thesis that an object's place is the inner surface of the body containing it. This thesis was in a way among the least influential of his doctrines, and in a way among the most influential. On the one hand, few philosophers in antiquity ever thought that Aristotle was right. The dissent started immediately with Aristotle's second successor Strato, and continued almost uninterrupted through the entire Aristotelian exegetical tradition.[1] On the one hand, the crucially important work on the concept of space carried out by Epicurus, Strato, and numerous Neoplatonists[2] took its initial impetus from Aristotle's challenge. Even the Sceptical treatment of place, as conveyed to us by Sextus,[3] takes Aristotle's list of the four possible characterisations of place as its starting point.

Philoponus – like Epicurus, Strato, and many Platonists – defends the view of space as an extension, especially in the Corollary on Place which forms a digression within his commentary on Aristotle's *Physics*.[4] He adopts an at first sight surprising characterisation of this extension – 'a space (*chôra*) which according to its own definition (*logos*) is void, although it is always filled with body'.[5] How can something be described as 'void' when it is never empty? Interestingly, Philoponus' description of occupied space as 'void' had well

[1] Simplicius *in Phys* 601,19-20, attributes the Aristotelian view to the entire Peripatos, which, even allowing for exaggeration, must include at least Eudemus (see fr 80 Wehrli) and Theophrastus (whose *aporiai* – puzzles – about Aristotelian place at ibid. 604, 5ff and 639,13ff I therefore prefer to take as purely exploratory). For Strato's dissent see fr 60 Wehrli.

[2] For Epicurus, see my 'Two conceptions of vacuum', *Phronesis* 27, 1982, 175-93. For Strato, D.J. Furley, 'Strato's theory of the void', in J. Wiesner, ed, *Aristotles Werk und Wirkung*, Berlin and N.Y. 1985, 594-609. For the Neoplatonists, S. Samburksy (1982), and I. Mueller, 'Space and place', forthcoming.

[3] Sextus Empiricus *Adversus mathematicos* 10.24-36.

[4] Philoponus *in Phys* 557,8-585,4, in H. Vitelli's edition (*CAG*, Berlin 1887-8). Contrast his probably earlier *in Cat*, e.g. 33,22ff, 87,7ff, where he appears to accept Aristotle's view.

[5] e.g. Philoponus *in Phys* 563,23; 569,8-10; 687,16-17; 689,50-2.

established antecedents. Something like it was already envisaged by Aristotle, though not of course as his own view;[6] it has parallels in Epicurus;[7] and it was, according to Simplicius, adopted by Strato and by the majority of Platonists.[8] Nevertheless, Philoponus goes beyond the general run of Platonists in actually adopting 'void' as his *name* for space. He defends the usage with the analogy of prime matter, which in its own definition is devoid of all form, yet always possesses form.[9] *If* we could isolate space from its occupants, he might say, it would be an empty extension. To be occupied is not part of what it is to be space.

Are these considerations enough to excuse the usage? Not obviously so. Certainly to be occupied is not part of what it is to be space, but it does not follow that to be *un*occupied *is* part of what it is to be space. Yet that implication is hard to disown when space is described as 'void according to its own definition'. Why not stick to the alternative, much less misleading description '*incorporeal* according to its own definition, although always filled with body'?

The paradoxical nature of this usage is highlighted by the fact that by 'void' Philoponus can mean two quite different things. Sometimes the term designates this container space, void in its own nature but always occupied. At other times – even in a neighbouring sentence and without apology for the shift[10] – it designates what Philoponus more accurately calls 'separated void', that is to say, vacuum, actually empty space, which, he concedes to Aristotle, does not actually exist.

A second puzzling feature is that this concession to Aristotle does not prevent Philoponus from devoting a good deal of effort to rebutting Aristotle's individual arguments against vacuum in *Physics* 4.8. In the Corollary on Void, a further digression in the *Physics* commentary, he elaborately counters Aristotle's arguments for the conclusion that, if there were a vacuum, motion through it would be impossible.[11]

There is no doubt that Philoponus' replies to Aristotle's arguments against both vacuum and space are outstanding contributions in their own right, and major landmarks in the history of science. However, beyond their individual worth there are two unresolved questions. What possible merit can there be in bracketing vacuum and space together with the single label 'void'? And why should Philoponus devote so much of his energy to defending against Aristotle a concept, that of vacuum, which on his own admission corresponds to nothing in reality?

To show how utterly different the arguments about vacuum are from those for space, the following samples should suffice. First vacuum.

In *Physics* 4.8 Aristotle argues from the supposition that the speed of a

[6] Aristotle *Phys* 4.8, 216a26-b2.

[7] For Epicurus' treatment of place and vacuum as essentially identical, cf my 'Two conceptions of vacuum'.

[8] Simplicius *in Phys* 618,20ff.

[9] e.g. Philoponus *in Phys* 569,10-11, and Text I in Part 3 below.

[10] e.g. ibid. 690,25-34, and 693,28ff (translated in Part 1 below).

[11] ibid 675,12-695,8. Contrast his earlier acceptance of Aristotle's arguments at *in Cat* 86,23ff.

moving object is determined by the ratio of its weight to the density of the corporeal medium through which it travels. Assuming that motion through a vacuum, if there is such a thing, will take time, he argues that the time taken by an object in moving through a vacuum will stand in some proportion or other to the time it would take in moving the same distance through a medium. Hence the density of the vacuum itself will stand in that same ratio to the density of the medium. But there is no such ratio, since vacuum has no density at all. (Philoponus tends to interpret this as an argument that motion through vacuum would, absurdly, take no time at all.) Aristotle also argues that heavy and light objects will stand in the same ratio to the density of the vacuum, and so will, unacceptably, have to move at equal speeds. Philoponus' answer,[12] in brief, is that what determines an object's speed is primarily its own inherent 'impetus' (*rhopê*), of which the heavier object will have more. The effect of passing through a medium like air or water is not to cause the motion to take time, but to add *extra* time to its journey. For all motion is in time, even in a vacuum.

When we turn to Philoponus' arguments for void *space*, we are confronted with a very different set of arguments. The following will serve as an example:

> It has then, I think, been adequately shown that even if there were void nothing would prevent there being motion – and motion in time, for there is no timeless motion. But that it is absolutely impossible for motion to occur without the existence of void you can learn in the following way. [Note the characteristic ease with which the designation of 'void' has shifted from vacuum to space.] If when motion occurs through a body, such as air, it is absolutely necessary that the parts of the air swap places with the moving object, and that a volume of air equal to the moving object be displaced, and if there is no void extension occupied by the body which is displaced reciprocally with the moving object and which by yielding to the moving object has given it room to move, but the air reciprocally displaced has left behind no extension, what need was there for reciprocal displacement? So that *what* may yield to the moving object? Not a mere surface. For the moving object is not a surface, to be yielded to by a matching surface; and there could be no match between a body and a surface. For if the moving object always takes over in exchange an extension equal to itself, and this extension is necessarily either void or body, and body cannot pass through body, the extension which it is always coming to occupy must be void. For if the moving object is three-dimensional, and the whole of it changes place as a whole, it will clearly need an amount of space equal to itself. Otherwise it would not be necessary for a volume of air equal to it to be displaced reciprocally with it. Say it is itself one cubic foot in volume: that, then, is also the volume of space that it will need, i.e. of void. Hence it is impossible for motion to occur without there being void.[13]

Philoponus has here made use of the widely acknowledged phenomenon of *antiperistasis* or *antimetastasis*, the 'reciprocal displacement' that takes place between a moving object and its medium. Such place-swapping would indeed,

---

[12] Philoponus *in Phys* 677,9-686,29.

[13] ibid. 693,28-694,12. At 694,2 I have felt compelled to emend to *epiphaneia monê*, where the *MSS* have the accusative form.

as he says, be hard to explain without the supposition of a portion of *space* needing to be filled as soon as the object leaves it. But what, in view of Philoponus' defence of vacuum, is particularly striking about the argument is its incongruous appeal to a phenomenon on which the *opponents* of vacuum had relied heavily in the past. For the old atomist argument that motion requires vacuum had been effectively countered with the alternative model of motion by redistribution within a plenum.[14] Thus while Philoponus' argument helps the cause of space, it also serves as an uncomfortable reminder that the possibility of vacuum is a quite separate thesis, and one which, although it entails the existence of space, may require very different argumentation.

2. The force of void

In seeking to explain Philoponus' incongruous bracketing of space with vacuum, we might hope to make some progress by examining his appeals to the 'force (*bia*) of void', the phenomenon which subsequently became familiar under the label '*horror vacui*', and which like his predecessor Hero of Alexandria he describes with the examples of such suction devices as the water-carrying clepsydra and the siphon.[15] This notion was widely assumed in antiquity, but rarely discussed. At least two quite different positions can be distinguished:

Position 1: Vacuum is a logical impossibility. It is therefore a matter of logical necessity that matter close up any gaps before they can form.
Position 2: Vacuum could occur, perhaps even does occur, but constitutes an unnatural state, which nature acts to avert or rectify with all speed.

The logical incoherence of void (which, being the non-existent, by definition cannot exist) was a fifth-century B.C. Eleatic thesis. Its consequences for cosmology were accepted by, for example, Empedocles, who must therefore be classed under Position 1. Another adherent of Position 1 is clearly Aristotle, although for non-Eleatic reasons.

A version of Position 2 can be attributed to the third-century B.C. Peripatetic Strato, and to Hero of Alexandria in the first century A.D., the latter's *Pneumatica* being plausibly held to reflect Strato's influence.[16] In their view, not only does vacuum already exist in minute interstitial portions between particles of matter, but in a case of *horror vacui* like the sucking of wine up a pipe an extended vacuum is first created above the wine and then proceeds to draw the wine upwards after itself.[17] It seems a fair guess that

[14] The most famous discussion of this thesis is at Lucretius 1.370-83; see also J. Barnes, *The Presocratic Philosophers*, London 1979, vol 2, 95-100.
[15] Philoponus *in Phys* 569,18-572,6.
[16] See especially Furley, 'Strato's theory of the void'. The main texts are Hero *Pneum* I introd. and 1-4, translated in M.R. Cohen and I.E. Drabkin (1958) 242-54. Strictly I should say that for Strato and Hero it is an *extended* vacuum that constitutes an unnatural state. Also adherents of Position 2 are the Stoics, for whom vacuum never occurs in the cosmos, but is intrinsically possible, since it does exist outside the cosmos.
[17] This is clearly implicit at Hero *Pneum* I 2.

this latter thesis had no empirical basis, but resulted from the understandable feeling that motive force could hardly be attributed to a vacuum which did not even exist. It is, after all, likewise a little disconcerting to be told that there is no actual vacuum in a vacuum cleaner.

Philoponus, by contrast, regularly insists that no vacuum ever occurs.[18] He has no doubt seen that no actual vacuum need intervene in the story, its mere threat being enough. Could we then have here one reason for his bracketing of vacuum with space? If there is no vacuum to exert a force, perhaps the 'void' which exerts it is space itself, precisely in virtue of its inability to become vacuum. Certainly one text could be read this way:

> Perhaps this is the force of void – the fact that this kind of quantity [*sc.* space] is never separated from substance.[19]

On the other hand, another text points in just the opposite direction:

> Again, the force of void will establish both points, for anyone who looks to the truth. I mean (a) that there exists a distinct extension apart from the entering bodies, and (b) that this extension is never without body. For why is it that in clepsydras (or 'snatchers', as most people call them here), if the vessel is full of water and there are numerous holes in the base, whenever we block the opening at its mouth with our finger the water does not escape through the holes in the base, despite being heavy and having so many exit routes, but rests on the air at the holes and, contrary to nature, stays up, yet when we unblock the upper opening the water gushes out through the holes in the base? The reason for this is nothing other than the force of the void. For *since it is not possible for there to be void*, if when the upper opening is blocked the water were to escape downwards, the air would have no route for being reciprocally displaced into the interior while the water is coming out ... Thus with the water travelling downwards and the air pushing against it, *the natural force of void in the world* produces immobility.[20]

Here the italicised words, if taken in conjunction, leave little doubt that the force of 'void' is that exerted by vacuum, in virtue of its inability to exist. We might still hope to make the two alternative analyses of 'the force of void' coalesce satisfactorily with the observation that the vacuum to which the force belongs is *potential* vacuum, and that potential vacuum is simply identical with space. Unfortunately so weak a notion of potentiality seems too dangerous to import here without explicit evidence. Philoponus, like Aristotle, would surely baulk at a potentiality, *dunamis*, which is never actualised. His own stated policy, indeed, is to confine *dunameis* to natural capacities or powers which are actively exercised, and he explicitly denies that space, being pure vacuity without qualities, can have any.[21] It is therefore safer to assign no

---

[18] e.g. Philoponus *in Phys* 569,10; 689,30-2, and many of the texts translated below.

[19] ibid. 579,14-15.

[20] ibid. 569,18-31; 570,3-4.

[21] ibid. 579,27-580,3; 632,4ff; only at 572,3-4 does he speak of *horror vacui* as a *dunamis*, and he may there perhaps be thinking of the natural power of *body* to forestall vacuum. For Philoponus' treatment of *dunamis*, cf ib. 342,15ff; Simplicius *in Phys* 1131,2-1140,8; and M. Wolff (1971) 92ff.

physical powers to space as such, and to take the force (*bia*) in question to be of only a very attenuated kind. This will be rather easier if we take it to be the purely notional force of a non-existent vacuum.

There remains the hope that a solution to our second problem may be offered by a later paragraph in Philoponus' discussion of the force of void:

> If then there were not within the clepsydra or the pipe of which we have spoken an extension distinct from the occupying body, void according to its own definition, whyever does nature either in the clepsydra prevent the water from moving with its natural motion when the upper mouth is blocked, or in the pipes make it travel upwards contrary to nature? The standard explanation says 'So that a void should not be created'. Well, what void was *going* to be created, my friend, if there is no extension between the edges of the vessel distinct from the extension of the water? It would follow that when the water has escaped nothing but the hollow surface of the vessel is left. So what was going to become void? The surface? I don't think anyone in his right mind would say that the surface is full or void. For if surface, *qua* surface, is by its nature filled with body, what is the quantity – the volume, that is – of the air filling the surface of the earth? You won't be able to specify. What is the quantity of air filling the surface of this cube? ...
>
> If, then, place is a surface, and there is no extension void in its own definition, what is this power in nature which often moves bodies in the contrary-to-nature direction in order that void should not arise, supposing that the extension threatening to become void does not exist in the world?[22]

Although Philoponus' primary purpose here is to underline the indispensability of space, in direct opposition to Aristotle, he does at the same time betray a concern that if *horror vacui* is to be intelligible the threat of vacuum should be a genuine threat, in other words that we should have a clear understanding of what it *would* be. Aristotle's view of place as surface, with its accompanying denial of an independent spatial extension, makes vacuum a logical nonsense, leaving it unclear why *any* physical process need intervene to prevent it becoming a reality. A similar concern on Philoponus' part may be evident in another passage:

> But I have often observed that this void, even though I say that it has independent existence in its own definition (*logos*) and is occupied by different bodies at different times while itself remaining motionless, nevertheless is at no time left without body, according to my account – just as matter is at no time left without form, or body without quality, despite having individual existence in their own definition. Thus even though the void has individual existence, it is never in fact without body. For the air, being fluid and mobile, never allows a void space to be left, but is displaced reciprocally with the bodies too quickly for void to be left.[23]

Philoponus' wording here appears to leave the non-occurrence of vacuum as no more than a contingent fact. Although it *does* not occur, here as often he

---

[22] Philoponus *in Phys* 571,9-23; 572,2-6; and cf 576,10-12.
[23] ibid. 694,19-27.

prefers to avoid the modal assertion that it cannot occur. And even the reason given for its non-occurrence looks much more physical than conceptual – the extreme mobility of air. It might seem a not implausible inference that Philoponus is an adherent of position 2, and that his elaborate defence of the notion of vacuum against Aristotle's criticisms is motivated at least in part by the need to show that, despite the total absence of vacuum in the world, the threat of it is an entirely real one, whose aversion requires physical measures – the instant refilling of all vacated spaces – and not just an effortless reliance on the laws of logic.

3. *The comparison of space with matter*

It would, however, be unwise to acquiesce in this solution without considering the following objection. Despite the evidence presented above, Philoponus must consider the non-occurrence of vacuum very much more than a contingent physical fact. For he regularly compares it to the non-occurrence of prime matter without form, and this latter fact, even if Philoponus chooses to express it non-modally, is surely rather more than the avoidance of an unnatural state. Of course, comparisons need not necessarily apply exhaustively to every detail, and it might be wondered whether to approach the question in this way is not to squeeze the analogy harder than its author intended. But we will see shortly that by consigning both dependence-relations to precisely the same area of his metaphysics Philoponus effectively precludes the possibility of any such disclaimer.

Two passages from the *Physics* commentary throw some light on the issue. The first is Philoponus' answer to the following objection. Space, if it exists, will be pure quantity. But quantity has a categorical dependence on substance: it can only exist as the quantity of some substance, never in isolation. Philoponus' reply includes the following remarks:

Text I

(i) Then again one can add that all the categories subsist not without reciprocal interweaving. For you cannot find one category without the others interwoven, not even substance, which is actually said to be able to subsist by itself. (ii) For both matter and second substrate, by which I mean three-dimensional body-without-qualities, can so far as depends on them subsist by themselves, yet never do subsist without qualities. (iii) Therefore spatial extension too, even if it could, so far as depends on it, have subsisted by itself (for what was there to prevent a space being void of body, as we said, if we conceive the jar as containing no body?), in fact never stays without body by itself, (iv) but just as in the case of matter when one form perishes another immediately succeeds it, so too in this case the reciprocal displacement of bodies never leaves the space void, but at the same time as one body leaves it another enters to replace it. (v) Thus this kind of quantity too can never be found without substance. And perhaps this is the force of void – the fact that this kind of quantity is never separated from substance. This would save what seems to be uncontroversial, so continually and habitually is it said, namely that quantity could not exist by itself without substance. For void can never exist in separation from body.[24]

The other potentially illuminating passage comes from Philoponus' reply to the following argument of Aristotle's. If there is void space coextensive with a body, it will be indistinguishable from the body's own extension, which we can abstract in thought by stripping away all its other properties. And if two identical extensions can coincide, then (this seems to be Philoponus' own addition to Aristotle's argument) why not two bodies? And indeed, why not *any* number of extensions? Philoponus' retort includes the following:

Text II

(i) Also, he is at fault in conflating void and body. For not even if you eliminate all the quality of the body will the bodily extension be identical with the void. For even if we remove all quality from the body, the volume of matter will be left, and the body-without-qualities, which consists of matter plus quantitative form. (ii) But void does not consist of matter plus form. For it is not even a body, but incorporeal and immaterial – merely the space of a body. (iii) If then when the qualities have been removed from the body that which is left is no less a body, and void is not a body, it will never result that a body is in a body, if body is in void as its place. (iv) 'Yes it will', comes the reply, 'for if it is *qua* extension that the extension is in a place, let its matter, which being by its own definition incorporeal is not even in a place, be removed too. In that way the extension of the body will not differ at all from the void.' (v) But such a person should know that he is asking the impossible. For the body, which we say is in a place and cannot pass through another body, is nothing other than that which consists of matter plus form. Therefore, once the matter is subtracted, the body's form vanishes too, since it is in matter that it has its being. (vi) Thus just as when we call white 'that which dilates vision' we mean that this only belongs to it when it is in body as substrate (and when we say that opposites cannot coexist in the same thing we mean those opposites which are in bodies as substrates, such as the cold and the hot in a body, and if we are going to conceive them outside body we *eo ipso* immediately also remove this kind of being from them; for if we are going to conceive opposites not in a body as substrate, they are not even in conflict with each other and there is nothing to prevent their existing together: for it has been shown that nothing prevents the genus from containing actual opposites, and one and the same soul possesses their definitions together; indeed, the knowledge of opposites is single), and neither will white dilate vision nor black contract it if they are not in a body, (vii) so too also, if you conceive a corporeal extension without matter, neither will it any longer be in a place, being no longer physical, nor could one say of a thing like that that it does not, <or does>, pass through a body. (viii) Nor indeed will it be in any way extended, except in definition; or rather, no such body in any way subsists, unless one is speaking of the paradigm or definitional account, which are irrelevant to the present issue, since our investigation is about physics. (ix) It is therefore absolutely necessary that the body which we speak of as being in a place be none but the physical body, i.e. the one composed of matter plus form. (x) Thus from the fact that this is in void as its place it would not follow that there is a body in a body (for void is not even body), nor that void is redundant, if it is the place of

[24] ibid. 578,32-579,18.

bodies. (xi) Nor indeed will it be *qua* extension that the body is in another extension, but *qua* bodily extension in spatial extension. Hence there is no necessity that void too be in a further extension, if it is not *qua* extension but *qua* body that the body is in a spatial extension.[25]

Apart from their intrinsic interest, these passages reveal the bare bones of the metaphysical theory with which Philoponus was working at the date of the *Physics* commentary. We can start with the following schema:

(a) void, or spatial extension

(b) matter
(c) quantity, or bodily extension
(d) quality
[(e) other categories – not relevant to present discussion]

This can be supplemented with the following equivalences:

b = 'first substrate'
b + c = 'second substrate', or 'body'
c, d = 'form'

Below the line stands the analysis of substance. Now under this heading we are told, in Text I (i)-(ii), that matter, quantity and quality are mutually dependent, i.e. that no one of them is ever found in separation from the others. However we also learn, from Text II, that this interdependence is asymmetrical. On the one hand, no item on the list is conceivable independently of those placed above it. Thus, in the example discussed in Text II, bodily quantity or extension is totally dependent on matter. Although in pure thought you can abstract it, the price is its disqualification as part of the furniture of the physical world, as explained in Text II (v)-(ix). A conceptual stripping operation (of the kind envisaged by Aristotle at *Metaphysics* Z 3 to isolate matter) could never get down to pure bodily extension, since to do so would require us to peel off matter, which is prior, while leaving form, which is posterior.

On the other hand, when we reverse the operation no such obstacle is encountered. Matter *can* exist without form – 'so far as depends on it', in Text I (ii), or 'according to its own definition (*logos*)' in other texts which we have met. Likewise 'body' (= matter plus extension) can exist without qualities.

What kind of separability is intended here? Presumably we are to think once again of the stripping operation. Whereas weight or colour which are not the weight or colour of some body are inconceivable in the physical world, a body stripped of both weight and colour *is* conceivable (something like air, perhaps?). Extend this stripping operation to all its other qualities, and the original body has still not forfeited its existence – it must remain as that to which the qualities could be reassigned when we reverse the process. Another

[25] ibid. 687,29-688,33. I read *te* for *de* at 688,21, and add <*ê chôrein*> after *chôrein* in 22.

way to put this is that its definition, say 'matter with extension', makes no mention of qualities. Similar, if more puzzling and controversial, is the further removal of extension. This still leaves something, matter, which is both ontologically and definitionally prior in that it is that to which the extension belongs. Thus matter is separable from form 'so far as depends on it'. Yet it *is* never without form (Text I (i)-(ii)).

The significance of this latter ontological relation is that it is precisely the one in which space stands to body. Returning to our schema, we can note that as matter stands to the categories below it, so space, above the line, stands to substance, below the line. Space could 'according to its own definition' exist without body. (Whether body could according to its own definition exist without location is not discussed.)[26] Nevertheless, space never *is* without body.

Philoponus' specific argument in Text II relies on our appreciating the great ontological difference between the two kinds of quantity, (a) spatial extension and (c) bodily extension, the two items which Aristotle wished to collapse into one. For the latter's existence is parasitic on that of matter (for a bodily extension to exist just *is* for it to belong to some portion of matter), whereas the former's is not. Yet this point about the independence of spatial extension can be, and is, made without the slightest suggestion that the actual separation of space from body is a real possibility. Indeed, although the relative weakness of the ontological dependence of space on body, or of matter on form, is emphasised by Philoponus' preference for saying just that the one *does not* exist without the other, he is fully entitled to the stronger assertion that it *cannot*. This stronger claim is scarcely deniable in the case of matter and form: pure prime matter could not even have a spatial location, according to the principle implicitly accepted by Philoponus at Text II (iv). And in connection with space and body he does indeed permit himself – hardly by accident – to express the point modally in the crucial last sentence of Text I (where the non-modal version would leave it quite unclear what the 'force' of void consisted in).

This same curiously ambivalent relation of dependence is one which elsewhere Philoponus calls that of being an 'inseparable accident'. Quantity, he says, is an inseparable accident of substance, meaning by this that quantity is definitionally posterior to substance but that every substance necessarily has some quantity.[27] Tentatively, then, we could attribute to

---

[26] Matter (b in the schema) does *not* presuppose space (a), since in itself it is unextended. What, if anything might be held to presuppose space is body (extended matter, = a + b), and Philoponus could insist that bodily extension is only conceivable in space. However, I know of nowhere where he does so. In his *Categories* commentary he explicitly denies it (33,15-31), but this was written before his development of his own characteristic theory of space (see n 4 above).

[27] Philoponus *in Phys* 561,3-24. The notion of an 'inseparable accident' occurs at 11 and 23 (where G's *all' achôriston* should surely be read, rather than *alla chôriston*, which Vitelli retains from the other *MSS*). Philoponus' discussion raises rather startling difficulties. In order to show, very much as in Text II above, that the coincidence of two extensions would not entail the coincidence of two bodies, he argues that mere three-dimensionality does not entail body because quantity is not definitional to body but its 'inseparable accident'. In support of this last claim he cites Aristotle *Phys* 1, 185a32-b3. But (a) what Aristotle there calls accidental to a

Philoponus the thesis that space has occupants, not by definition, but as its *inseparable accidents*. This expression throws immediate light on one question, why Philoponus should prefer to say that space *is* never without body, but still feel entitled when the need arises to add that it *can* never be without body. The former locuaion does justice to the notion of an accident, the latter to that of inseparability.

We can conclude that the impossibility of there being space without body, although no doubt conceptually weaker than the impossibility of there being a body without a place, is still an ontological impossibility comparable to the impossibility of there being prime matter devoid of form. Returning to the evidence examined in the previous section, in which *horror vacui* seemed to result from nothing more than the extreme fluidity of air, we can now see that this impression was, at best, misleading. The fluidity of air explains *how* vacuum is always forestalled. But to know *why*, we must turn from physics to metaphysics. Hence (a further consequence) when we look back to the two main versions of *horror vacui*, (1) vacuum as a logical impossibility, and (2) vacuum as an unnatural state, it now becomes hard not to relegate Philoponus from position 2 to position 1. (It could be added here that far from identifying *horror vacui* with the avoidance of an unnatural state, the description '*force* of void' deliberately emphasises its ability to move things in a contrary-to-nature direction.)[28] Thus we seem no nearer a solution to our second puzzle, why Philoponus should so energetically defend the notion of vacuum while denying that there is any such thing.

4. The solution

Nevertheless, the last section has offered us a positive way forward. It has revealed Philoponus' own predilection for thought experiments, provided only that they are properly conducted, and in particular for those which lay bare some underlying metaphysical item by peeling away all the layers which are definitionally posterior to it. We have also seen a hint, at Text I (iii), that Philoponus may have envisaged some such thought experiment for isolating void. The experiment is more fully described elsewhere. For example:

> Alternatively, if you imagine the air reciprocally displaced with it [sc. the moving object] as having itself been displaced from its own place, but neither

substance must be not the determinable, quantity, but any determinate quantity. And (b) Philoponus *does* normally regard quantity as definitional to body, since it is precisely what differentiates it from matter: he only dodges this obvious challenge by first running the argument for substance, then substituting body for substance as definitionally independent of quantity; that no doubt raises problems about what his precise conception of substance is, but it is hardly enough to warrant the parallel treatment of body without further justification. It is these deficiencies that have led me to accord the passage less emphasis than it would otherwise deserve.

For a further probable use of the same notion in a later work, see his words quoted by Simplicius *in Phys* 1333,4ff: if a body is indivisible this is not by its own *logos* but because of an extra, accidental (*episumban*, ib. 9) form; here once again Philoponus invokes the comparison of the matter-form and body-quality relations.

[28] For Philoponus' use of *bia* in the familiar Aristotelian sense of a counter-natural force, see e.g. the reports of Simplicius *in Phys* 1135,11-12, *in Cael* 158,13ff, and Philoponus *in Phys* 571,2-13.

the moving object having come to be in the same place nor the surrounding air having collapsed in, what else could that which is contained within the surrounding air be except a void extension into which the moving object passes? (A broader discussion of this can be found in my treatise on place). From this too, then, it is clear that the moving object occupies one void space after another in succession.[29]

In the broader discussion to which Philoponus parenthetically refers here,[30] he conducts the more ambitious experiment of imagining the entire region between the earth and the heavens empty of body, and points out that a line drawn from the centre to the circumference would have to be described as passing *through* something, which could only be a void three-dimensional extension.

What is immediately illuminating about these experiments is that by isolating space from its occupants they provide a context in which the distinction between void as space and void as vacuum is erased. Although space is ontologically prior to vacuum, in order of understanding it is not: the most effective way to get to the notion of space is *through* that of vacuum. To call space 'void according to its own definition' is thus not to imply that it is its nature to be unoccupied, but to observe that you can only appreciate its intrinsic nature by subtracting its occupants. Similarly, to say 'Man is in his own nature naked' would not be to advocate nudism, but simply to observe that clothing is not intrinsic to human nature, that to appreciate the nature of man you must, if only in thought, subtract his clothing.

Moreover, there is an excellent reason for Philoponus to have felt compelled to isolate space in just this way. According to his theory of motion, at which we took a brief glance in part 1 of this chapter, a moving object's speed is primarily determined by its own intrinsic 'impetus', while the resistance of a corporeal medium is a secondary factor which slows it down.[31] Now in order to distinguish these two factors in any given case, it would be necessary to isolate the first by asking what the object's speed would be *if it were moving in a vacuum.* For to ask that is no different from asking what is its speed *through space as such.* Once this has been answered, the resistance of the corporeal medium can be entered into the calculation as a secondary factor.

We have here, then, answers to both our initial problems. 'Void' designates both space and vacuum because the notion of pure space *can* be effectively isolated by the mental operation of subtracting all the body from it, and *must* be isolated in this way if we are to understand the respective contributions to motion (a) of the object's inherent tendency through space *per se* and (b) of the resistance of the bodily medium occupying the space. And this explanation already has built into it the answer to our second problem. Philoponus' commitment to defending the conceptual coherence of vacuum, with special emphasis on the possibility of *motion* through a vacuum, is essential to this whole enterprise. For motion through space *per se* is only intelligible if

[29] ibid. 694,12-19.
[30] ibid. 574,13-575,20.
[31] See especially ibid. 384,11-385,11; 641,13-642,20. Also M. Wolff (1971) and (1978).

conceived as motion through a vacuum.

This solution carries with it implications about Philoponus' own philosophical priorities. From the defence of void summarised in Part 1 above, it might have understandably been inferred that his true goal was the rehabilitation of vacuum, and that to this end he elaborated his impetus theory, adding the distinction between the basic time taken by a moving object and the extra time due to the resistance of the medium, in order to ward off Aristotle's charge that speed in a vacuum would be infinite. Now, however, we can see that it is precisely the other way about. The distinction between basic time and extra time is an integral component of the impetus theory, and lies closest to Philoponus' heart. The entire defence of the logical coherence of vacuum is a strictly ancillary operation in its support.

To end, it is worth drawing attention to an outstandingly ingenious argument in which Philoponus leaves his reader seriously wondering whether there is *any* proper distinction to be drawn between space and vacuum.

That even if there were a void extension separated from body there would be nothing to prevent bodies from moving through it in time, but that there would be different degrees of speed in motion, has already been demonstrated by the refutation of the arguments which purported to establish this [sc. Aristotle's] view, but we must try to establish it directly as well.

First, if there is circular motion, and that not a singular motion, but differentiated (for the different spheres move in different ways, one faster, another slower); and if things moving with a circular motion do not move through some body (for it is not even by dividing different bodies at different times that they move in this way, but they rotate within themselves without cutting through any body, and the sphere of the fixed stars does not even have contact with anything external); if, then, the spheres, although cutting through no body, nevertheless move both in time and at different speeds, it is not after all the fact that motion is through a body that is responsible either for time being spent in motion or for there being different degrees of speed in motion. Rather, different degrees of speed result from differences of power in the moving objects, and the reason why time is always spent in motion, and would be, they think, even if the motion were supremely rapid like that of the sphere of the fixed stars, is the very nature of motion – I mean the facts that every motion is from somewhere to somewhere, and that it is impossible for the same thing at one and the same moment to be in different places. If therefore the fastest-moving thing there is, which in its motion divides no body, nevertheless owing to the very nature of motion moves in time, why should not the same also be true of objects moving with rectilinear motion, even if they are moving through void? Thus this very fact of moving is the reason for their moving in time, while it will be because of their differing innate impetus that they have different degrees of speed. For if in the spheres all motion is in time even though there is no medium to be divided, and the difference in degrees of speed exists through no other cause than the powers innate in them, presumably all the more in generable and perishable bodies, even if they move through void, will it follow both that they move in time and that they have different degrees of speed.[32]

[32] Philoponus *in Phys* 689,32-690,27. I read *kan hôs* with *G* at 690,14, in place of *kalôs*, and add <*mallon*> after *ge* in 25 (cf 32). In 27, which I take to mark the transition from this passage to a

The principal insight of this passage is that in the case of spherical rotation there is no proper distinction to be made between motion through occupied space and motion through a vacuum. A sphere rotates through pure space, with no obstructive medium to push apart, and yet that space is entirely filled at all times. It would be hard to find more effective confirmation of Philoponus' position, that motion as such is through space as such, and that space's own contribution to motion in no way depends on the presence in it of any body other than the moving object itself.[33]

separate, although closely parallel, argument, I would suggest emending *de ei* to *epei* (for a simple *allôs te* introducing an alternative argument, cf 694,12).

[33] In thinking out these problems I was lucky enough to have access to David Furley's invaluable summaries of Philoponus' two Corollaries (see above, Chapter 6). I am also deeply indebted, for expert criticism and advice, to Henry Blumenthal, Geoffrey Lloyd, Henry Mendell, Ian Mueller, Richard Sorabji, and Christian Wildberg, none of whom, however, should be assumed to endorse all the views expressed in this chapter.

CHAPTER EIGHT

Philoponus on Self-Awareness

Wolfgang Bernard

In modern philosophy awareness and self-awareness are usually considered to be central to psychology and epistemology. It is well-known, however, that awareness in no way holds a key position in Platonic and Aristotelian descriptions of the soul. It is, in fact, hardly discussed at all. H.-R. Schwyzer collected a number of passages bearing on the subject from Plotinus and many other ancient philosophers.[1] He did not mention a most instructive passage in Philoponus' commentary on the *de Anima*, which adds a new term to the list of Greek expressions roughly equivalent in meaning to '(self-)awareness'. My attention has been drawn to this passage by Arbogast Schmitt, who will be explaining its epistemological significance in a forthcoming book.[2] I believe the passage to be by Philoponus, despite the verdict of the editor of this commentary, Hayduck.[3]

[1] H.-R. Schwyzer, ' "Bewusst" und "Unbewusst" bei Plotin', *Les Sources de Plotin (entretien sur l'antiquité classique*, Tome V) (Fondation Hardt), 1957, 343-78 (including a short bibliography on 377f). Cf also A.C. Lloyd, 'Nosce teipsum and conscientia', *Archiv für Geschichte der Philosophie* 46, 1964, 188-200 and R. Mossé-Bastide, *Bergson et Plotin*, Paris 1959, ch 2.

[2] Arbogast Schmitt, *Subjektivität und Innerlichkeit. Zur unterschiedlichen Begründung und Darstellung der Selbständigkeit des Erkennens und Handelns in Antike und Neuzeit (Saecula Spiritalia* Band 6) Baden Baden 1987.

[3] It has occasionally been disputed that Philoponus is the author of the Greek commentary on *de Anima* 3 that is printed in *CAG* XV. The editor, M. Hayduck, maintains in his preface (p v) that in the commentary on book 1 and 2 Philoponus' 'verbose labouring' (*verbosa industria*) is to be found everywhere, whereas 'a certain jejune and scanty brevity of interpretation' is characteristic of book 3, 'which is quite removed from his [Philoponus'] character and custom'. Hayduck therefore suggests with some diffidence ('even though I dare not say anything with certainty') that book 3 might be by Stephanus (codex Parisinus 1914 (twelfth-century) has an adscript by a later hand saying 'Third book from the voice (*apo phônês*) of Stephanus'; the same appears in the late (fifteenth-century) codex Estensis III F 8; but 'from the voice of Stephanus', if correct at all, would imply that the book was written by a disciple, rather than by Stephanus himself). In addition to that, Hayduck takes 543,9 'as we learnt in *de Interpretatione*' to be a direct reference to Stephanus' commentary (*CAG* XVIII,3), 8,32ff; but how can we be sure that Philoponus did not lecture on *de Interpretatione*? There is at least one other commentary on *de Anima* 3 attributed to Philoponus, part of which (on 3.4-7) survives in a Latin translation by William of Moerbeke (G. Verbeke (1966)). There is, however, no difficulty in assuming that Philoponus held lectures on the *de Anima* in different places at different times and that several collections of his comments existed side by side in his day (cf also O. Schissel von Fleschenberg (1932)). Hayduck also points out that the Greek commentary on *de Anima* 3 is divided into *praxeis*, lectures, whereas books 1

At the beginning of *de Anima* 3.2 Aristotle raises the question of how we perceive that we perceive.[4] He first shows (425b12-16) that to postulate an additional faculty for perception of perception leads to logical difficulties. Consequently, the perceptive faculty itself must be assumed to perceive both perceptibles (*aisthêta*) and its act of perceiving them (b17). This, in turn, is not quite unproblematic, for if, for example, the object of vision is colour, the actualised perceptive faculty[5] would have to be coloured so as to enable vision to perceive the act of seeing (425b17-20). Aristotle solves this problem by drawing attention to the fact that vision discriminates not only colours, but also light and darkness, though in a different way (b20-22). He does not elaborate on this but he obviously means that we discriminate darkness from light by perceiving that visual perception is impaired or impossible.

Aristotle's second point is that one might indeed say that the actualised perceptive faculty is coloured, in a manner of speaking.[6] After all, the sense organ takes in the perceptible form (*aisthêton eidos*) without matter (*hulê*). And this enables it to retain perceptions and representations (*phantasiai*) in itself after the *aisthêta* have gone (425b22-25). That is not to say that the actualised perceptive faculty (or the sense organ *qua* sense organ) is a coloured object, something coloured. Rather it is the 'discriminative holding' of the perceptible form.[7] Hence the form (e.g. colour) is present, though not in association with matter, so that one can speak of the actualised perceptive faculty as being coloured, though not in the ordinary sense. Perception of perception could therefore be explained as the cognition of the presence of the perceptible form in dissociation from matter, which is the immediate result of the discriminatory activity of sense perception.

I have tried to show elsewhere[8] that this is the best interpretation of the Aristotelian text but, even if this is not granted, it is certainly based on Philoponus' exposition of the problem.[9]

---

and 2 are not. The most reasonable explanation of all this seems to be that the commentary on books 1 and 2 was planned and written as a whole and that the commentary on book 3 either was left uncompleted by Philoponus for some reason or may have been lost fairly early (though one part of it may have survived till the time of William of Moerbeke, only to disappear shortly afterwards). At any rate, a collection of comments on *de Anima* 3 by Philoponus might then have been used to supplement the now mutilated original commentary. This would explain the slightly different structure of book 3 (Hayduck definitely overstates the matter, the passage to be treated in this article (and book 3 in general) can hardly be described as 'jejune'). In any case, it certainly takes more hard evidence seriously to dispute someone's well-attested authorship of a work.

[4] I cannot explain the reasons for my interpretation of the passage in the present context. For a full discussion (and bibliography) cf W. Bernard (forthcoming) chs III 1 and II 9.

[5] Literally 'that which sees', reading *to horôn* twice at 425b19, which is better attested and reappears at b22.

[6] *estin hôs kechrômatistai* (425b22f). The *estin hôs* should not be overlooked.

[7] Cf Philoponus *in DA* 470,1-12; see also S.O. Brennan, 'Sensing and the sensitive mean in Aristotle', *The New Scholasticism* 47, 1973, 279-310, who speaks of 'the holding of this form in disassociation from the sentient's own material constitution' (306).

[8] In Bernard (forthcoming).

[9] Philoponus differs in taking Aristotle's 'Therefore, even when the perceptibles are gone, there are still perceptions and representations (*phantasiai*) in the sense organs' (425b24f) to be a reference to *marmarugai*, 'seeing stars' (*in DA* 464,14-17) a phenomenon which he rightly rejects

So, on Philoponus' interpretation, Aristotle is saying that every single sense perceives both its proper object (*idion aisthêton*) and its own act of perceiving that proper object.[10] Philoponus then mentions Alexander's view that it is the common sense that perceives acts of perception[11] and Plutarch's doctrine according to which it is a part of the logical soul that apprehends acts of perception, namely judgment (*doxa*).[12]

And now we come to a most interesting passage. For Philoponus goes on to say:

> More recent exponents, however, neither respect Alexander's brow, nor do they follow Plutarch; indeed, they even reject Aristotle himself and they have devised instead a new exposition. For they say that it is the task of the attentive part (*tou prosektikou merous*) of the rational soul to apprehend the acts of the senses. For the rational soul has not only five faculties, noetic thought, discursive thought, judgment (*doxa*), will and choice – according to them – they add an additional sixth faculty to the rational soul, which they call the attentive. This attentive part (*prosektikon*), they say, attends to (*ephistanei*) what is going on in the person and says 'I have thought noetically', 'I have thought discursively', 'I have judged (*egô edoxasa*)', 'I have been angry', 'I have desired', and, in general, this attentive part of the rational soul goes through all the faculties, the rational, the non-rational and the vegetative ones. So, if the attentive is to go through all [the parts of the soul], then let it pass through the senses as well and let it say 'I have seen', 'I have heard'. So, if it is the attentive that makes such statements, it itself is that which apprehends the acts of the senses. For that which apprehends all [i.e. all the different acts of the soul] must be *one*, since the person is also *one*. For if one [part] apprehended this, and another that, that would be equivalent – as he [i.e. Aristotle] himself says in a different context[13] – to your perceiving this and my perceiving that. Hence the attentive must be *one*. For this attentive [faculty] goes through all the faculties, both the cognitive and the vital ones. But when it goes through the cognitive ones, it is called the attentive (*to prosektikon*) – which is why when we wish to rebuke someone who is dilly-dallying (*rembomenôi tini*) with respect to the cognitive acts we say 'watch yourself' (*proseche sautôi*) – whereas when it goes through the vital [faculties] it is called conscience (*suneidos*) – hence tragedy says 'the conscious knowledge (*hê sunesis*) that terrible deeds are on my conscience' (*hoti sunoida emautôi deina eirgasmenôi*).[14] So it is the attentive which apprehends the acts of the senses. (464,30-465,17)

---

as irrelevant to the problem Aristotle is trying to solve. Philoponus is, however, engaged in refuting at least part of Aristotle's doctrine in this passage, as we shall see, and his interpretation of Aristotle's words is not very charitable. Aristotle seems to be talking of the transition from perception to representation (*phantasia*), and how this retention of percepts implies that the percept is immaterial and therefore stable (cf Bernard (forthcoming) ch II 9).

[10] Cf Philoponus 464,18-20.

[11] Cf ibid. 464,20-23.

[12] Cf ibid. 464,23-30. At 465,23-25, however, Philoponus points out that he cannot see how the doctrine came to be attributed to Plutarch. He has not found any passage to support that claim, rather, he has found Plutarch to be in agreement with Alexander on this question.

[13] 426b19ff; Aristotle shows that ultimately there must be *one* faculty of sense perception, i.e. the common sense, because there must be *one* faculty capable of discriminating between e.g. 'white' and 'sweet', since we know they are different, which neither vision nor taste can tell us.

[14] Cf Euripides *Orestes* 396.

The view attributed to certain 'more recent exponents' is striking in its resemblance to modern positions. The phenomenon of self-awareness, which they term '*prosektikon*', is to be explained – according to these 'more recent exponents' – by assuming there is a special faculty of the soul cognising the acts of the soul. And since it is one and the same person who thinks noetically and discursively, judges, perceives and wills, etc., they argue that there must be *one* cognising agent which apprehends all the different acts of the soul, that is, they derive the unity of self-awareness from the unity of the person. It should not escape our attention, though, that Philoponus does not seem to agree with their theory. He inserts 'they say' at regular intervals and, in particular, he says that 'according to them' the rational soul has six parts instead of five and that they 'add an additional one'.

As we read on we shall find that Philoponus – while in no way disputing the phenomenon described by the term '*prosektikon*' – rejects the idea that it constitutes a *faculty* of the soul. Rather, it is an *ability* possessed by the highest cognitive faculty, reason (*logos*), to reflect on its own acts of cognition. We shall also see that Philoponus' position entails a different way of deriving the unity of self-consciousness.[15]

Plutarch is criticised by the advocates of the *prosektikon*-theory for making judgment (*doxa*) the faculty which apprehends the acts of the soul. For, while judgment could apprehend the acts of the faculties lower down on the scale it could not possibly say 'I have thought noetically', that is, it could not contemplate the acts of the faculties higher than itself. Hence we should have to introduce an additional faculty to apprehend the acts of the higher parts of the soul, but, as their argument goes, that which apprehends all acts of the soul must be *one* (465,17-22).

Philoponus moves on to give his own view now. He says he has not found any passage in Plutarch claiming that it is judgment (*doxa*) that apprehends the acts of the senses,[16] he takes Plutarch to be in agreement with Alexander in attributing this function to the common sense (465,22-25). But they must both be wrong, because the common sense apprehends nothing but perceptibles, and all perceptibles are in a body (*en sômati*), that is, they subsist as compounds (*suntheta*) together with matter (*hulê*), whereas the act of perception is not in a body (*en sômati*) and therefore is not a perceptible. So it does not belong to the common sense (nor to any other kind of perception) (465,26-31).

Notice that Philoponus – unlike the advocates of the *prosektikon*-theory – does not refute Plutarch and Alexander by claiming that, since the person is *one*, that which apprehends the person's acts must also be *one*, and that perception could not apprehend the acts of the higher faculties, Philoponus refutes Plutarch's and Alexander's theory by demonstrating that the common sense could not possibly apprehend *any* cognitive act, not even its own acts of perception, because the common sense – like perception in general – is directed towards external compounds (*suntheta*), from which it

[15] For this problem cf Bernard (forthcoming) ch III 1.
[16] Cf n 12 above.

disassociates the perceptible forms (*aisthêta eidê*), so the common sense cannot reflect on its own act or any other cognitive act since such acts are not compounds.

Returning to Aristotle's argument, Philoponus accepts his point that there is no additional sixth sense to perceive the act of perception (465,31-35) but he rejects the idea that the sense itself both perceives and cognises that it perceives (465,35-466,1). Philoponus then refutes Aristotle's first argument in support of that theory, which said that vision discriminates not only colours, but also light and darkness ('darkness' meaning 'vision impaired' or 'vision impossible'). So why should it not detect the act of vision? Philoponus replies: only colour is perceived in itself by vision, light and darkness are known to it only by negation (*kata apophasin*), that is, by not being colour. But nobody knows his own act in the same way as things known by negation (466,2-6).

Philoponus goes on to criticise Aristotle's statement, that the act of vision (the actualised visual faculty) is coloured, in a manner of speaking. After all, the act is incorporeal, so how can it be coloured? And, since vision sees white and black at the same time, it would then have to be white and black at the same time.[17] Maybe one could argue that the organ is coloured, not the act. But even that would not help Aristotle, since his theory requires the act to be coloured[18] (466,7-18).

Philoponus is not treating Aristotle fairly in this somewhat polemic piece of criticism, since Aristotle has pointed out himself at 425b23f that the sense organ *qua* sense organ takes in the perceptible form without matter, which according to Philoponus' own interpretation[19] means holding the perceptible form in dissociation from any matter whatever (not just in dissociation from the matter of the external compound), so that actual perception is the cognitive apprehension of the perceptible form dissociated from all matter. Thus in saying that the actualised perceptive faculty is coloured, in a manner of speaking, Aristotle can hardly be taken by Philoponus to mean that it is a

---

[17] This is polemic; Aristotle makes it quite plain in 3.2, 426b8-427a14, that the distinction between e.g. 'white' and 'black' cannot be made by anything corporeal since this can only be either white or black at any given time or in one part white and in another black, whereas that which says they are different must be *one* and undivided in actuality (cf Bernard (forthcoming) ch II 9). This is Philoponus' own interpretation (cf 477,21-482,7).

[18] Philoponus omits the 'in a manner of speaking' (*estin hôs*, 425b23). He is giving Aristotle short shrift in this passage. But cf below.

[19] Philoponus *in DA* 437f (on 2.12, 424a17ff); 'to take in the perceptible form without matter' means not just 'to receive the perceptible form without *its* matter' (this is often taken to be Themistius' interpretation, *de Anima Paraphrasis* (*CAG* V,3) 77,33-78,3; 36-38; but cf 78,7ff), but 'to receive the perceptible form without any matter being involved in the process'. As Philoponus says (309,19-25): 'For, just as we say the wax is potentially like the signet, because when affected by it [the wax] gets to be as [the signet] is actually, having received not its matter, but only its form, in this way perception is also affected by perceptibles and receives an incorporeal imprint (*asômatôs anamattetai*) of their forms (*eidê*). There is, however, a difference in that the wax gets to be itself the matter of the form in the signet, whereas perception does not get to be the matter of the perceptible, rather it cognitively receives an imprint (*gnôstikôs ... ekmattetai*) of its form'. It should be noted, however, that this interpretation did not – as is often assumed – originate with Philoponus. Alexander says exactly the same in his *de Anima* (Suppl. Arist. II,I), 83,13-84,6, so Philoponus' interpretation is the traditional one. Cf Bernard (forthcoming) ch II 9.

coloured compound involving matter. Philoponus does show a little later[20] that even if one takes 'coloured' to refer to the mere presence of a perceptible form without any matter, Aristotle's view cannot be vindicated, since the perceptible form is the substratum (*hupokeimenon*) of perception, whereas the act is the discriminatory apprehension of the substratum.

There is a further reason, Philoponus goes on to say, why it does not make sense to speak of perception apprehending its own act. For the apprehension of one's own acts is the same as reflecting on oneself (*pros heauto epistrephein*). And what is able to reflect upon itself is able to act in separation from the body (*chôristês estin energeias*) and must therefore be eternal and incorporeal[21] (in the strong sense of being completely dissociated from body). But the perceptive faculty is neither eternal nor incorporeal (in the strong sense). Therefore, it could not possibly be able to reflect and apprehend its own acts (466,18-27).

So, Aristotle is wrong, and it is indeed the task of the *prosektikon* to cognise the acts of the senses, Philoponus says (466,27-29). From what we have read so far we might almost be led to assume that Philoponus himself agrees with the 'more recent exponents' who put forward the *prosektikon*-theory. But let us see how he continues: that it is the *prosektikon* that apprehends the acts of perception can be seen from experience.

> For when reason (*logos*) is occupied with something, even though vision is seeing, we do not know that it has seen, because reason is occupied. And later, after reason has returned to itself, even though it is not now seeing our friend, it now says 'I saw him', as if it had taken in a weak impression of that which was seen, even though it was occupied,[22] and now that it had returned to its senses (*ananêpsas*) it was saying 'I saw'. So it is the task of the *logos* to say 'I saw'.[23] (466,30-35)

The example Philoponus uses is most instructive.[24] You are walking home,

---

[20] 470,18-471,10; this is Alexander's interpretation, cf his *Aporiai kai Luseis* III,7, 92,27-93,22 (Suppl. Arist. II,II).

[21] Even though Philoponus uses Neoplatonist terminology in this passage it can, I think, be shown that he is not deviating from Aristotelian doctrine (see Bernard (forthcoming) ch III 1). Aristotle makes it plain in 3.5 that only intellect (*nous*) in the pregnant sense is separable from body. And in 2.5 he points out that perception is dependent on external stimuli (417a2ff.; cf Bernard (forthcoming) ch II 5). In 3.4 Aristotle says that one difference between intellect (*nous*) and perception is that the latter is associated with the body (429a24ff; cf Bernard (forthcoming) ch II 10).

[22] Reading *ei kai êscholêto, nun ananêpsas* ... in 466,34. Hayduck's edition has *ei kai êscholêtai nun, ananêpsas* ... I do not see how this could be made sense of. 'Even though <the *logos*> is occupied now'? How does that square with *ananêpsas*, 'having returned to his senses, having sobered up'? It seems better to place the comma before *nun*. But then we need the pluperfect *êscholêto* rather than the perfect, since it must be a reference to the *logos*' former state of concentration on something else. MS A has the corrupt reading *êscholêtoinau*, which seems to be a jumbled version of *êscholêto nun*.

[23] Or 'I have seen'. The aorist *eidon* can mean both, of course.

[24] Plutarch uses a similar example to show how perception is attended to and guided by *nous*, in *de Sollertia Animalium* 961A, saying: 'we often run through pieces of writing with our eyes, and words reach our ear, yet they remain unknown to us and escape us because our attention (our

let us say, and your mind is set on a difficult problem; you are completely lost in thought, as one says. You arrive at your house. After a while you have solved the problem you were pondering on, or else your concentration slips just a little. And all of a sudden the thought crosses your mind 'Wait a minute! Didn't I see my friend Peter on the way home?' So you must have perceived[25] him as you were walking past, that is, you apprehended an image, which was stored in your memory and returned when your concentration went. But you did not recognise your friend at the time and, consequently, you were not aware of your act of seeing either. Philoponus has chosen this example to show that self-awareness is dependent on cognition – 'cognition' in the Aristotelian sense of 'immediate discriminatory apprehension of a form' (either perceptible (*aisthêton*) or intelligible (*noêton*)), not in the sense of 'awareness of something' – first you have to cognise something and only afterwards can you be aware of yourself as the cognising subject. This applies not only to perceptive self-awareness, but also to self-awareness of, for example, thoughts or imaginings. The act of thinking (or imagining) comes first and the unity of this act gives the possible subsequent act of self-awareness its unity and its definiteness. You are aware of yourself as the cognising subject perceiving your friend, for example, or creating a mental image of your friend or thinking of your friend.

Philoponus uses this example taken from the field of perceptual self-awareness because in order to make his point he wants a case where the primary activity of the soul (perceiving, imagining, thinking, willing, etc.) is delayed, so as to be able to show that self-awareness is also delayed in such a case, proving that it is secondary to and thus dependent on the primary acts of the soul. Since the soul is not self-sufficient as regards perception, but directed towards external objects, it is easy to devise an example that has the necessary feature (primary act delayed) by referring to perceptual self-awareness. Of course, other activities of the soul are not in the same way dependent on external objects, this is peculiar to perception (and, indirectly, to self-awareness of perception, since this depends on a previous act of perception), but all self-awareness is dependent on a primary act of the soul (thinking, perceiving, etc.), which precedes it – and this is Philoponus' point.

Thus, according to Philoponus, self-awareness is neither self-contained, nor a faculty. It is dependent on a primary (usually cognitive) act and it is the ability of the ultimate cognising faculty, reason (*logos*), to reflect on its own acts. It is only after one has performed a definite psychical act (e.g. after one has cognised something, without that 'something' having to be external to the soul) that one can also be self-aware, that is, aware of oneself as the (cognising) agent. Thus the unity and the definiteness of self-awareness is derived from the unity and the definiteness of the act one is aware one is performing (or, as in Philoponus' example, has performed). And the unity of

---

nous) is directed to other things. Then it returns, collects everything it has missed and pursues and hunts it'.

[25] Throughout this chapter I use 'perceive', 'perception', etc. as the equivalent of *aisthanesthai*, *aisthêsis*, etc. Therefore 'perception' does not carry its modern meaning as opposed to 'sensation' here.

these primary acts is bestowed by **reason** (*logos*).

Philoponus goes on to answer an objection he anticipates. Reason is usually thought to deal with universals, yet Philoponus is saying that it acts together with the particular senses. The answer is that when reason is using the body as its organ it acts with respect to particulars, whereas when it is not using the body as an organ, reason is concerned with universals. So, when reason attends to and guides perceptive cognition, which does involve the use of bodily organs, reason deals with particulars (466,35-467,3).

> So [says Philoponus] we are not paying Aristotle his salary, rather we consider his theory to be wrong. At any rate we say that he agrees with us. For, when he says that perception knows the fact of its acting (*to hoti energei*) he says it knows this not in its capacity as perception but in its capacity as an organ of reason. For he said above[26] that vision knows 'sweet', not in its capacity as vision but in its capacity as the common sense, and he made his present remarks in the same spirit. Nor is he the only one to do this; common usage is in the habit of this when it says 'the adze made the throne'. And note how the work is generously attributed to the instrument (*organon*), not because it itself did [the work], but because the craftsman who used it did. (467,4-12)

This is the end of Philoponus' summary and general discussion of *de Anima* 425a12-25. He follows it up, as he always does, with an analysis of the text in detail.

What has Philoponus told us about self-awareness? Is he justified in claiming that his account is in agreement with Aristotelian doctrine? Philoponus shows that self-awareness is not to be regarded as a faculty of the soul and he criticises the advocates of the *prosektikon*-theory for making it one. Self-awareness depends on previous acts of primary cognition, which shows that it is not itself a cognitive faculty, but an ability possessed by reason to reflect upon its own acts. And thus the unity of self-awareness cannot be derived from the unity of the person because on that assumption one would have to be self-sufficient as regards self-awareness and should not depend on previous cognition of anything. In truth, the unity of self-awareness derives from the unity of the primary act of the soul which one is aware one is performing. Self-awareness is an awareness of oneself as the agent performing a definite psychical act.[27]

Is this viewpoint Aristotelian? The answer you give to that will obviously depend on your interpretation of Aristotle's doctrine. He definitely rejects the idea that there could be an additional sense to perceive the acts of the other senses. Such an assumption leads to an infinite regress, he says (425b16). The same argument would also apply to the postulation of an additional faculty to perceive the acts of the other faculties. Why – on that assumption – do we not need another faculty to make us aware of the act of the faculty we added? It is

---

[26] 425a22.

[27] On the question of the unity of cognition in Platonic and Aristotelian epistemology as compared with modern philosophical theories and on the related question of spontaneity see Arbogast Schmitt's forthcoming *Subjektivität und Innerlichkeit*. On the unity of perceptive cognition cf Bernard (forthcoming) ch II 9 and part III.

plain that Aristotle is of the opinion that a phenomenon like self-awareness must be an act of reflection by the primary (cognising) agent on its own acts (in the case of perceptual self-awareness: a perception itself of itself (*autê hautês*) (425b15)). To postulate a faculty of self-awareness is to misconceive the nature of self-awareness. It must be an ability of the cognising agent to reflect upon its own acts, which explains why one can only be aware of oneself as the agent performing a definite psychical act. In this way the unity of self-awareness presupposes the unity of the primary psychical act of which one can then be aware as the agent performing it.[28]

Thus Philoponus' treatment of the problem is, indeed, in agreement with fundamental Aristotelian doctrine in that both philosophers insist that self-awareness is a reflecting of the cognising agent on its own acts, which means that self-awareness is not a faculty of the soul. This makes self-awareness secondary to the primary acts of the soul, which are performed by separate faculties.

There is, however, an objection some modern readers might want to raise against Philoponus' claim that his description of self-awareness is Aristotelian. Is it not Aristotle's position that there cannot be two faculties which apprehend, for example, colour? And does Philoponus' solution of the problem not conflict with this by making both vision and reason (*logos*) apprehend colour? The answer is that Philoponus does not say that it is reason as opposed to vision, that is able to reflect consciously on acts of perception. He points out that reason is the principle of all cognition, perceptive and other; all acts of discrimination are performed by reason (*logos* or *nous*, as Philoponus might equally well have said), the difference between perception and thought being that perceptive discrimination – which is 'rational' insofar as it is discrimination – is bound to the particular and apprehends only perceptible forms ('white', 'hard', etc.), which are accidental to the real determinateness (noetic form) of a thing. Philoponus is not talking about the proper acts of reason supervening on perception, he is referring to ordinary acts of perception and specifying the real agent of perceptive discrimination, which is reason, though bound to the deficiencies and limitations of the mode of cognition which characterises perception. And it is because of this, says Philoponus, that there can be such a phenomenon as perceptual self-awareness. Since reason was involved from the start – even while the soul was performing the most basic perceptive discriminations – it can then reflect upon its own 'bound' act of perceptive discrimination (a reflective act which is outside the scope of the limitations characteristic of perception).

So, on Philoponus' account there are not two faculties apprehending, for example, colour, there is only one, reason in its bound activity of seeing, and reason in this activity can equally be called 'vision', which is what Aristotle does call it (425b12ff). Normally, it is perfectly all right to speak of 'vision', 'perception', etc.; there is no need continually to emphasise that there is really only one principle of discrimination and cognition cognising in different modes. But in a description of perceptual self-awareness the slight inaccuracy

---

[28] Cf Bernard (forthcoming) part III (and see above, n 27).

of the ordinary phrase leads to the dilemma that you have to make vision reflect consciously upon its own act, which does not fit in well with the limitations of visual (and all other perceptive) cognition (i.e. that it is dependent on external stimuli and directed towards what is outside the soul (cf. *de Anima* 417a2ff and 417b20ff)). Thus, in dealing with this particular problem, it is advisable to specify the active principle of visual (like all other) cognition and refer perceptual self-awareness to reason (*logos* or *nous*) in as much as it is present in all acts of perceptive discrimination. Vision is nothing more than 'the adze which makes the throne' (Philoponus 467,9-12), reason is the craftsman who uses his tool, vision, to see.

The question now is whether Aristotle shared the view expressed by Philoponus that reason (*logos*) is the active principle of all cognition. It is, of course, impossible for me to discuss this problem adequately in the present context and I must refer the reader to my interpretation of the *de Anima* for a more detailed discussion.[29] It should be remembered, however, that Aristotle at 432a30f speaks of 'the faculty of perception (*to aisthêtikon*), which one could not easily classify as either non-rational (*alogon*) or rational (possessing *logos*, *logon echon*)'. Reason is the principle which gives unity to all the psychical acts (430b5f: *to de hen poioun, touto ho nous hekaston*). I regard this as the correct interpretation of Aristotle's *de Anima*, but even if the reader disagrees with that, it has to be admitted that this is Philoponus' interpretation and that he bases it on a detailed examination of the Aristotelian text in its entirety.

But if Philoponus' treatment of the problem is really in agreement with Aristotelian theory and he is merely bringing out the active principle of perception and perceptual awareness, why does he begin by saying Aristotle is wrong in this matter? He even repeats this at the end, before showing that Aristotle's solution is ultimately the same as his own and that Aristotle has merely failed to make a necessary qualification. Well, even in this somewhat harsh way of dealing with the theories of one's predecessors Philoponus seems to be following Aristotle's own example. In the *Nicomachean Ethics*, for instance, in his discussion of lack of self-control (*akrasia*) Aristotle begins by mentioning Socrates' view that nobody does what he believes to be bad and that it is through ignorance that we do wrong (*NE* 7.2, 1145b25ff), but adds that this theory is manifestly at variance with facts (*amphisbêtei tois phainomenois enargôs*), which sounds like a straightforward rejection of the whole theory. Yet Aristotle goes on to show that our wrong-doing is due to a particular kind of ignorance and concludes his analysis by saying 'and thus there seems to follow what Socrates sought to prove' (*NE* 7.3, 1147b15).

Philoponus certainly treats Aristotle's doctrine in a rather similar way in our passage. An apparently straightforward rejection of Aristotle's theory is the starting point for a most instructive discussion of the problem, setting out in detail (and with the necessary reservations) how it should be explained. Yet in conclusion Philoponus shows that – provided you make one necessary qualification – Aristotle's solution is correct.

[29] Cf Bernard (forthcoming) ch II 9, II 10, part III and passim.

CHAPTER NINE

Infinity and the Creation

Richard Sorabji

Greek philosophy lasted over a thousand years, from 500 B.C. to A.D. 550, but
the recognition of Anglo-Saxon philosophers has been confined to its early
period – the period down to, and including, Aristotle. There has been a
renaissance in the last five years of studies in the next period, the period of
Stoics, Epicureans and Sceptics down to A.D. 200. But the period of
Neoplatonism and of its interaction with Christianity from A.D. 250 to 550 has
been almost totally neglected by contemporary philosophers. It is this period
that I want to present as a philosophically exciting one. I want to recommend
that it should be studied by philosophers – not indeed *instead* of the early
period, but *as well*.

In A.D. 529, the Byzantine Emperor Justinian forebade pagans to hold
public teaching posts anywhere in the Empire, and also closed the private
Neoplatonist school of philosophy in Athens. There has been discussion
inspired by a former professor of this college, Alan Cameron, of the sense in
which he closed the Athenian school. He forebade the Athenian philosophers
to continue teaching, but was it possible for them to continue researching, at
their own expense? In this year of 1982, when schools of philosophy are being
closed elsewhere, we must not allow Justinian to escape the ignominy, even
though we may understand that his motive, a defence of Christianity, was
loftier than any motive discernible nowadays. None the less, there has been a
persistent defence of Justinian; he did not kill Neoplatonist philosophy, it is
said. That philosophy was already dead; he merely buried the corpse.

This view is practically universal, not only among philosophers, but also
among historians. Alan Cameron was one of the very few to challenge it. I
should like to quote a representative statement, not because I think it will be
easy to rebut, but because it is a particularly good statement of an almost
universal view. The onus will be on me to see if I can persuade you to take a
different view from the orthodox one. Typically, the orthodox view disparages
not just the late Neoplatonism of the Athenians, but late Neoplatonism as a
whole.

* Reprinted from an inaugural lecture given at King's College, London, on 16 March 1982,
with minor alterations.

It is clear that, in all this, Neoplatonism had become scholastic. After this time there are no major figures, and when Justinian closed the philosophical schools in A.D. 529, there was no fruitful philosophy to which to put a stop. The spirit of Greek philosophical thought, which, it is clear, had steadily become weaker during this period, had finally died.[1]

The most direct way to challenge this view would be to talk about the last Athenian Neoplatonists, Damascius and Simplicius. But that is something I did recently elsewhere.[2] Today I want to choose a Christian who studied in the other great Neoplatonist school in Alexandria. His name is John Philoponus. It may be thought wrong of me to choose a Christian, since what the conventional view attacks is Neoplatonist, not Christian, philosophy. But that would involve a false distinction. Philoponus was both a pupil and an editor of the Neoplatonist Ammonius in Alexandria.[3] A large part of his output consisted of editions of Ammonius' lecture courses on Aristotle, and at least one modern writer finds comparatively little Christian influence in these.[4] At any rate, the philosophy Philoponus breathed was that of Neoplatonist circles, and when he did argue for Christianity against paganism, he still used Neoplatonist assumptions in doing so. Often there was no need for argument because there was so much overlap between Christian and Neoplatonist views in this period.[5]

The inextricable link between Platonists and Christians is illustrated in another great contemporary, Boethius of Rome (died *c.* 524/5). It used to be asked whether Boethius was a Neoplatonist or a Christian. As those of you will know who recently heard Henry Chadwick in this room, scholarship now tends to the view that he was both.[6] So inextricable is the link that a slur on the philosophy of the Neoplatonists of this period can only reasonably be taken as a slur on the philosophy of their closest Christian associates. I would admit that the most brilliant philosophers in these circles were both Christians, Philoponus and Boethius. None the less I would be perfectly willing, on another occasion, to argue for the merits of Philoponus' teacher Ammonius.

The arguments of Philoponus on which I want to focus concern the Christian view that the universe had a beginning. But here already I must draw a distinction. For in talking of the *universe* beginning, I am not talking merely of the present orderly arrangement of the earth, sun, moon and stars. Many pagans would have accepted that the *present* arrangement of matter had a beginning. What, with very few exceptions, they all thought absurd was

---

[1] D.W. Hamlyn, 'Greek Philosophy after Aristotle', in D.J. O'Connor, ed, *A Critical History of Western Philosophy*, London 1964, 78.

[2] 'Is time real? Responses to an unageing paradox', Dawes Hicks Lecture 1982, published in the *Proceedings of the British Academy* and as a separate lecture.

[3] 'Editor' and 'edition' need to be understood in the rather loose sense explained above in Chapter 1.

[4] H.J. Blumenthal (1982) 54-63; 244-6.

[5] For example, some Neoplatonists agreed with Philoponus (*in Phys.* 189,10-26), in ascribing to Aristotle the view that the universe depends on God for its (beginningless) existence (Ammonius ap Simplicium *in Phys* 1363,8-24; Simplicius *in Phys* 256,16-25; Simplicius and Ammonius ap Simplicium *in Cael* 271,13-21). For other rapprochements, see H.J. Blumenthal (1982) 58-60.

[6] See Henry Chadwick, *Boethius*, Oxford 1981.

that matter *itself* should have had a beginning. Indeed, Jews and Christians themselves were embarrassed about this doctrine, and were by no means unanimous in accepting it. It has been suggested that the oldest references to creation in the Old Testament come in Job, and that there God is envisaged as imposing order on pre-existing matter, not as creating matter itself.[7] It has further been doubted whether there is any clear statement in the Bible of creation out of nothing.[8] The opinion of Philo the Jew, in the first century A.D., is a matter of controversy,[9] but I believe that he takes different sides in different works.[10] A little later, Hermogenes and others[11] offered a surprising reason for denying matter a beginning. They pointed to the use of the word 'was' in the opening of Genesis, where it is said that the earth *was* without form and void, and they took the use of the past tense to show that earth, or matter, was already in existence, when the Creator began work. It is often held, although I am not inclined to agree myself, that Boethius endorsed the Neoplatonist view of a beginningless universe at the end of his *Consolation of Philosophy*.[12] What I would acknowledge is that other Christians in these centuries, such as Synesius and Elias, did deny the universe a beginning or end under the influence of Platonism.[13] If we skip to the thirteenth century, we find Thomas Aquinas, and his teacher Albert the Great, saying that it cannot be established by philosophy one way or the other whether the universe had a beginning. It is only Scripture which reveals that it did.[14] Two slightly younger contemporaries in Paris went a step further, indeed, a step too far. Boethius of Dacia (the Dane, not the sixth century Roman) and Siger of Brabant maintained that philosophical argument showed the universe to be beginningless, but that none the less reason must

[7] Job 28 and 38. See Robert M. Grant, *Miracle and Natural Law in Graeco-Roman and Early Christian Thought*, Amsterdam 1952, ch 10.

[8] See H.A. Wolfson, *Philo*, vol I, Cambridge Mass. 1948, 300-12; David Winston, 'The *Book of Wisdom's* theory of cosmogony', in *History of Religions* 11, 1971-2, 185-202; Gerhard May, *Schöpfung aus dem Nichts*, Stuttgart 1980, 1-21.

[9] See Wolfson loc. cit., May loc. cit., Winston, *Philo of Alexandria, The Contemplative Life, The Giants and Selections*, London 1981, 7-21.

[10] I believe that the fullest statement of the orthodox view that matter had a beginning comes in *de Providentia* 1 6-7 (Aucher), but contrast *de Aeternitate Mundi* 2. 5 and *de Opificio Mundi* 5.21. My view is now explained in Richard Sorabji (1983) 203-9.

[11] Tertullian *Adversus Hermogenem* XXIII 1, XXVII; Similarly Basil of Casesarea *in Hexaemeron* 2.2; Ambrose *Hexaemeron* 1.7; Augustine *de Genesi contra Manichaeos* I.3.5.

[12] Boethius *Consolation of Philosophy* 5.6. So P. Courcelle, *Late Latin Writers and their Greek Sources*, Cambridge Mass. 1969, translated from the French of 1948, ch 6, 316; 322; Philip Merlan, 'Ammonius Hermiae, Zacharias Scholasticus and Boethius', *Greek, Roman and Byzantine Studies* 9, 1968, 193-203; J. de Blic, 'Les arguments de Saint Augustin contre l'éternité du monde', *Mélanges de science religieuse*, Lille, vol 2, 1945, 33-44.

[13] Synesius *Epistle* 105 makes the universe endless. Elias *in Cat* 187, 6-7, treats the heavens as indestructible.

[14] Albertus Magnus *Summa Theologiae* II, tr 1,q 4,a 5, partic 3; *Liber VIII Physicorum*, Tract.I, ch 11. For Thomas Aquinas, see e.g. *de Aeternitate Mundi; Scriptum Super Libros Sententiarum Magistri Petri Lombardi*, in II, dist 1,q 1,a 2 solutio; and *Summa Theologiae* I,q 46, a 1 and 2. John F. Wippel, 'Did Thomas Aquinas defend the possibility of an eternally created world?', *Journal of the History of Philosophy* 19, 1981, 21-37, argues that, in works up to the *de Aeternitate Mundi*, Thomas treats a beginningless universe as not disproved or disprovable, while in the *de Aeternitate Mundi* he treats it as actually possible.

bow to revelation.[15] They had to flee Paris in the condemnation of 1277, and there is a tradition that Siger was murdered.

Up to A.D. 529, Christians were on the defensive. They argued that a beginning of the universe was not impossible. In 529, Philoponus swung round into the attack. He argued that a beginning of the universe was actually mandatory, and mandatory on the pagans' own principles. The most exciting set of arguments, the one I shall be considering, has to do with the concept of infinity.

529 was an annus mirabilis for Christianity. St. Benedict, on the usual dating, founded the monastery at Monte Cassino, the Council of Orange settled outstanding matters on free will, Justinian closed the Neoplatonist school at Athens, and Philoponus produced his book of eighteen arguments *On the Eternity of the World (de Aeternitate Mundi)* against the Neoplatonist Proclus. Simplicius, one of the victims of the closure in Athens, survived to write his replies to Philoponus. He speaks of him with contempt, and refers to him dismissively as 'the grammarian'.[16]

Philoponus included the arguments which I shall be considering in two of his commentaries, those on Aristotle's *Physics*[17] and *Meteorology*. Here the arguments are not very prominent, but the *de Aeternitate Mundi contra Proclum* of 529 is the first of four or more whole books which he devoted to the subject.[18] Some leading scholars of the mediaeval West have created the impression that these arguments were invented by Bonaventure in the thirteenth century.[19] In fact, Bonaventure is repeating Philoponus' arguments, and even using the very same examples. He was able to do so because the arguments had been retailed innumerable times, with elaborations, by Islamic philosophers. The facts of transmission have been set out in a magisterial article by H.A. Davidson,[20] who makes it clear what an impact Philoponus' arguments had on Islamic thought, while the same arguments had nearly as much impact when they were rediscovered by the Latin West. Philoponus, coming at the end of Ancient Greek philosophy, has not always got the credit he deserves. And many philosophers nowadays, if they know of the infinity arguments at all, will know of them from Kant. But Kant's version in the *Critique of Pure Reason* is only a faint echo of the ancient originals.

---

[15] Boethius of Dacia *Tractatus de Aeternitate Mundi*, ed. G. Sajó, Berlin 1964, 47-8; 60-2; Siger of Brabant, *de Aeternitate Mundi*, ed W.J. Dwyer, Louvain 1937. Translations of relevant works of Aquinas, Siger and Bonaventure, are conveniently assembled, with useful introductions in Cyril Vollert, L.H. Kendzierski, P.M. Byrne, eds, *St. Thomas Aquinas, Siger of Brabant, St. Bonaventure, On the Eternity of the World*, Milwaukee Wisconsin, 1964.

[16] For further aspects of this name, see above, Chapters 1 and 3.

[17] For the date of the *Physics* commentary, see Chapter 1 above p 37 and n 260a.

[18] The evidence is given by H.A. Davidson (1969).

[19] This is the impression, intentional or otherwise, created by E. Gilson, *La Philosophie de Saint Bonaventure*, Paris 1924, 184-8; John Murdoch, 'William of Ockham and the logic of infinity and continuity', in Norman Kretzmann, ed, *Infinity and Continuity in Ancient and Mediaeval Thought*, Ithaca New York 1982, 166. See also G.J. Whitrow, 'On the impossibility of an infinite past', *British Journal for the Philosophy of Science* 29, 1978,40 n 1.

[20] op. cit.

There is an opportunity here because Philoponus' major work, the *de Aeternitate Mundi contra Proclum*, has never been translated into a modern European language. Its summaries of the eighteen arguments of Proclus were translated into English in 1825, but that translation is no longer in print;[21] and snippets of Philoponus' counter-arguments from one work or another have been translated into various European languages.[22] This situation is being remedied in part by the work of Christian Wildberg, who is preparing a translation, with commentary, of the fragments of Philoponus' treatise against Aristotle, and who is including a new fragment which he has discovered in Syriac. None the less, the *de Aeternitate Mundi contra Proclum* still awaits translation.[23]

In order to understand Philoponus' arguments about infinity, we must first understand Aristotle's analysis of infinity. For what Philoponus does is to exploit that analysis in order to turn Aristotle against himself, and to show that the pagans must admit a beginning of the universe. Aristotle's treatment of infinity in *Physics* 3.4-8 was very original. He complained that his predecessors had got things the wrong way round. They thought of infinity as something which is so all-embracing that it has nothing outside it. But the very opposite is the case: infinity is what always has something outside it.[24] For it should be thought of as an *extendible finitude*. However large a finite number you have taken, you can take more. It is the reference to the possibility of taking more that guarantees this infinity will always have something outside it. The analysis is summed up in the following words:

> For in general infinity exists through one thing always being taken after another, what is taken being always finite, but ever other and other.[25]

That is why I describe Aristotle's infinity as an extendible finitude.

This conception of infinity would have a lot of appeal nowadays. For example, the modern idea of *approaching a limit* is very much in the spirit of Aristotle. It is said that the series $\frac{1}{2} + \frac{1}{4} + \frac{1}{8}$ etc. approaches 1 as a limit, and this way of talking enables everyone to avoid the naughty word 'infinity'. The idea is that you can get *as close as you like* to 1 by adding more fractions to a *finite* collection. And this talk of getting as close as you like by a *finite* operation

[21] Thomas Taylor, *The Fragments that Remain of the Lost Writings of Proclus, Surnamed the Platonic Successor*, London 1825. Where the Greek (*ms*) is deficient, omitting Proclus' first argument, the want is supplied by an Arabic translation of Proclus himself, now available in French: G.C. Anawati (1956) 21-5. The Arabic text of Proclus' first nine arguments has been published by A. Badawi, *Neoplatonici apud Arabes, Islamica* 16, Cairo 1954.

[22] There is a selection from Philoponus' work available in German translation: W. Böhm (1967); but this contains only twelve pages on the eternity of the world. Some brief extracts from the Commentaries on *Physics* and *Meteorologica* are translated into English by S. Sambursky (1972) and by Robert B. Todd (1980). A few other extracts on the subject are translated in Sorabji (1983). A tiny fragment is preserved in Arabic of Philoponus' book against Aristotle, and this has been translated into English by Joel L. Kraemer (1965). Finally, there is a translation by S. Pines of an Arabic summary of what may be a distinct work on the subject: Pines (1972).

[23] This translation has now been undertaken, along with others.

[24] *Phys* 3.6,206b33-207a2; cf 206b17-18; 207a7-8.

[25] *Phys* 3.5, 206a27-9.

is very much Aristotle's own. But Aristotle is closer still to those modern finitist mathematicians who deny that there is ever more than a finite number of points in a line: the points are brought into existence only as they are marked off.[26] I say that Aristotle is closer to this, but in order to bring this out, I shall have to draw attention to an ambiguity in the notion of an extendible finitude.

The question is whether Aristotle thinks that you can always go on *creating* more divisions in a line, or that you can always go on *recognising* more of the divisions which exist already. If they exist already, their number is greater than any finite number. But I believe that Aristotle is committed to what I might call the *finitist* view that there is never more than a *finite* number of divisions in existence. Thus he repeatedly says, for example, that infinity exists through a process of one thing *coming into being* after another (*Phys* 3.6, 206a21-3; 30-3; 3.7, 207b14). And whatever his point about divisions he will hold the *same* view about numbers for he thinks that there is infinite number only in the sense in which there is an infinite number of divisions (3.6, 206b3-12).

Certainly, Aristotle would allow only a finite number of *actually* existing divisions in a line. An *actually* existing division is one that has been marked out in some way, either physically or mentally. But the situation is more complex, when we ask how many *potentially* existing divisions there are. The very question is ambiguous, for potentially existing divisions might be understood as divisions that could be made, although they have not yet been made, or they might be understood as points at any finite number of which divisions can be made. Certainly, the *Physics* passage envisages that there are divisions which could be made over and above those which actually exist: why else should it insist that infinity always has something outside it? But here and in *On Generation and Corruption* 1.2, 316a10-317a12, it is also implied that the number of makable divisions is not more than finite. And the *GC* passage finishes (317a9-12) by implying that the number of points in a line is not more than finite either. Otherwise (Aristotle evidently fears) the line would, absurdly, be made up of sizeless points. I can go further, for Aristotle cannot afford to admit any collections which are more than finite, if his analysis of infinity is to surmount the problems which it is intended to surmount.

One such problem is whether infinity does not have *parts* which are infinite. Aristotle complains that his predecessor Anaxagoras is committed to this.[27] Modern theory recognises that in a sense it is possible: the whole numbers contain an infinite sub-set of odd numbers. But Aristotle declares that the same thing cannot be many infinities, and his analysis of infinity is taken to avoid this outcome.[28] If it is to do so, the finitist interpretation needs to be applied to all collections, even to collections of potentially existing divisions.

If Aristotle's view of infinity is finitist, I believe it will be perfectly adapted for some cases, but inadequate for others. If, for example, we take pairs of

[26] D. Hilbert and P. Bernays, *Grundlagen der Mathematik*, Berlin 1934, 15-17.

[27] *Phys* 1.4, 188a2-5.

[28] *Phys* 3.5, 204a20-6.

whole numbers, they will never be separated from each other by an all-embracing infinity of whole numbers. The most we can do, by selecting pairs still further apart, is to obtain an extendible finitude of Aristotle's sort. On the other hand, my view would be that the totality of whole numbers is an all-embracing infinity of an un-Aristotelian type. I shall later suggest that there may be an asymmetry between the past and the future in respect of years traversed: that the series of future years traversed, starting from now, may best be viewed as an extendible finitude in Aristotle's manner, while the series of past years, if there was no beginning, should be viewed as more than finite.

There are two consequences which Aristotle draws from his conception of infinity as an extendible finitude, and which are relevant to Philoponus' argument. One is that infinity is merely *potential* and *never actual*;[29] the other that it can never be *traversed*, that is, gone right through.[30] Both points require a word of discussion. The new application of the word 'actual' is different from its application to divisions; it is because infinity is never more than finite, and yet is extendible, that it should be thought of as existing potentially, and never actually. An actual infinity, as I understand it, would be *more* than a finitude.

The other principle, that infinity is never *traversed*, undergoes an important qualification, when Aristotle comes to discuss Zeno's paradox of the half-distances. According to that paradox, those of you who hope to leave the hall at the end of this lecture will be disappointed. For even to reach the door, you would first have to go half way, then half the remaining distance, and so on, *ad infinitum*. You would have to traverse an infinity of sub-distances, and how can you do that? In his final discussion, Aristotle offers the following solution. Although we cannot traverse an infinity of *actually* existing divisions (*entelecheiâi onta*), we can traverse an infinity of potentially existing ones (*dunamei* sc. *onta*).[31]

This last concession is startling: how can Aristotle think himself free to allow more than a finite number of potential divisions? That they are more than finite is implied by the contrast with the finite number of actual divisions. To allow such an infinite collection would be a possible policy decision, but it would conflict with that of *Physics* 3 and *On Generation and Corruption* 1.2, where makable divisions and points are not more than finite, and, worse, it would debar Aristotle from solving the problems of infinite sub-sets and of lines made up of unextended points. So the decision could be taken only at a cost. If this cost is to be avoided, it is not enough that the divisions should *exist* potentially (*dunamei onta*); their *infinity* ought to be potential as well (*dunamei apeira*), and that in the sense which I earlier defined of not being more than finite. Aristotle's unexpected concession will prove relevant to subsequent controversy.

Philoponus' attack on the pagans can now be explained. He offers a whole

---

[29] *Phys* 3.6, 206a14-23, with a qualification at 206b13.

[30] *Phys* 3.5, 204b9; 6.2, 233a22; 6.7, 238a33; 8.8, 263a6; b4; b9; 265a20; *Cael* 1.5, 272a3; a29; 3.2, 300b5; *Metaph* 2.2,994b30; *An Post* 1.3, 72b11; 1.22, 82b39; 83b6.

[31] *Phys* 8.8, 263b5-6.

battery of infinity arguments, but I shall confine myself to the two most spectacular.[32] The first is that the universe must have had a beginning, or it would by now have *traversed*, or gone right through, an infinity of years. And that infinity would be *actual*, and not a mere extendible finitude. To make matters worse, the second argument says that the infinity would have to be *increased*. For if there has already been an actual infinity of years by 1982, how many will there have been by 1983? Infinity plus one. And how many days will there have been? Infinity times 365. But to talk of something larger than infinity is absurd.

Simplicius, still suffering from the ban on teaching, was none the less free to write, and he penned his replies to John, the Grammarian.[33] Aristotle had already anticipated Philoponus' objections, he said, and answered them in advance. For he had pointed out that the past years do not stay, but have perished, and this implies that you do not get an infinity of them existing. Nor need we fear that next year will increase an infinity, for there is not an infinity there.

At first sight, Simplicius' point seems arresting: what made it possible for an infinity of divisions to be a merely *potential* infinity was that the divisions not yet made did not *exist*. Has not Simplicius shown that years are like divisions, in that no more than a finite number exists? Unfortunately, we have seen that the situation is more complex, because Aristotle is willing to think of points and potential divisions as entities of a sort, capable of forming collections. And he would have to allow the same for past years – all the more so because the sense in which past years no longer exist is only the rather weak sense of no longer being present. They are still entities enough to form a collection, and Aristotle ought therefore to avoid their forming an actually infinite one, just as he does in *Physics* 3 and *GC* 1.2 for potential divisions and points. Otherwise, he will be back with the problems which he hoped to avoid of infinite sub-sets and of lines composed of points. Admittedly, we have seen him allowing an actual infinity of potential divisions in *Physics* 8, but first that was argued to be an inconsistency, and secondly potential divisions have less claim on existence than past years and so hardly set a precedent for them.

Things will be no better if Simplicius' point is transposed and phrased in terms of actuality instead of existence, so as to say that past years are no longer *actual*. We should then have had at least three senses of actuality: divisions are actual when *marked out*, years when *present* and infinities when they are *more than a finitude*. But the answer would be that the status of the years does not settle the status of their infinity. Even though the past years in a beginningless universe are not actual, their infinity must be. And for some purposes Aristotle needs to avoid any actual infinity.

It seems to me that Philoponus has trapped the pagans: since they follow

---

[32] Philoponus *de Aeternitate Mundi contra Proclum*, ed Rabe, pp 9-11 and 619; *in Phys* 428, 14-430,10; 467,5-468,4; *in Meteor* 16,36ff; *contra Aristotelem* ap Simplicium *in Phys* 1179,12-26. It is the last passage which most graphically illustrates the multiplication of infinity, using the examples subsequently repeated by Bonaventure. I have adapted his examples in the text.

[33] Simplicius *in Phys* 506,3-18; cf 1180,29-31; relying on Aristotle *Phys* 3.6, 206a33-b3 and 3.8, 208a20-1.

Aristotle in denying actual infinities, they ought to admit that the universe must have had a beginning. So now I want to turn to a new task, and to consider what response *ought* to be made to Philoponus, at the cost, if need be, of abandoning the Aristotelian view of infinity. I shall start with Philoponus' argument about the impossibility of an infinity being *increased*.

I want to distinguish a sense in which infinity can be increased and a sense in which it cannot. I shall do so without entering at all into the complication of transfinite numbers. Let me take as an example the infinity of *whole* numbers as against the infinity of *odd* numbers, and let me first explain the sense in which one of these infinities is *not* larger than the other. I would like you to imagine the column of whole numbers stretching away from your left eye infinitely far into the distance, and parallel to it, stretching away from your right eye, the column of odd numbers, also stretching infinitely far. The two columns should be aligned at the near end, and the members of the two columns should be matched against each other one to one. 1, 2, 3, 4 in the one column should be matched against 1, 3, 5, 7 in the other. I can now explain the sense in which the column of whole numbers is not larger than the column of odd numbers: it will not *stick out beyond the far end of* that column, because neither column has a far end.

Now for the sense in which the column of whole numbers *is* larger than the column of odd numbers. This can be brought out by saying that it contains all the same members as the column of odd numbers *and more besides*; whereas the column of odd numbers does not contain more besides. The two key words in the explanation are *beyond* and *besides*. One column does not go *beyond* the other, but it does have members *besides*.

I have deliberately chosen the simplest possible example. If I were to choose some other examples, I would have to switch to something more complicated than the idea of *besides*, in order to explain how one infinity can be larger. But the simplest example will suffice for present purposes. One interesting fact is that these ideas seem to have been worked out for the first time only in the fourteenth century. I am relying on an account given by the historian of science John Murdoch, who describes the work of Henry of Harclay, William of Alnwick, William of Ockham and Gregory of Rimini.[34] I may be over-simplifying in picking out from their discussions the two words that interest me. But there is no doubt that they used the two Latin words *ultra* and *preter* (elsewhere *praeter*), beyond and besides, to explain the sense in which one infinity can be greater than another, and the sense in which it cannot.

One of Philoponus' arguments can now be answered as follows. If there has been an infinity of years, it is true that that infinity will be increased next year. But it will be increased only in the innocuous sense, of containing one year *besides*. It will not be increased in the objectionable sense, which would

---

[34] John E. Murdoch, 'Mathesis in Philosophiam Scholasticam introducta: the rise and development of the application of mathematics in fourteenth-century philosophy and theology', *Arts libéraux et philosophie au moyen âge, actes du quatrième congrès de philosophie médiévale*, Paris 1969, 222-3; 'The "equality" of infinites in the Middle Ages', *Actes du XIe congrès international d'histoire des sciences*, Warsaw-Cracow 1968, vol 3, 171-4.

arise from the two columns of years being aligned at the *near* end, in such a way that one stuck out *beyond* the other at the *far* end. Consequently, there will not have been an increase in any *objectionable* sense. The same point could be made about the infinity of *years* and the infinity of *days*, although I should have to say something more complicated, in order to bring out the sense in which one infinity *is* larger that the other here. But the important point is that it is *not* larger than the other in the objectionable sense of being liable to stick out beyond the far end.

So much for Philoponus' objection about increasing infinity. But what now about the other objection, that an infinity cannot be actual, or traversed? This consideration is still found compelling by many people, and it has been defended in recent years by our colleagues Pamela Huby, William Lane Craig and G.J. Whitrow.[35] What I want to say is that an actual infinity of past years can perfectly well have been traversed. But there are many reasons why this view may seem to be mistaken. I expect that it will seem to be mistaken to some of you. And I am therefore going to consider eight sources of temptation which may lead to its being resisted. My hope will be, by countering the temptations, to break that resistance down.

(i) One source of resistance may be the idea that, if an infinity of days had to pass before the arrival of today, then today would never arrive. This would certainly be so, if there was a first day, and then an infinity of days to cram in before today. But of course no first day is envisaged by those who postulate a beginningless universe, so there is ample room for a preceding infinity.

(ii) The first mistaken objection is related to a second one about counting. We might try to imagine that the years have always been subjected to counting, as they arrived. If the universe had no beginning, then earlier than any year we care to name, the count should already have reached infinity. But, the objection goes, it is absurd to suppose that this infinite count could be completed. What this objection overlooks is that counting differs from traversing in a crucial respect, for counting involves taking a *starting* number. This is, indeed, part of the reason why it would be so difficult to complete an infinite count. We will not be able to complete it, unless we can accelerate in our counting in Zenonian fashion, taking half as much time for each successive act of numbering. There are no such obstacles, however, to completing an infinite lapse of past years, precisely because it involves no *starting* year. Moreover, this difference, the absence of a *starting* year, has a second consequence. For it means that after all we cannot imagine that the beginningless series of past years has been subjected to counting in any straightforward way; for it has no first member to match the first number used in counting. The counting argument, then, must fail, although it has

[35] Pamela Huby, 'Kant or Cantor? That the universe, if real, must be finite in both space and time', *Philosophy* 46, 1971, 121-3; and 48, 1973, 186-7. W.L. Craig (1979) esp 83-7; 97-9. G.J. Whitrow, *The Natural Philosophy of Time* 1961, 2nd edition Oxford 1980, 27-33; 'On the impossibility of an infinite past', *British Journal for the Philosophy of Science* 29, 1978, 39-45; Review of Craig, *British Journal for the Philosophy of Science*, 31, 1980, 408-11.

been very popular. It was frequently used by Philoponus,[36] and it has been repeated recently.[37]

But now I must face an objection. If the only obstacle to completing an infinite count is that conventional counting takes a *starting* number, what about counting in a *backwards* direction? Ought it not to be possible for a beginningless being to count off the years, descending from the higher numbers and finishing, say, in this century with the years, four, three, two, one and zero? A backwards-counting angel might then sigh with relief in 1982 and say 'Thank heavens, I've reached year zero, and I've finished counting infinity.' If this is *not* possible, then how can the traversal of an infinity of years be possible?[38]

My answer to this is that something like the backwards count would indeed be possible *in principle*. I am not at all sure that it ought to be called *counting*, but it is conceptually possible that God should have included a beginningless meter in his beginningless universe, to record how many years remained until some important event, say, until the incarnation of his Son. At zero B.C., the meter would register zero, but the counting would never have been begun. Rather, for every earlier year, the meter would have displayed a higher number. Whether or not this should be called *counting*, there is no logical barrier to it, I believe; and therefore no logical barrier has been exhibited to the traversal of an infinity of past years.

(iii) The counting objection, in its original form, is akin to Kant's argument in the first antinomy, which has also been endorsed recently,[39] that the universe must have a beginning because an infinite series can never be completed by a successive synthesis. Admittedly, if the number of years is infinite, this will not be the result of completion by successive synthesis, that is, of adding to a finite collection. For that would suggest that mere addition (without any Zenonian tricks of acceleration) could take you from a finite number of years to an infinite. But the hypothesis of an infinite past never envisaged that infinity was reached by that particular method. On the contrary, the hypothesis was not that the number of years had ever passed from being finite to being infinite, but that it had always been infinite.

(iv) A further objection used by Bonaventure has reappeared in modern times:[40] if we think backwards from the present, we will not find an infinitieth year. But this objection, like the last, represents an *ignoratio elenchi*. For those who believe that there has been an infinity of years do not mean that one of them occurred an infinite number of years ago, just as someone

[36] Philoponus *aet* 9-10, as above; *in Phys* 428-9; ap. Simplicium *in Phys* 1178; in Arabic summary 3rd treatise, translated by S. Pines (1972).

[37] Pamela Huby op. cit. 1971, 128.

[38] I would thank Norman Kretzmann for this objection. I do not accept Fred Dretske's complaint that, however far back you go, a backwards counter will have *finished* his task. He will, it is true, have counted infinitely many numbers, but infinitely many does not imply all (Fred I. Dretske, 'Counting to infinity', *Analysis* 25, suppl 1965, 99-101).

[39] W.L. Craig (1979) 103; 109; 189.

[40] Bonaventure, *in IV Libros Sententiarum Magistri Petri Lombardi*, tomus II, in librum II, dist 1, p 1, q 2 argument (3) (pp 107-8 in Byrne); Richard Bentley, Sixth Boyle Sermon, 1692; Pamela Huby op. cit. 127; W.L. Craig (1979) 98; 200-1.

who says that a crowd is large does not mean that any of the members is large: it may be a crowd of midgets. Infinity is a property of the collection as a whole, not of one member. If a collection could not be infinite, without one of its members being the infinitieth member, we would get the absurd result that the set of whole numbers is not infinite.

A variant on Bonaventure's argument has also reappeared.[41] It is conceded that the pagans are not committed to a *first* year which would have been the infinitieth year ago. But it is alleged that they are committed to there being many past years separated from the present by an infinite gap. I do not believe that they are so committed, for once again the infinity of the whole number series does not involve there being *any* whole numbers separated by an infinite gap from 1. But as to why this misconception should have arisen, I shall have more to say shortly.

(v) An objection discussed in the Islamic world was that the infinite, by definition, cannot come to an end, and so cannot be completed.[42] The *simplest* answer to this is that an infinite series can easily have one end, as, for example, the series of positive whole numbers does at zero. And if the infinity of past years is regarded as ending at the present, this will only give the series one end.

(vi) A more subtle objection has been put to me in discussion by Pamela Huby, and it is also suggested by the work of G.J. Whitrow, who is in turn inspired by the Boyle sermons of Richard Bentley, delivered in 1692.[43] This objection points to the fact that an infinity of *future* years starting from now would always remain potential, and never be completed. So should we not in consistency say the same about the infinity of *past* years starting from now? The same objection could be raised about *any* set of years that runs forward from a given date, say, from 1700. But, for convenience, I shall consider the set of years running forwards from the *present*.

In reply it is important to bring out why there is a disanalogy between the past and future. Past years do not start from now. If anything, it will be our *thoughts* about past years which start from now, if we choose to think of them in reverse order, but the years themselves do not. Indeed, on the pagan view, they do not have a start at all. And this is important, because it means that when we say that they *have* been gone through, and therefore assign them a *finish*, we are not thereby assigning them *two* termini. This is what leaves us free to think of them as forming a traversed series which is more than an extendible finitude. Contrast a series of future years to which we assign a start, say, at the present. Whenever we think of such a series as *having* been gone through, we will automatically be assigning it a *second* terminus, a finish as well as a start. And this is what prevents the future series of traversed years

---

[41] G.J. Whitrow op. cit.

[42] So the opponents envisaged by Avicenna in chapter 4 of an unpublished treatise, summarised in English by Pines, in the appendix to Pines (1972) (see 348); also ascribed to others by Averroes *Tahāfut al-Tahāfut* (Bouyges), 19-22 and 31, translated by S. van den Bergh, London 1969, vol I, pp 10-11 and 17.

[43] Richard Bentley, Sixth Boyle Sermon. I am grateful to Nicholas Denyer for drawing my attention to this Sermon.

from being more than finite. The asymmetry to which I am pointing between past and future series depends crucially on the occurrence of the perfect tense: '*having* been gone through'. It is not future years *as such* which have a different infinity from past years, but future years which *have* been traversed, for the traversed ones will have *two* termini.

I think that failure to appreciate the nature and extent of the disanalogy between past and future series of traversed years has provided one motive for the view in (iv) above that an infinite past would involve events infinitely far removed from the present.[44] Admittedly, a *future* set of years which started from now would become actually infinite only if, *per impossibile*, it attained to a year that was infinitely far removed. But the same ought not to be said about the *past*.

(vii) A very ingenious argument against an infinite past has been built by Craig out of a suggestion originally made by Bertrand Russell and endorsed by Whitrow.[45] It involves Tristram Shandy, who is to be imagined as keeping a diary, but as recording his life at the snail's pace of one day recorded for every year lived. If there had been an infinity of years, it is alleged, then we would get the absurd result that he could catch up, for he would have had time for an infinite number of entries. And the absurdity of this is intended to cast doubt on the conceivability of an infinite past.

This argument has several things wrong with it, I believe. First, it confuses the idea of *infinitely many* with the idea of *all*. There would have been time for an infinity of entries, but not for all. These are not the same: for example, there is an infinity of odd numbers, but these are not all the whole numbers. Secondly, the truth of Russell's original claim, that no part of the biography will remain unwritten, depends on how Tristram Shandy behaves. For what if (contrary to Russell's hypothesis) he records only every January 1st, and then skips to the next January 1st, leaving permanent gaps? He must avoid such gaps, if he is to bring it about that no part of the biography remains unwritten. Moreover, it is hard to see how he *can* avoid gaps, if with Craig we diverge from Russell's story by adding the idea that he has lived without beginning in the past, so that the diary has no beginning. For he then has no possibility of filling up the diary in a systematic way from its opening pages – there being no opening. There is yet a further thing wrong with the argument. For we can admit that, if the diary has a beginning, and if it is kept without gaps, and kept for ever, then no day will remain for ever unrecorded. But it does not at all follow from this that a time will eventually arrive when *all* the days *have* been recorded. Nor does it follow that they will *already* have been recorded, if Tristram Shandy has *already* lived an infinity of years.

(viii) I now come to a final kind of argument, which is perhaps even more ingenious – ingenious, but desperate. It seeks to discredit the possibility of an *actual* infinity, that is of an infinity which is more than an extendible finitude,

[44] G.J. Whitrow, op. cit.

[45] Bertrand Russell, in 'Mathematics and the metaphysicians', in *Mysticism and Logic*, London 1917 (p 70 of the 1963 edition), revised from an article written in 1901, and published in *The International Monthly*; also in *Principles of Mathematics* 1903, 2nd edition, London 1937, 358-9. W.L. Craig (1979) 98. G.J. Whitrow op. cit. 1978-80.

by alleging that all actual infinities must lead to absurdity. The most spectacular example, used by two recent authors,[46] concerns the case of Hilbert's hotel. The hotel, named after the notable mathematician D. Hilbert, contains infinitely many rooms, but every single one is full. Along comes a late traveller and says, 'I know your hotel is full, but cannot you fit me in somewhere?' 'Certainly', says the manager, 'I can accommodate you in room number one.' And then, in a loud voice, he declares, 'Will the occupant of room number one step into room number two? Will the occupant of room number two step into room number three? And so on *ad infinitum*'. You may have a nasty feeling that some unfortunate resident at the far end of the hotel will drop off into space. But there *is* no far end. It is like the column of odd numbers and the column of whole numbers which we considered before: the line of residents will not *stick out beyond the far end of* the line of rooms.

Once it is seen like this, the outcome should no longer seem an absurdity which can discredit the idea of an actual infinity. It should instead be seen as an explicable truth about infinity. It may be a surprising truth, even an exhilarating and delightful truth, but a truth for which we can perfectly well see the reasons, when we reflect on the idea of sticking out beyond the far end.

The other puzzles which have been pressed, in an attempt to discredit actual infinities, can all, I think, be shown explicable, in the light of principles which have already been explained.[47]

I want now to return to Philoponus. I have considered eight reasons why you might be inclined to agree with his claim that the universe cannot have gone through an actual infinity of past years; but I have argued that all the reasons are wanting. This means that I do not think Philoponus' arguments succeed absolutely: it cannot be proved that the universe must have had a beginning. I would leave the question open whether it began or not. None the less, I do think that Philoponus' arguments succeed *ad homines* against the pagans. He has found a contradiction at the heart of paganism, a contradiction between their concept of infinity and their denial of a beginning. This contradiction had gone unnoticed for 850 years. Moreover,

[46] Used by Pamela Huby op. cit. 1971, 128, though not directly in connection with time, and by W.L. Craig (1979) 84-5.

[47] W.L. Craig (1979) 83-7 envisages a library with infinitely many books, and is surprised that you can make various additions, subtractions and subcollections without getting any collection of a different size from the original. What remains unrecognised is the reason, namely, that none of the resulting collections would stick out beyond the far end of any other. And this also provides the reason why certain other subtractions would after all reduce the size of the collection. Craig further assimilates 'infinitely many' and 'all', when he supposes that the numbers on the spines of an infinitely numerous book collection would exhaust *all* the whole numbers. Why should not just the *odd* numbers be used? Even if all integers had been used initially, it does not follow that it would be impossible to accommodate a new book with a number of its own. One would simply re-number the original collection, perhaps using odd numbers and giving the new book an even number. Finally, more analysis is required for the claim that if you close the gaps after subtracting books, the shelves will remain full. If 'full' merely means that there will be no gaps, this is a tautology. But it is false, if it means that there will be no room for reinsertion: that would apply only to a full shelf with two ends. To the general literature on the subject I would now add Robert Bunn, 'Quantitative relations between infinite sets', *Annals of Science* 34, 1977, 177-91.

the materials for beginning to answer Philoponus' puzzle about increasing infinity were not even assembled until Henry of Harclay and others, some 800 years later. We can therefore see Philoponus as being at the centre of a 1600-year period. For the first time, he put Christianity on the offensive in the debate on whether the universe has a beginning. This might well be called a turning point in the history of philosophy.

I should like to remind you now of the almost universal view with which I started this lecture, that the Neoplatonist philosophy from which Philoponus arose was by now dead. Edward Gibbon was speaking of the *Athenian* Neoplatonists in particular, when he said, in *The Decline and Fall of the Roman Empire*, that Plato would have blushed to acknowledge them, and that, with their interest in mystery religions, they exhibited the second childhood of reason.[48] But my interest is in late Neoplatonism quite generally, and it is time for me to ask for your verdict. Admittedly, in choosing Philoponus I have decided to talk about a Christian. But he was a Christian nourished by late Neoplatonism, who wrote and thought within the tradition of late Neoplatonism. Are you persuaded by the received view that by 529 this tradition was dead?[49]

[48] Edward Gibbon, *The Decline and Fall of the Roman Empire*, ch 40, §VI.

[49] I should like to thank the many people who have helped me by offering comments. Only some are acknowledged in the text and notes; others are acknowledged in Sorabji (1983).

CHAPTER TEN

God or Nature? Philoponus on Generability and Perishability

Lindsay Judson

Philoponus deploys many arguments, in a number of works, against the Aristotelian and Neoplatonic doctrine that the cosmos is eternal. Most of his positive arguments, and his replies to objections, have been charted by others,[1] and I shall look instead at a major theme of the *de Aeternitate Mundi contra Proclum* which is hard to classify as either: his appeal to the authority of Plato for the view that the cosmos had a beginning. Even if Philoponus' interpretation of Plato's doctrine is correct, his appeal to it is hardly an argument (except, of course, *ad hominem* against Proclus), and any persuasive force it has should be resisted by even the most ardent Platonist; nevertheless, the highly controversial nature of this interpretation of Plato leads Philoponus to formulate a number of arguments and theses which are worth examining.

In the *de Aeternitate*[2] Philoponus wishes to defend the view that Plato meant his creationist account of the cosmos in the *Timaeus* to be taken literally. He disagrees with Plato's belief that the cosmos will never *end*, but is anxious to claim Plato's authority for its having a beginning. In so doing, Philoponus flies in the face of the orthodox Neoplatonic reconciliation of the cosmologies of Plato and Aristotle, which took both to believe in an eternal cosmos, and took creation in the *Timaeus* to be a mere device for expository purposes. Thus the Proclan view, for instance, was that Plato's cosmos is literally 'generated' only in the senses of 'having its being in coming to be (*to gignesthai*)' and of having an external cause in the Demiurge.[3] In his commentary on Aristotle's *Physics* Philoponus himself endorses the *non*-literal interpretation of the *Timaeus*: he thinks that the prior existence of the disorderly matter in

---

[1] See especially Herbert A. Davidson (1969) S. Sambursky (1972): Robert B. Todd (1980); Gérard Verbeke (1982); Richard Sorabji (1983), ch 14.

[2] Hereafter abbreviated to *aet*; references are to the Teubner text, edited by Hugo Rabe (Leipzig 1899).

[3] Philoponus reports and argues against other interpretations as well (see *aet* 6, especially chs 7-26). For modern discussions of whether the *Timaeus* should be taken literally, see Gregory Vlastos, 'The disorderly motion in the *Timaeus*' and 'Creation in the *Timaeus*: is it a fiction?', in R.E. Allen, ed, *Studies in Plato's Metaphysics* (London 1965), and Sorabji (1983) 268-75.

separation from the forms imposed by the Demiurge is impossible, and that this separation is made by Plato 'in thought' only, in order to show how much of the actual cosmos is due to matter itself (*in Phys* 575,7-11). Philoponus' interpretation here is entirely in line with that of the orthodoxy which he rejects in *aet*; but he adopts this non-literal reading only on the issue of whether the coming into existence of the ordered cosmos *from a pre-existing chaos* is to be understood literally;[4] he thinks that Plato is in this respect to be interpreted metaphorically, since (a) this squares with his own view and (b) it squares better with his belief that Plato made *time* begin simultaneously with the cosmos.[5] On the issue as to whether the cosmos had a beginning at all, however, he sides with Aristotle in taking Plato literally.[6]

I

Proclus thinks he can show that to hold the view which the literal interpretation ascribes to him would commit Plato to serious inconsistency: a major part of Philoponus' defence of this interpretation is an attempt to refute this charge and to demonstrate the internal consistency of Platonic creationism. Here is Proclus' main argument, quoted by Philoponus at the start of *aet* 6:

> Proclus: If the Demiurge alone <bound together> the cosmos, he <alone> would dissolve it; for it is in every way indissoluble, <Plato> says, except for the one who bound it together ... But the Demiurge would not dissolve the cosmos; for it is he himself who says, 'it is evil to wish to dissolve what has been well fitted together and is in a good state'. It is impossible that the one who is truly good should become evil; impossible that the cosmos be dissolved ... Either, therefore, he did not fit it together well, and is not the best demiurge, or he did fit it together well and will not dissolve it unless he has become evil, which is impossible. Thus the universe is indissoluble, and consequently imperishable. And if it is imperishable, <it is ungenerable>. 'Destruction belongs to everything that has come into existence,' says Socrates in the first of the Timaean works [*Republic* 546a2] – not speaking in his own part, but saying that the muses are speaking. And, I presume, Timaeus does not immediately take the muses' claim as an eccentricity, and posit something which has come into existence but is imperishable. If this is true, therefore, then that to which no destruction belongs is ungenerable. Destruction does not belong to the cosmos; so it is ungenerable.

---

[4] Cf Sorabji (1983) 268 n 22.

[5] *Tim* 37d-e; cf 38b6, 'Time has come into existence together with the heavens'. Philoponus too thinks that time began along with the cosmos (*aet* 5.4.117,14-118,15; cf 6.7.140,12-142,1), and he defends non-temporal usages of temporal words to describe the beginning of time (5. 1-4; cf *in Phys* 456,17-458,16, ap. Simplicium *in Phys* 1163,25ff). Elsewhere (*aet* 9.9-11; cf 11.11-12, *in Phys* 54,8ff, ap. Simplicium *in Phys* 1141,5ff), he argues on both his own behalf and Plato's that creation *ex nihilo* is possible (and indeed is a sign of the superiority of God's creative power over that of art and nature: *aet* 9.9.339,25-341,23); this does away with the need for the pre-existence of chaotic matter out of which the orderly cosmos has to be generated.

[6] *de Cael* 1.10-12. But Aristotle (in *de Cael*; contrast *Phys* 8.1, 251b17-19) and Philoponus disagree over whether time also had a beginning in Plato's theory. The question as to whether Aristotle is really interested in Plato's *actual* position is a moot one, however; see my 'Eternity and necessity in *De Caelo* I.12', *OSAP* I (1983) 217-55, at 238.

Thus the cosmos is eternal, if it is ungenerable and imperishable (6.119,14-120,14, with gaps; cf 15.549,7-550,24).

This argument for the eternity of Plato's cosmos has two stages. In the first, Proclus understands Plato to be claiming that the cosmos is imperishable, and offers an argument based on Platonic premises to show that Plato is in fact committed to this; in the second, he argues that, if the cosmos is, in Plato's view, imperishable, he is committed to its being ungenerable – and hence ungenerated – as well; for Proclus holds that everything which is imperishable is also ungenerable.[7]

Philoponus attacks some of Proclus' other moves in this argument, and in particular he denies the claim that the destruction of something well-ordered cannot be consonant with God's goodness;[8] he does not, however, attack the dubious idea that indestructibility or imperishability entails ungenerability. This is surprising: although Philoponus' claim that a good God might readily destroy the cosmos will suffice to defend his own position, it will not save the consistency of Plato's, as we shall see later. We might expect him to attack this seemingly major weak point in Proclus' argument in any case – and all the more so since Aristotle too had used the closely related doctrine that imperishability and ungenerability entail each other to demonstrate that a Timaean cosmos is impossible in *de Caelo* 1.12. Aristotle, moreover, claims that his argument also rules out even the possibility of a *finite* cosmos – the sort of cosmos in which Philoponus himself believes.[9] Despite all this, Philoponus does not question Proclus' use of the 'imperishability entails ungenerability' thesis at all.

Why is this? One reason is that Plato himself appears to endorse a thesis

[7] There is some obscurity in Proclus' presentation of this second stage: he does not make it clear whether he intends the quotation from the *Republic* to support the preceding claim that imperishability entails ungenerability, or to introduce a fresh argument. In other words, does he employ one thesis – 'imperishability entails ungenerability' – or two – 'imperishability entails ungenerability' *and* 'generation entails destruction'? (For a more precise formulation of these, and of related theses, see below, p 182.) For the moment I shall just consider the argument which appeals to 'imperishability entails ungenerability'. Philoponus, however, is aware that there is another argument in the offing, whether or not Proclus is; and since he designs his reply to meet both, we shall return to the 'generation entails destruction' argument later (pp 182-5). This obscurity arises partly because Proclus does not make it clear whether he takes the phrase 'destruction belongs to *x*' to mean '*x* will perish', or merely '*x* is perishable' (see below, p 185); consequently he leaves it quite vague as to how inferences such as 'destruction does not belong to the cosmos; so it is ungenerable' are meant to go through. For this reason it is also unclear exactly what thesis Proclus means to ascribe to Plato, and in particular whether he appeals to 'imperishability entails ungenerability' as something *he* takes to be true, or as a Platonic doctrine.

[8] *aet* 6.4.128,1-131,25. Proclus' argument threatens Philoponus' own view that the cosmos is finite, and hence is allowed by God to perish.

[9] Aristotle is wrong about this. Even if we granted the validity of his argument in *Cael* 1.10-12, he has still not refuted the possibility of a finite Democritean cosmos (which will not succumb to the move he deploys against Empedocles' *regular* series of finite cosmoi); and it still leaves open the possibility of the *absence* of natural explanation (see my 'Eternity and necessity in *De Caelo* I.12', 236-9). Philoponus does not acknowledge the threat posed by Aristotle here, but he avoids having to face it, in effect, by allowing (as Aristotle will not) a non-natural feature – God – into his account of the formation and destruction of the cosmos.

equivalent to 'imperishability entails ungenerability' (even though Proclus' own presentation of the argument leaves this rather unclear); for the Demiurge affirms that everything which has been bound may also be unbound (*Tim* 41a8-b1). Thus if Philoponus wishes to defend the internal consistency of a creationist Plato, he cannot afford to do it at the expense of this thesis. The other reason is that, with suitable qualifications, Philoponus himself believes the thesis to be true.[10] More precisely, since he thinks that the converse entailment holds as well, he can be said to hold the following:

(1A) Necessarily, if a thing is able to come into existence, it is also able to cease to exist. (For short: if it is generable, it is also perishable.)

(1B) Necessarily, if a thing is able to cease to exist, it is also able to come into existence. (For short: if it is perishable, it is also generable.)

(1A) and (1B) are straightforwardly equivalent to the theses that imperishability entails ungenerability and *vice versa*. Philoponus also subscribes to two even stranger beliefs which will be of importance later; they are analogous to (1A) and (1B), but are concerned with *actual* coming into existence and ceasing to exist. These beliefs are (again with an important qualification):

(2A) Necessarily, if a thing comes into existence at some time, it also ceases to exist at some later time.

(2B) Necessarily, if a thing ceases to exist at some time, it also comes into existence at some earlier time.[11]

II

What are Philoponus' grounds for these doctrines? Proclus too believes (1 A&B), as the fifteenth argument quoted by Philoponus shows,[12] and has an

---

[10] And thus is also ready to cite Plato's authority in support of such doctrines in contexts where he is not engaged in defending Plato's consistency (see *de Opificio Mundi* (hereafter *Opif*) 5.9.

[11] (1A): *aet* 6.28.226,10-14, 17.2.596,2-6. (1B): *aet* 6.28.230,21-2, 29.241,26-7, 7.13.272,11-16, 8.1.304,8-9, 11.8.441,18. (2A): *aet* 9.17.380,7-9 (this passage could, less naturally, be read as expressing (1A) instead), 11.12.458,8-15, ap. Simplicium *in Phys* 1145,25; cf *aet* 8.4.311,8-9. (2B): *aet* 4.14.94,19-20, 11.8.441,16-17; note also 6.13.163,1-2, where, expounding a passage from the *Laws*, Philoponus rewrites Plato's words, 'no beginning and no end' (162,10-11), as 'no beginning *and therefore* no end'. I shall label the relevant biconditionals (1 A&B) and (2 A&B). (Whether (2 A&B) entails (1 A&B) depends on what sorts of modality are involved in the necessity operator and in the 'abilities' mentioned in (1 A&B): if these were different (e.g. metaphysical necessity in the first case, but, say, natural abilities or capacities of some sort in the second), then the entailment would look dubious.)

[12] *aet* 6.29.239,9-11 (= 8.1.298,24-299,1 = 18.5.626,18-19) may support the ascription of (2A) to him as well: ' ... the whole coming-into-being thing, then, is, in respect of itself taken by itself, perishing also in every way ...' But (a) Proclus asserts no *connection* here between coming into being and perishing; (b) from the context it seems that 'is perishing' may mean 'having its being in ceasing to be (*to phtheiresthai*)' here, since 'coming-into-being' may be used here in the sense of 'having its being in coming to be (*to gignesthai*)'.

argument for (2A) which would serve as well (if at all) for (2B). He thinks that everything which comes into existence does so out of some pre-existing matter which it 'masters'; eventually the original matter will regain the mastery and destroy the generated object (see, for example, *aet* 9.313,17-22). But this rather unconvincing argument is in any case not available to Philoponus, as he believes in the possibility of creation *ex nihilo* (see above, n 5). Philoponus himself has another argument to hand, which starts from the Aristotelian doctrine that no finite body has more than a finite *dunamis*, or power; since the cosmos is – as Aristotle himself agrees – a finite body, it will have a finite *dunamis* to maintain its orderly existence:

> If, therefore, nothing is added to it from outside, and the natural *dunamis* of the subject for preservation is itself unable to be sufficient always, because it is limited, it thereafter allows its subject to travel to destruction, and perishes with it – like a shipwrecked helmsman [in the sea] who is unable to endure, when he loses hold of the rudder-piece, not because he is overcome by rough seas, but because his *dunamis* does not suffice for him, because he grows weary in the course of time (*aet* 8.1.303,17-25).[13]

Thus one could conclude that any finite body can only last for a finite time, and hence must both come into existence and perish.

But to base the belief in even Philoponus' importantly qualified versions of (1 A&B) and (2 A&B) upon the finite *dunamis* argument would require a further move, namely showing that a thing is generable or generated (or is perishable or going to perish), if and only if it is a finite body. For his present purpose, of course, Philoponus only needs these principles to be true of corporeal objects – of finite (and infinite) bodies. He does not, however, restrict their truth to such objects, since he applies (1A) to the Platonic Forms (*aet* 6.28.226,10-14), and presumably would think them applicable to souls as well.[14] Aristotle's doctrine that no finite body possess infinite power will not suffice to ground these applications of the principles. Moreover, it is striking that Philoponus himself does not use this argument to demonstrate that finite bodies must be *generated*, nor that they are generable, but only that they are perishable (and must perish); thus, for instance, at 6.29.242,11-15 Philoponus uses the finite *dunamis* argument only to show that the cosmos is perishable,

[13] Cf *aet* 6.29.241,22-242,22, 8.4.312,16-26 (quoted below, n 30), ap. Simplicium *in Phys* 1327,11ff, 1333,20, and ap. Simplicium *in Cael* 142,22-5: 'If Aristotle's proof that every body has a limited *dunamis* is sound, and the heaven is a body, then it is clear that it is also receptive of destruction, having a *logos* of destruction' (*logos* = 'formula' or 'definition'); see also, in a different connection, *aet* 1.2. 2,10-14 (quoted below, p 185). Proclus too uses something like this argument (6.29.238,6-240,9 = 8.1.297,24-300,2; part of this series of quotations is repeated again at 18.5.626,11-627,20); but it may have a different orientation from Philoponus' (see above, n 12).

[14] Philoponus thinks that the rational (part of the) soul is ungenerable and imperishable, and that the unreasoning soul, being inseparable from the body, is generated and destroyed (*in DA* 9.3-18,33, ap. Simplicium *in Cael* 129,19-20, *Opif* V.13). At *Opif* VI.23-4 he does speak of the generation (*genesis*) of the rational soul; but *genesis* here probably refers to its *incarnation* (Philoponus himself ascribes this usage to Plato when trying to reconcile the ungenerability of the world-soul with its apparent *genesis* in the *Timaeus*: *aet* 6.24.195,7-200,3).

and then uses (1B) to show that it is also generable. This sits ill with the idea that the finite *dunamis* argument is the *basis* of (1B) and the rest.[15] In fact, it is hard to resist the impression that Philoponus takes versions of these doctrines to be self-evident truths. This impression is perhaps confirmed by a passage where Philoponus quotes with approval *Galen's* claim that (1 A&B) is (when suitably qualified), an undemonstrable, primary and self-warranting axiom (*aet* 17.5.600,9-10).[16]

If he accepts these principles as undeniable and self-evident truths, then Philoponus seems hard put to it to defend the coherence of Plato's creationist account; for in that account, according to Proclus, the cosmos is both generated and imperishable, in violation of (1A). One way round the problem might be this. Since it is the *imperishability* of the cosmos which is causing the difficulty, Philoponus could stress the fact that Plato seems to think that the Demiurge *can* dissolve the cosmos (*Tim* 41a7-b3); so he could argue that Plato (like Leibniz) must deny the soundness of Proclus' inference that if destroying the cosmos is evil, then a good Demiurge *cannot* do it, and must in fact regard the cosmos as *perishable* (though of course unperishing). Thus Plato could be represented as consistently holding that the cosmos is generable (because generated), and also that, though perishable in consequence of (1A), it will not actually perish. This way out would involve Philoponus in ascribing beliefs to Plato which he himself thinks are false;[17] but Philoponus is elsewhere happy to do this, since, as I said earlier, he is concerned to defend the consistency, not the truth, of Plato's position.[18]

This line of approach is not so far from the one Philoponus actually adopts,[19] but as it stands, it faces two major difficulties. (i) Since there is no natural force outside the cosmos to affect its workings,[20] and since the

---

[15] The sole exception, as far as I am aware, is at *aet* 1.2.2,10-14 (quoted below, p 185). This might seem, if anything, to prove the rule: for in this case Philoponus is not arguing for the generation of the cosmos, but is deploying a *reductio* argument against the idea that God's power is limited, and his goal seems to be to show that if it were limited, God the creator of time would himself be *in* time (in Aristotle's sense). I discuss this issue further on pp 194-5.

[16] *Anapodeikton te kai prôton kai ex hautou piston axiôma* (for the first two adjectives, cf *Posterior Analytics* A.2, especially 71b26ff). Philoponus is, of course, far from being the only ancient thinker to subscribe to such doctrines – indeed, according to Galen, *everyone* accepts (1B) (and perhaps a suitably qualified version of (1A)). For later Near Eastern thinkers who subscribed to (2 A&B) as well, and who take it to be an *a priori* truth, see Davidson (1969) 378 and nn 169, 173, 174. We shall return to this appeal to Galen (which Richard Sorabji first pointed out to me) below, p 196.

[17] Philoponus does not appear to question Proclus' claim that, since goodness is a necessary property of God, he lacks the freedom to perform actions which are evil; and I suspect that he actually agrees that the inference is sound. This suspicion is perhaps strengthened by the argument at *aet* 6.5, apparently designed to show that God *must* exercise all his *dunameis* (discussed below, n 38).

[18] He argues explicitly against Plato that it need not be evil to destroy something good (6.4), and that the cosmos will have an end (6.5-6); he disputes various other Platonic beliefs in book 9.

[19] Thus Philoponus does stress the point that Plato's Demiurge *can* destroy the cosmos, and merely chooses not to (*aet* 6.28.225,14-226,19, quoted below, p 189; 8.4.311,10-15); and on his interpretation, there is a clear sense even beyond this in which Plato's cosmos is not imperishable.

[20] Cf Proclus' eighth argument for the indestructibility of the cosmos (*aet* 8.294,2ff.), and Philoponus' reply.

cosmos is, on this view, perishable, it might seem to be a matter of *accident* whether it endures for ever or not;[21] this consequence would be fine for a Democritean, but seems hard to square with the general tenor of the *Timaeus*, especially as viewed both by Philoponus and by his Neoplatonist opponents.[22] This point is not found explicitly in Philoponus' discussion, but is, I think, reflected in his desire to see Plato's cosmos as, in a sense, imperishable, and not merely unperishing. (ii) This solution may be successful against Proclus' objection as outlined earlier; but there is a second objection of a similar sort which it does not avoid. Plato appears not only to commit himself to (1A), but also to (2A): 'Destruction belongs to everything that has come to be.' A cosmos which is generated but unperishing violates this principle even if it is not imperishable; a creationist Plato still contradicts himself.[23]

As I have said, it is not clear whether Proclus himself is aware of this second objection – or, indeed, whether he is at all careful to distinguish '*x* is generable' and '*x* is perishable' from their non-modal counterparts (see above, n 7). Philoponus, on the other hand, is alive to the distinction, and to the new objection. His solution is designed to meet both Proclus' original objection to Platonic creationism and the new one sketched in (ii); it also avoids the unfortunate consequence (i), by providing a way in which the cosmos' unperishingness is not accidental.

III

Philoponus achieves all this by drawing a distinction between those properties of a subject, *S*, which are 'in accordance with the *logos* of *S*'s own nature' (*kata ton tês idias phuseôs logon*) and those which arise from something outside its nature. This idea was already hinted at in a different context in *aet* 1.2; speaking about God, Philoponus says: 'For if his *dunamis* is limited, it is unable to be sufficient always; therefore either he will have both a beginning and an end because of the limitedness of his *dunamis*, or something else will be responsible for his existence and his eternity' (1.2.2,10-14). Here is his account of the distinction as he applies it to imperishability in *aet* 17:

> 'Deathless' and 'imperishable' have two senses: <signifying> (1) that which has deathlessness and imperishability in accordance with the *logos* of its own nature, nor having a *logos* of death or destruction; (2) that which is receptive of destruction and dissolution, but has gained deathlessness in addition as a

---

[21] Thus Galen in the passage mentioned earlier, who ascribes to Plato the reflection that the Spartan state, though generated, might none the less *chance* to be unperishing (*aet* 17.5.601,5-11).

[22] See, for instance, the debate in *aet* 15 between Proclus and Philoponus as to which features of the cosmos are intended by Plato to constitute an image of the sort of existence enjoyed by the unchanging Forms.

[23] One could avoid this difficulty by taking Plato's words here as the weaker claim that generated things are *perishable*, and not the claim – (2A) – that they will actually perish (cf n 7). But this is the less natural reading, and to that extent offers a weak defence of Plato. In any case, Philoponus himself does take Plato to believe (a qualified version of) (2A): see *aet* 8.4.311,8-9.

further acquisition [*epiktêton prosktômenon*] (17.2.594,18-23).[24]

The idea of a contrast between properties in accordance with *S*'s nature and properties which are further acquisitions might suggest:

> *S*'s *F*-ness is a further acquisition if and only if an adequate explanation of *S*'s being *F* would require the citing of something outside *S*'s nature.

This is supported by *aet* 8.1.301,21-6, where Philoponus argues that if the cosmos' being unperishing is not due to something outside it, then it must be 'in accordance with its nature'. (Philoponus assumes throughout his discussion that such properties always have an adequate explanation.) On the other hand, the examples which he uses to illustrate the distinction in 17.2 contrast the heat of fire with that of heated water or iron, and the blackness of ebony with that of dyed wool (594,23-595,4). These examples suggest that:

> *S*'s *F*-ness is in accordance with the *logos* of *S*'s nature if and only if being *F* is part of the *essence* of *S* (or: *S*'s being *F* is *necessitated* by *S*'s nature).

This in turn seems to be confirmed by some of the terminology used in the chapter. Philoponus employs a number of different phrases to express this idea of 'in accordance with *S*'s nature'; while most are simply stylistic variants,[25] one is strikingly different, and gives an important clue to Philoponus' meaning. At 594,25-6 things which are in accordance with *S*'s nature are said to be 'essentially joined to it in its essence (or: in what it is)' (*en têi ousiai sunousiômenon tou echontos*); so also in the recapitulation following the illustrations mentioned above:

> Just as in these cases, each of the things mentioned has two senses, (1) joined essentially to the subject from <its> nature, and (2) having come to it as a further acquisition, in this way it is necessary for 'imperishable' to have a double sense, either (1) belonging to something in accordance with <its> *ousia*, or (2) having come to something as a further acquisition and not from the *logos* of its proper nature (595,13-19).

Now we might well think that when these two characterisations are put together, they fail to constitute an *exhaustive* distinction, as Philoponus clearly takes his distinction to be. For some properties of *S* might be explicable wholly in terms of its nature, while not being *required* or necessitated by that nature. If our conception of explanation in terms of a thing's nature were of an Aristotelian sort, however, which demanded necessitation by the *explanans*, the possibility of such a *tertium quid* would be ruled out. I take it that

---

[24] Note that *phusis* ('nature') as Philoponus uses it has a wider sense than it has in Aristotle, since Philoponus is ready to use it of the unchanging Forms (*aet* 6.28.226,10-14, cited above, p 182).

[25] In 17.2 alone we find: 'in accordance with its nature'; 'by nature'; 'has a *logos* of'; 'in accordance with the *logos* of its nature'; 'from the *logos* of its specific nature (*oikeias phuseôs*)'; 'in accordance with the *logos* of its own nature (*idias phuseôs*)'.

Philoponus is employing some such conception.

There are two ways in which *S*'s *F*-ness can fail to be required by its nature, and hence be a 'further acquisition' (on the assumption that it is adequately explicable at all). The first way is available when *S*'s nature neither requires nor forbids *S*'s being *F*. Then its being *F* can be due to the natural workings of other natural objects, as with the water's being hot and the wool's being black. The second comes into play when being *F* is *forbidden* by *S*'s nature; for Philoponus thinks that the natures of things can be overruled by God.[26] Philoponus holds that *S*'s being *F* in respect of its nature is *compatible* with being not-*F* by reason of God's further intervention: this is crucial for his response to Proclus. In the case of the putative imperishability of the cosmos, the first of these two ways is ruled out since Philoponus and Proclus agree that there is no natural power outside the cosmos (see above, n 20). Thus Philoponus tends to speak of the cosmos' imperishability as either in accordance with the *logos* of its nature, or as an acquisition due to God's wish.[27]

Both (1 A&B) and (2 A&B) now stand in need of clarification: are they true in whatever sense imperishability, etc., belong to the object in question, or only in some restricted way? Philoponus thinks that they are only true if all the relevant properties are due to the *logos* of the object's nature. Thus if *S* is generable by nature, then it follows that it is perishable by nature. But it does not follow that it is perishable *tout court*; for its nature could be overridden by God's wish, and in that case *S* would be perishable in respect of its nature, but imperishable as a further acquisition by reason of God's intervention:

> What kind of necessity is there that ungenerability should follow from deathlessness which is a further acquisition and not in accordance with a

---

[26] *aet* 6.29.242,15-22, ap. Simplicium *in Phys* 1331,24-5; cf *aet* 8.4.311,8-12. Thus if we take 'being *F* *simpliciter*' to mean 'either being *F* by nature or being *F* through acquisition', then being *F* *simpliciter* is compatible with being not-*F* *simpliciter* (see, for example, 6.28.226,5-7, quoted below, p 189, and 29.242,15-22). This means that for a precise characterisation of a property's being in accordance with the *logos* of *S*'s nature we need a modification of the normal account of essential or necessary properties:

> *S*'s *F*-ness is in accordance with the *logos* of *S*'s nature if and only if *S* could not (exist and) be not-*F* *simpliciter* without a violation of a law of nature.

(Cf David Lewis' characterisation of determinism in 'Causation', *Journal of Philosophy* 73 (1970) 556-67, at 559.)

Philoponus does not say what limits there are on God's ability to overrule essential natures in this way: could God make fire cold, or trees rational, for instance? He is, however, committed to at least one restriction. God cannot bestow ungenerability as a further acquisition on something which is by nature generable (see below, pp 189-95).

[27] Thus, for example, *aet* 6.27.216,4-6: 'But [Proclus] does not draw a distinction between the imperishability of the cosmos due to God's wish and its being imperishable by its own nature' (cf 7.13.272,11-13, ap. Simplicium *in Phys* 1333,23-7). This distinction between what is due to nature and what is due to God's further wish is to be found in both Proclus and Simplicius as well (*aet* 6.29.238,6-240,9, cf Proclus, *in Tim* (Diehl) 1.276,27-29, translated in n 35; Simplicius *in Cael* 105,32-107,24). (Thus in the remark just quoted, Philoponus means only that Proclus fails to show awareness of the distinction *here*.) Like Proclus and Simplicius, Philoponus believes that *everything* depends causally on God (cf *aet* 6.18); but this does not, of course, vitiate the distinction between nature and further acquisitions from God.

thing's nature? For the opposite is altogether necessary: things which have something as a further acquisition are among the things opposite to those which are receptive of it in accordance with the *logos* of their nature (*aet* 17.2.596,14-18; cf 6.28.225,22-4, 8.1.304,5-8, 17.3.598,6-9).

Similarly with (2 A&B): if *S*'s nature requires that it have a beginning (because it has a finite *dunamis*, for example, or because its nature requires it to *develop* into a mature organism), then its nature will require *S* also to have an end; but once again, God can intervene.[28]

Within the assumption that (1 A&B) and (2 A&B) are true in some sense, Philoponus' restriction seems entirely reasonable: if *S*'s possession of, say, imperishability or generation is due to some additional choice by God, then there is no guarantee that God will make the further additional choice to bestow the corresponding ungenerability or destruction as well.[29]

Philoponus now uses his nature/acquisition distinction to show that Proclus misunderstands the sense in which Plato would mean his cosmos to be generated but imperishable, if he intended this literally:

> If the cosmos is perishable by nature, as Proclus shows in the argument which has been stated, and is for this reason generable also, the philosopher is revealed as intentionally misleading us here with faulty reasoning, when he conveys the manner in which Plato says that the cosmos is imperishable (*aet* 6.29.240,23-8).

He takes the words of the Demiurge at *Tim* 41a7-b3 to mean that the cosmos is perishable in its nature, and thinks that Plato is right to regard it as such, because of the finite *dunamis* argument (*aet* 6.29.242,11-14).[30] If it is by nature perishable, then its imperishability must be a further acquisition from the Demiurge, who constantly restores its waning power (17.3.598,1-9); and of course to support his claim that this is what Plato himself thinks too, Philoponus can cite *Tim* 41b3-6: '...you shall not be dissolved nor have death as your portion, having instead my wish as a greater and more sovereign bond than those with which you were bound together when you came into

---

[28] *aet* 8.4.311,8-12; cf 7.13.271,23-272,26. 'Having an end by nature' must be taken to mean 'being bound to perish, unless there is a violation of a law of nature', if we are to make sense of the compatibility of 'having an end by nature' and 'not having an end thanks to God's further intervention'; similarly 'having a beginning by nature' must mean 'being bound to have a beginning (if the thing exists at all), unless there is a violation of a law of nature' (cf n 26 above).

[29] Likewise, if (*per impossibile*) the imperishability were due to the *per accidens* intervention of some other natural agent, there would be no guarantee that it or any other such agent would intervene to bestow ungenerability as well.

[30] So also 8.4.312,16-26: 'But Plato has not fallen away from right thinking here – not because he thinks that the cosmos is imperishable, but because having thought it necessary that it be imperishable he does not say that the infinite *dunamis* of the nature of the cosmos is the cause, but the wish of the Demiurge. For quite generally, if it were necessary for some body to endure for ever, it could not be imperishable otherwise than if the *dunamis* of the Demiurge supplied it with its perpetual endurance, since no body has perpetual existence in so far as the *logos* of its nature goes.' Philoponus ascribes the finite *dunamis* argument to Plato himself at 6.29.235,4-19.

existence.'[31] Thus Proclus is wrong to infer that a Timaean cosmos would have to be ungenerable because of its imperishability; this would only be the case if that imperishability were part of its essential nature:

> We should consider whether Plato contradicts himself; for if his view is that destruction belongs of necessity to everything which has come to be, and by conversion with the negation it follows from this that everything which does not perish has not come to be either, how can he take one and the same cosmos to be at once both imperishable and generable? To this the reply is brief. For if what is ungenerable in respect of time is also imperishable in accordance with its nature, then if <something> is not imperishable in accordance with its nature it is not ungenerable in respect of time either; but Plato did not take the cosmos to be imperishable in accordance with its nature – but on the contrary says that it is perishable by nature. For he says that everything which has been bound is dissoluble... If, therefore, the cosmos in Plato's view exists as mortal and dissoluble by nature, but indissolubility <belongs> to it because of the wish of the Demiurge (for he says that they are indissoluble not by nature but 'at my wish'), it is clear that it would also be generable in respect of time. Thus Plato drew the right conclusions from his own principles (*aet* 6.28.225,14-226,19, with one gap, in which Philoponus quotes *Tim* 41b2-6).[32]

In this way Philoponus uses his qualified version of (1A) to show that a Timaean cosmos can be generated *and* imperishable (though perishable by nature), as an acquisition from God – and so he out-manoeuvres Proclus, who had tried to use (1A) to demonstrate the inconsistency of the Timaean account if taken literally.

IV

I do not hold a brief for the truth of any version of (1 A&B), and still less for the truth of (2 A&B), but this reply to Proclus is to Philoponus' credit: within the terms of the debate, he succeeds in defusing Proclus' objection to a creationist interpretation of Plato. His use of (1A) to turn the tables on Proclus is deft, and reflects his characteristic ability to utilise his opponents' own arguments against them. His reading of the *Timaeus*, moreover, achieves a commendable degree of success in according with the text (especially given that no interpretation is likely to command the support of everything which Plato says). Philoponus gets into unexpected difficulty, however, when we combine this defence of literal creation in the *Timaeus* with what he says in response to another Neoplatonist argument, developed this time by Porphyry.

> Porphyry says that Plato's proofs concerning the cosmos' having come to be are not appropriate for the cosmos' having come to be in a *temporal* sense, from which

---

[31] He can also quote *Politicus* 269e7-270a7 (*aet* 6.28.229,21-230,6) – although Simplicius interprets this passage differently: *in Cael* 143,20-31. Philoponus cites *Tim* 41a7-b6 as a case of Plato deploying the nature/acquisition distinction, but Plato makes no mention there of *nature*. For ascriptions of Philoponus' interpretation to Plato, see, for example, 7.13.272,22-6, 13.7.494,5-11, 14.3.548,3-10, 18.9.638,6-14.

[32] Cf 15.3.559,15-21 and 17.2.596,1-18.

he draws the conclusion that Plato does not say that it has come to be temporally ... (*aet* 6.25.200,4-8).

Porphyry argues that the grounds which Plato offers for the 'generation' of the cosmos – namely that it is visible, tangible, and corporeal (*Tim* 28b7-c2) – do not entail that the cosmos came into existence; but they do entail that it is *composite*, which is how Porphyry understands 'generated' (*aet* 6.10; Porphyry also takes 'generated' to mean 'having an external cause': 6.17). Since the idea that corporeality and the rest entail temporal generation is absurd, Porphyry claims, Plato cannot be talking about temporal generation, but about generation in some other sense. Philoponus for his part defends the disputed entailment. At 201,22-203,19 he argues that, since all the *elements* are generable and perishable in a temporal sense, anything made of the elements is also generable and perishable in that sense. Anything which is visible and corporeal is made of the elements, and so Plato is perfectly correct to infer temporal generability from visibility and corporeality. (He then goes on to give another defence of Plato's inference.)

This argument is unimpressive.[33] The inference from 'every part is generable and perishable' to 'the whole is generable and perishable' is highly dubious, and certainly to assume that it is sound is to beg the question against Porphyry.[34] But I am not concerned with the strength or weakness in general of Philoponus' response here, but with the fact that his distinction between properties due to nature and those due to God's intervention creates an inevitable gap in his argument against Porphyry, no matter how strong that argument might otherwise be.

Although Philoponus usually speaks in this chapter in terms of Plato's showing the cosmos to be *generable*, the inference which Plato makes from the cosmos' corporeality is that it was actually generated – *gegonen*, as Timaeus tersely says at 28b7. This is also the inference which Porphyry attacks: 'What could be more absurd than these things? For what is the necessity of the entailment, if <the cosmos> is visible and tangible and has body, that it has come to be from a temporal beginning?' (200,21-3). Now all that Philoponus' – or Plato's – arguments from the cosmos' corporeal *nature* can hope to achieve is to show that the cosmos is generable *in accordance with the logos of its nature*. As we saw earlier, Philoponus' defence against Proclus depended on the claim that a thing can be perishable by the *logos* of its nature, and yet be unperishing if God intervenes; why, then, cannot the cosmos be generable by the *logos* of its nature, and yet be ungenerated thanks to God's overruling

---

[33] Although it shares its faults with an argument of Aristotle's designed to show almost the opposite conclusion; see my 'Eternity and necessity in *De Caelo* I.12', 254.

[34] Philoponus offers as a fresh argument something which might be taken to support this inference at 203,20-5: 'It would appear to be unreasonable (*alogon*), that while the elements of the universe are always taking a temporal beginning and coming to be numerically different at different times, the heaven composed of these has never had a beginning of existence nor will have an end, and does not in between undergo some reversal and alteration in respect of its *ousia*'. But this argument merely claims that the onus of proof is on those who think that the inference is unsound. Note that, like the first argument, it is explicitly symmetrical with respect to generation and destruction: see below.

intervention?[35] Nor does it help if we grant Philoponus a stronger argument, which would show that the corporeal nature of the cosmos requires it not only to be generable, but actually to have a beginning.[36] For Philoponus' rebuttal of Proclus also involved the claim that 'required by one's nature to perish', does not entail 'actually perishes', since nature can again be overruled by God.[37] The argument for this claim is quite general, so it appears to apply to generation as much as to destruction. Thus Philoponus is committed to denying the very entailment he must here defend – namely from 'required by one's nature to be generated' to 'actually generated'.

Philoponus gets half-way to seeing this problem; for he acknowledges that his defence of Plato against Porphyry also shows that the cosmos is perishable 'as far as the *logos* of its nature goes' (203,17-19). He does not, however, seem to notice that if generation and destruction are symmetrical, then either his argument proves too much – by showing that corporeality entails actual perishing – or too little – by failing to justify Plato's inference to actual generation. Philoponus needs to show that if something is generable, then it must have actually come to be (if it exists at all), and that not even God can overrule this. But he needs in addition to break the symmetry between generability and perishability: the argument must not also show that if something is perishable, it will actually perish.

V

A number of passages suggest that Philoponus takes it for granted that '(exists and) is generable' entails 'is generated' in the way that his defence against Porphyry requires. If not even God can bestow ungenerability on something which is generable by nature, then nothing can possess ungenerability as a further acquisition; but the same will not be true of imperishability. So it would not be surprising if Philoponus drops the 'by nature' qualifier when talking of ungenerability, but retains it for imperishability; the version of (1A) which he regards as true would then be:

> If a thing is ungenerable, it must also be imperishable by nature.

We have in fact had an example of this already, at 6.28.225,22-24 (quoted above, p 189), and there is another example at 6.27.216,1-4: 'if it were ungenerable, it would have the cause and origin of its imperishability in itself, since he agrees that what is ungenerable is imperishable by its own nature.' In other passages, he takes the entailment of 'generated' by 'generable' for granted in a more straightforward fashion. Thus at 11.8.441,16-18 he argues

---

[35] That the cosmos is generable by nature, but is granted ungenerability as a further acquisition from God, is perhaps the view which *Proclus* actually holds – at least in his commentary on the *Timaeus*: 'Having shown that the cosmos is generable only in accordance with its body, [Plato] grants ungenerability to it in respect of something else' (*in Tim* 1.276,27-9).

[36] The finite *dunamis* argument would do the job as adequately as it can be done (though see above, pp 183-4 and n 15).

[37] See above, pp 187-8 and n 28.

from 'perishable' to 'has come into existence', but appeals only to the principle that 'perishable' entails 'generable', and makes no comment on the gap left in the argument. Again, at 6.29.237,7-11 he writes without further comment: 'If, therefore, the cosmos is not deathless by nature, but receives deathlessness from outside as a further, restorative acquisition, then *by nature* it is mortal; and if it is by nature mortal, in Plato's view, then he is right to give it a beginning of existence.'

Philoponus' belief that 'generable' entails 'actually generated' might suggest that he subscribes to the so-called 'Principle of Plenitude' – that is, to the notion that (in an infinite time) every genuine possibility is eventually realised; for then, if *S* is generable, it will actually be generated sooner or later. Philoponus' attitude to modal ideas of this sort is in fact quite difficult to pin down, and it is hard to make everything he says on the subject consistent.[38] In any case, Philoponus cannot afford to rely on such a principle here, since it fails to provide the required asymmetry between generability and perishability: if the former necessitates actual generation because *all* possibilities are eventually realised, then the latter will likewise necessitate actual destruction.[39]

In a quite different context, however, Philoponus offers two arguments to show that what is generable is necessarily generated. His opponent here is

---

[38] (1) *aet* 16.4.578,3-580,18: Richard Sorabji has canvassed the suggestion (in correspondence) that Philoponus is here assuming the truth of the Principle of Plenitude. But all that Philoponus need be saying, I think, is that the 'conjunction' of the planets is possible *and* will of necessity happen (for purely astronomical reasons) – rather than that it is possible *and so* will of necessity happen (the crucial lines are 578,24-579,1). The same holds for the reference at 579,10-14 to there being the whole of time for this conjunction to come about.

(2) 6.5.131,26-132,28: here Philoponus argues that Plato is wrong to think that the cosmos will never cease to be. In the course of this argument Philoponus seems to commit himself to the view that God must exercise all his powers or potentialities at some time or other. For if he does not, Philoponus argues, then (a) he will be imperfect (*atelês*), since he has a potentiality which he never exercises (cf 4.15.100,23-26); (b) he will have the *dunamis* to no purpose (*matên*); (c) he will fall foul of Aristotle's doctrine that the potentialities of eternal things are of necessity actualised (cf 11.8.443,23-44,9, *in Phys* 406,1-16). (c) is perhaps suggestive of a highly restricted version of the Principle of Plenitude (i.e. a version true only of the possibilities attaching to eternal things). A similarly restricted commitment to this is made at 10.2.385,16-21: nothing remains in an unnatural condition for an infinite time (cf 10.3.390,17-26, where the reasons are theological and teleological). But if God is causally responsible for everything which happens (see above n 27), then the idea that he must exercise all his *dunameis* seems to commit Philoponus to a full-blooded version of the principle. (It is true that Philoponus himself does not believe that time is infinite, but arguments (a) and (b) are set in terms of 'the whole of time', and it makes no difference to them whether this is finite or infinite; thus Philoponus will still be committed to the belief that God will exercise all his powers sooner or later, and so to a 'finite-time' version of the full-blooded principle.)

(3) On the other hand, Philoponus insists that the full-blooded version of the principle is false: possibilities which attach to finite things can go unrealised forever (11.8.443,23-444,9). He may also be committed to the falsity of the highly restricted version as well. In a passage to be discussed below, he allows that a thing might be visible but never seen, because hidden from view by the earth (6.9.151,4ff); if parts of the earth itself are eternal (because they are too close to the centre to be affected by the cycles of elemental change), then this unrealised possibility could attach to something eternal.

[39] If the reply is that God could intervene to prevent the destruction, then we are back where we started: why cannot God intervene in the case of generation as well?

Taurus, who defends the non-creationist interpretation of the *Timaeus*, claiming that when Plato describes his cosmos as 'generable' he means that it is 'in the genus of the generables', but not actually generated. Philoponus replies that nothing can be in the genus of the generables without also being generated.

In his first argument (*aet* 6.9.151,5-23), he contrasts the cases of visibility and generability. *S* can be visible in its nature, and yet never be seen. If so, however, its failure to be seen is not (wholly) due to its own nature; for if it were, then the thing would not be visible by nature after all. Thus *S* may remain unseen because, for instance, it is on the (uninhabited) underside of the earth. In that case *S*'s own nature is not malfunctioning in any way as far as being seen goes, and its being unseen must be due instead to some limitation in *us*, the potential perceivers: 'if someone had the eyes of Lynx in the story, so that he could "without difficulty see even under the earth", he would also see the objects covered by the centre of the earth' (151,13-15).[40] In other words, if the unseen *S* *were* to have been seen, this would have required nothing to be done to *S* itself: all that would have been required is an improvement in us. Philoponus now draws the moral for generability:

> The result is that if the cosmos exists in actuality and is in the genus of generables, it is altogether necessary that it should also have come to be, notwithstanding that the origin of its being generated is not left clear to us – not because it is ungenerable by nature (for thus it would belong in the genus of ungenerables and not in that of generables) – but because our understanding [*gnôsis*] grasps only the things which are present (151,16-23).

I take this to hint at the following *reductio* argument. 'Suppose the cosmos were generable but ungenerated; then, as with *S*'s visibility, this would mean that if the ungenerated cosmos *were* to have been generated, this would have required nothing to be done to the cosmos itself – all that would have been required is an improvement in us. But this is absurd, since whether the cosmos had a beginning or not is quite independent of the state of our cognitive abilities. The most that could really be true would be that the cosmos *seems* ungenerated to us (thanks to our cognitive limitations), but in fact is not.'

This first argument is hardly persuasive. The contrast between generability and visibility, if I have construed it correctly, depends not on a special feature of being generated, but on one of being seen – namely that its being actualised depends on the activity of *another* causal agent, the perceiver. Let us grant that if the ability to be *F* is necessitated by *S*'s nature, then its failure ever to be *F* must (if it is explicable at all) involve the operation of some factor external to *S*. This external factor need not, however, operate upon some *third*

---

[40] It might look from the lines preceding this as if Philoponus' point is that *S*'s remaining unseen is due to *the intervention of the earth*. But his conclusion shows that he wants to concentrate on the idea of the limitations of the perceivers; it is crucial for this conclusion that the earth does not affect the natural functioning of *S*, but only the functioning of something else. This point, however, though it generates the conclusion, also generates the fallacy in the argument: see below pp 193-4.

party such as a perceiver: it can operate upon S itself. Take a helium balloon which never rises towards the heavens because it is sealed in a lead box; it is clearly not that the balloon really does rise into the heavens, but that the box (still on the ground) prevents us from perceiving that it does so. Indeed, the Demiurge's intervention to preserve the perishable world from destruction is of exactly the lead box type.[41]

In the second argument (151,23-152,14) Philoponus again starts from the point that something can be visible yet never seen (and repeats that this must involve some external intervention); this means that what it is for something to be visible is different from – and does not include – its being seen. By contrast,

> It is impossible for that which never has nor <ever> will come into existence to be something generable; for that which is generable possesses its generability by not only being of a nature to come into existence but also coming into existence in every way, in either the past or the future; for generation (*genesis*) is its road to existence. It is impossible for that which has never come into existence, nor is doing so now, nor is of a nature to come into existence at some time in the future to be or to be called generable (151,5-152,7-14).

When Philoponus claims that what it is to be generable by nature *includes* being generated, he is not simply trying to insist that whereas 'visible' means 'able to be seen', 'generable' means 'not only able to be generated, but actually being generated as well'. He gives an argument for this conclusion which makes no appeal to usage: 'generation is its road to existence'.

It is not clear what Philoponus' point is here. He may just be confused by the fact that, while 'S can be visible without ever being seen' does capture the idea that an existing thing can fail to realise some possibility which attaches to it, 'S can be generable without ever being generated' might seem to express a weaker claim – namely that S can be generable, but never succeed in existing at all. If this is the *only* sense in which a thing's generability can fail to issue in generation, then, given that the cosmos does exist, its generability has led to generation. But of course it is false that this is the only sense in which actual generation could fail to happen to something generable.

There is another possible argument in the offing here, however, which is both more interesting and more in line with Philoponus' thinking in the defence against Proclus and elsewhere (though it will, unfortunately, turn out to be no more sound than the previous one). If S is by nature able to be F, but fails ever to be F in actuality, this failure, as we have seen, requires the intervention of something external – call it E.[42] This intervention requires

---

[41] For this reason too, there seems no chance of any asymmetry here between generation and destruction, even if the argument were better. (Note that not even all cases of failure to be seen have to be as Philoponus suggests, if devices like the legendary Tarnhelm – a helmet which could render its wearer invisible – are possible, and work by affecting the wearer, not the perceivers.)

[42] Two qualifications must be made. (i) The *absence* of some (normally present) external factor would do equally well – the absence of tigers explains why the gazelle in my garden never runs very fast. (ii) S may fail to be F simply because its natural (finite) *dunamis* runs out, and it ceases to exist before it can become F (Philoponus himself makes this point at *aet* 11.8.443,225-444,8).

the causal action of E either upon some third party, as in the visibility case, or upon S itself. Now in the case where E prevents S from being generated, it will have to operate upon S itself, as we saw earlier. But E cannot act causally upon S, the argument continues, if S does not exist; once S does exist, however, it is too late for E to intervene to prevent its being generated. Thus if S exists and is generable by nature, then it must actually have been generated; for the external causal action which would be required to explain its failure to be generated cannot take place.

This line of thought has the crucial advantage of providing the required asymmetry between generability and perishability, since it does not show that what is by nature perishable cannot be acted upon so as not to perish. That it is congenial to Philoponus is perhaps suggested by his reluctance to use the 'finite *dunamis*' argument to show that the cosmos is by nature generable, despite the fact that he is happy to use it to demonstrate its perishability (see above, pp 183-4 and n 15): Philoponus may think that, since the cosmos' *dunamis* does not exist before the cosmos does, it cannot operate as an explanatory factor until the cosmos exists – but by that time, it is too late for it to be what explains the cosmos' generation, since that has already happened.

Be that as it may, the argument which I have sketched for the entailment of 'is actually generated' by '(exists and) is generable' rests on a fallacy. Although E cannot act upon S before the latter exists, S does not in turn need to exist *before* E can act upon it, if 'before' is taken in the temporal sense which the rest of the argument requires. It is true that S must exist *when E* acts upon it, but this is compatible with E's having *always* acted upon S to bestow ungenerability upon it as a further acquisition.

VI

I conclude that Philoponus does not succeed in defending his claim that if S is generable in accordance with the *logos* of its nature, it must actually be generated. Without this, his defence against Porphyry in 6.25 is undermined; but a wider issue is also at stake. Philoponus' commitment to this entailment means that in addition to holding the versions of (1 A&B) and (2 A&B) in which the properties are all stipulated as in accordance with the thing's *logos*, he also believes two further principles:

(3B) If a thing is ungenerable *simpliciter*, it must be imperishable in its nature.
(3A*) *It is not true that* if a thing is imperishable *simpliciter*, it must be ungenerable in its nature.

In effect, the arguments which I have been examining in Parts III-V of this chapter are arguments defending the rationality of holding *both* (3B) and

---

Neither of these possibilities arise, however, in the case of S's existing and being generable, but never generated.

(3A*). Since this hinges on the possibility of an argument for the generable/generated entailment which preserves an asymmetry with perishability, Philoponus' defence has turned out to be inadequate.

It can be said in mitigation that Philoponus is not alone in believing (3B) and (3A*). For in the passage mentioned earlier (*aet* 17.5) he quotes Galen as holding both principles to be indisputable truths. Galen, too, has the *Timaeus* in mind, and thinks that while imperishability in the sense of 'having no *logos* of destruction' entails ungenerability, imperishability bestowed as it is by the Demiurge does not: (3A*). As for (3B), Galen actually maintains the stronger claim that 'is ungenerable *simpliciter*' entails 'is imperishable *tout court*' (by this I mean 'is imperishable by nature *and* cannot be made to perish through God's intervention'): this in turn entails (3B). Philoponus' beliefs put him in very good company after all.

But this defence of Philoponus is, of course, only an appeal to authority.[43]

[43] I should like to thank Richard Sorabji for suggesting that I write this paper, and for all his help and encouragement in its preparation.

CHAPTER ELEVEN

Prolegomena to the Study of Philoponus'
contra Aristotelem

Christian Wildberg

Judging from the number and content of his commentaries, Philoponus was a thinker in the Aristotelian tradition. One of his major achievements lies in the fact that as a commentator he accepted and developed the heritage of his teacher Ammonius. For that reason alone it is remarkable that he composed a treatise which attacked vital topics of Aristotle's philosophy with little compromise. Although it is true that throughout Antiquity many philosophers ventured to criticise the great Aristotle, one may agree that Philoponus did so, as Cesare Cremonini put it in 1616, 'more sharply than anyone' (*acerrime omnium*).[1] Where does this attack fit into the context of Philoponus' doctrinal development? No doubt his outspoken critique of Aristotle in the *de Aeternitate Mundi contra Aristotelem* somehow swayed Philoponus to desert the philosophical and join the theological camp. But the story is probably more complex. The general point of dissent was, as the title indicates, the doctrine of the eternity of the world. Being a Christian, Philoponus perhaps possessed a particular motivation for launching his attack – as a feat of *praeparatio evangelica*. This fact has been sufficiently recognised and appreciated. Less appreciated and studied, however, has been the philosophical side, i.e. the actual argument and structure of the treatise in question. Since it has not survived the content must be reconstructed from a number of substantial fragments found mainly in the commentaries of Philoponus' adversary Simplicius. An adequate treatment of the double controversy Simplicius *v* Philoponus *v* Aristotle would fill a volume on its own and cannot be the subject of this chapter.[2] Instead, I will attempt to revise apparently firmly established views about the treatise, in particular its composition and date. This, it is hoped, may lead to a revised view of that treatise and at the same

---

[1] In his work *Apologia dictorum Aristotelis de quinta caeli substantia. Adversus Xenarcum, Ioannem Grammaticum, et alios*, Venice.

[2] The publication of a collection and translation of the surviving fragments as well as a commentary on Philoponus' criticism of Aristotle's theory of ether (dealing with the first five books of the *contra Aristotelem*) are in preparation by the author.

time encourage a more advanced study of Philoponus' doctrinal development in general.

I

In the first instance, it seems unanimously accepted that the *contra Aristotelem* consisted of only six major sections or books in which Philoponus attacks Aristotle's theory of ether as developed in the first chapters of the *de Caelo* (books 1-5), and the arguments for the eternity of motion and time in *Physics* 8 (book 6). This belief about the composition of the treatise is based on the fact that Simplicius discusses the arguments of six books in his commentaries *in de Caelo* and *in Physicorum*, and on the fact that Arabic bibliographers like Ibn Abī Uṣaybiʿa, Ibn al-Nadīm and Ibn al-Qifṭi know of six books only.[3] However, this picture of the *contra Aristotelem* seems to be itself a fragmentary impression of the original. Let us begin with a remark made by Simplicius which has puzzled modern scholars considerably. Having almost completed his unfavourable discussion of Philoponus' ingenious 'proofs' of the temporal finitude of the universe in book 6 of the *contra Aristotelem*, Simplicius says (*in Phys* 1177,38-1178,5):

> Having said these things <Philoponus> claims that he will show that the world does not change into absolute nothingness but into something different, greater and more divine. ...He declares that this world changes into another world which is more divine – a <proposition> he elaborates in the following books (*en tois hexês bibliois*) – not realising that this is not a destruction of the world but a perfection.

Two things are remarkable about this passage. First, Simplicius clearly speaks of additional books beyond the sixth, not just of further sections of the sixth book. And secondly, the content vaguely hinted at seems to be strikingly different in character. The idea that the whole world is transformed into a more divine entity has nothing to do with Aristotelian philosophy but is – in a Christian context – reminiscent of the New Testament topic of *kainê ktisis*.[4] Unfortunately, Simplicius has no interest in these ideas and provides no further information. When he again stoops to discuss the 'rotten arguments' of the Alexandrian grammarian in the final pages of his commentary he clearly deals with a different treatise, i.e. an apparently brief series of arguments showing that an infinite force cannot reside in a finite body.[5]

Simplicius' puzzling reference to further books, however, receives clarification from a passage in an anonymous Syriac manuscript of the seventh century in the British Museum Library. The manuscript (Add. 17

---

[3] See Steinschneider (1869). Cf also M. Mahdi (1972) 269 n1.

[4] Cf 2 Cor 5:17; Gal 6:15; Mk13:31.

[5] From which Philoponus infers that the world, being a finite body possessing a finite *dunamis*, cannot exist for an infinite amount of time. The treatise is written in a similar spirit to the *contra Aristotelem*, but is not part of it. See H.A. Davidson (1969) 358f; cf S. Pines (1972) 340f.

214) contains extracts from the writings of the Church Fathers,[6] and in the relevant passage[7] the author states that he cites from the eighth book of Philoponus' treatise. The passage may be translated as follows:

> Of John Grammaticus, the title of the second chapter of the eighth book of Against Aristotle:
> On <the proposition> that that which is resolved into not-being is not wicked on its own and by itself; and on <the proposition> that the world will not be resolved into not-being; this is what our discourse is about.
>
> From the second chapter:
> The world will not be resolved into not-being, for the words of God are not resolved into it either; for we clearly speak of new heavens and a new earth.

The translated passage contains two citations from the second chapter of book 8 of the *contra Aristotelem*, first the title of that chapter[8] and then what the excerptor supposedly regarded as a key proposition of the argument. The content and tenor of this fragment agrees remarkably well with the scanty information given by Simplicius. In both passages Philoponus is reported to endorse the belief in a new creation of the world after the destruction of the old aeon. In spite of the Arabic evidence which knows of only six books of the *contra Aristotelem*, it is very likely that the anonymous author of the Syriac manuscript has cited correctly. For in lines 19-29 of the same column, directly following the above passage, he adds a citation from the *de Opificio Mundi* and refers the reader correctly to the sixteenth chapter of its first book. There is trustworthy evidence, therefore, that Philoponus appended his repudiation of Aristotle's arguments for the eternity of the world with a positively Christian discourse showing that the world, though certainly coming to an end, is not simply annihilated but rather transformed into an eschatological universe with new heavens and a new earth.[9]

Someone who is familiar with the fragments of the *contra Aristotelem* found in Simplicius' commentaries could raise an objection against the authenticity of the above Syriac fragment on the grounds that it is theological in character. In contrast, all other fragments we possess discuss highly 'technical' points concerning Aristotle's natural philosophy. But this objection can be met. The impression that the arguments of the first six books are philosophical seems to be due, to a large extent, to Simplicius' selectivity. Plausible evidence for this may be drawn from two sides. First, there is one argument of the *contra Aristotelem* which survived independently both in Greek

[6] See W. Wright, *Catalogue of the Syriac Manuscripts in the British Museum acquired since the Year 1838*, London 1871, 915-17.

[7] Fol 72 v b 36-73 r a 19. The passage is also mentioned in Th. Hermann (1930) 214 n3.

[8] These titles of sections may well have been assembled in a way familiar from the earlier treatise against Proclus.

[9] In his polemic against Proclus, Philoponus disagreed with Plato's belief that the world will not come to an end; see above, Chapter 10. There also argued that the processes of generation and corruption occur out of and into not-being, see *contra Proclum* 9. In the *contra Aristotelem* he seems to have moved in Plato's direction in so far as he now posits that the universe will not be destroyed into not-being after all.

and in Arabic. The Arabic fragment, which has been published and analysed by J.L. Kraemer, occurs in an anonymous abstract of Abū Sulaimān as-Sijistānī's treatise *Ṣiwān al-Ḥikmah*.[10] If this fragment is indeed authentic and has not suffered severely from arabesque decorations, a comparison with Simplicius' version[11] clearly shows that the latter concerned himself almost exclusively with the philosophical backbone of the argument and ignored all blantantly Christian aspects. This impression may be supported by one of Simplicius' own remarks. He says at one point, *in de Caelo* 90,13-15:

> Due to his vain contentiousness it escaped <Philoponus'> notice that even this David, whom he honours so much, teaches the contrary.

There is no other evidence in Simplicius suggesting that Philoponus referred expressly to the Psalms of the Old Testament; yet, he must have done, otherwise Simplicius' ironical remark remains baseless.

The evidence shows, I think, that the *contra Aristotelem* consisted of more than six and of at least eight books, and that these last books differed considerably from the preceding ones. They were theological in character and no longer attacked Aristotelian philosophy directly. If this is true, one could plausibly explain not just the fact that Simplicius, commenting on Aristotle, dealt with the first six books only, but also the fact that the Arabic bibliographers spoke of six books *against Aristotle*. The whole treatise may well have been truncated by philosophers who were not interested in Philoponus' theological views. The Syrian Christians, on the other hand, were greatly interested in the monophysite theologian, and in fact most of Philoponus' theological treatises survive in Syriac only.[12] Hence there seem to be good reasons for accepting the genuiness of the Syriac fragment, and for revising the tenet of the six books of the *contra Aristotelem*.

II

It is worthwhile reconsidering the absolute and relative dating of the *contra Aristotelem*. First the absolute date, which does not seem to be determinable with precision. The *terminus post quem* is no doubt the date of completion of the *de Aeternitate Mundi contra Proclum*, which is the year 529, in fact, as can be shown, the second half of that year. Towards the end of that treatise, Philoponus remarks (579,14-17):

> For just now in our time, in the 245th year of Diocletian, the seven planets were in conjunction in the same sign of the zodiac, Taurus (*gegonasin en tôi autôi zôidiôi tôi taurôi*), although not all of them were in the same part.

---

[10] See J.L. Kraemer (1965). The Arabic text of the abstract has been edited by D.M. Dunlop, *The Muntakhab Ṣiwān al-Ḥikmah of Abū Sulaimān As-Sijistānī. Arabic Text, Introduction and Indices*, The Hague, Paris and New York 1979.

[11] See *in Cael* 141,11-19.

[12] See the valuable edition and translation into Latin by A. Šanda (1930).

In order to transform a date of the Diocletian calendar into a date of the Julian calendar one must add 284 years, and thus one arrives at the year A.D. 529. This date can be confirmed by means of astronomical tables. Philoponus speaks of a semi-conjunction of all seven planets, i.e. of the Moon, the Sun, Mercury, Venus, Mars, Jupiter, and Saturn, in Taurus. This conjunction occured on 21 May 529.

In checking Philoponus' report of the conjunction of the planets with the help of astronomical tables (ephemeries) it was supposed that by 'Taurus' Philoponus does not mean the constellation of that name but the section of longitude 30° – 60° measured eastward from the vernal equinox.[13] According to Tuckerman's calculations[14] Saturn was in Taurus from early May 527 to early June 529, Jupiter from the middle of May 529 to the end of May 530. Mercury entered into Taurus in early April 529, followed by the Sun in the third week of April. Mars followed in early May, Venus in the second week of that month. 'Conjunction' of the seven 'planets' occurred when the Moon entered into Taurus on 21 May 529. The sun, at that time, was about to pass on to Gemini. Philoponus acknowledges correctly that the planets were not in the same part of Taurus at any one time. Since this conjunction was unobservable, Philoponus' remark indicates that he followed the celestial movements closely by means of the astrolabe and astronomical tables.[15]

Since Philoponus' *contra Aristotelem* is criticised for the first time by Simplicius in his commentary on Aristotle's *de Caelo*, the date of this commentary may serve as a *terminus ante quem*. This date, however, is much more difficult to determine.[16] The commentary was written before the commentaries on the *Physics* and the *Categories*,[17] but it is only possible to establish a *terminus post quem* for the *Physics* commentary. This commentary was composed some time after 537.[18] If one supposes that the commentary on

[13] See O. Neugebauer, *A History of Ancient Mathematical Astronomy*, Berlin, Heidelberg and New York 1975, 1079 (vol III): 'In the usage of ancient and mediaeval astronomy, however, signs [i.e. of the zodiac] are nothing but names of longitudes, counted from a properly defined vernal point. It is a mistake, often made by modern historians, to interpret a sentence like "the planet entered Leo on this and that date" as an expression for the planet's position with respect to the constellation "Leo".' Even if one supposed, incorrectly according to Neugebauer, that Philoponus referred to the constellation 'Taurus' rather than the section 30°–60° on the ecliptic, Philoponus' remark remains true because in the sixth century 'Taurus' was in fact approximately in Taurus. Today, due to the precession of the equinoxes, the constellation of that name is approx. 20° further to the east.

[14] B. Tuckerman, *Planetary, Lunar and Solar Positions at Five-day and Ten-day Intervals*. Memoirs of the American Philosophical Society 56 and 59. Philadelphia 1962/1964, vol. 2, 281f.

[15] Philoponus was no doubt proficient in the use of the astrolabe, see J. Drecker (1928); his treatise on the astrolabe has been edited by A.P. Segonds (1981). On the instrument see D. Hill, *A History of Engineering in Classical and Medieval Times*, London and Sydney 1984, 190-5.

[16] In the most recent major publication on Simplicius, Madame Hadot carefully avoids assigning an absolute date to the commentary in question; see I. Hadot, *Le Problème du néoplatonisme alexandrin: Hiéroclès et Simplicius*, Paris 1978, 20-32.

[17] See ibid. 27ff.

[18] The commentary must have been written after the death of Damascius, see *in Phys* 795,11-17 and cf Hadot op.cit. 28f. Damascius lived probably until 537/38, see Hadot op.cit. and

the *de Caelo*, being the first of Simplicius' major commentaries, was written not long after the Athenian philosophers returned from their exile at the court of Chosreos in 532,[19] then the years 534-536 may be assumed as a rough estimate for the date of that commentary. If this is taken to be the *terminus ante quem* for the *contra Aristotelem*, and if one further considers that Simplicius suggests at one point that Philoponus had already won a certain reputation for having written the *contra Aristotelem*,[20] one may infer that the treatise against Aristotle was composed soon after the *contra Proclum* in the period between the years 530 and 533/4.

III

The question of the relative date of the treatise is considerably more complex but also more interesting. É. Evrard originally discussed the relative dating of Philoponus' later writings.[21] In his important article Evrard aimed to show that the *Meteorology* commentary is not an early work, and that the assumption that Philoponus was a convert to Christianity rather than a born Christian is unjustified.[22] Evrard further thought he was able to show that the *Meteorology* commentary was written after the treatise against Proclus but before the *contra Aristotelem*. The *contra Aristotelem* is viewed as a work which marked the final break with Aristotelian-Neoplatonic philosophy.[23] Evrard's arguments and his conclusion have found general acceptance and approval. The following discussion aims to show first that Evrard's arguments are inconclusive, and secondly that there is certain evidence suggesting that the treatise against Aristotle may well have preceded the composition of the *Meteorology* commentary by a short period.

Evrard's article is a learned exposition of numerous important passages of the relevant works. Without undue simplification one may reduce his reasoning to three arguments on the basis of which he establishes his thesis that the *Meteorology* commentary predates the *contra Aristotelem*.

1. In his commentary on Aristotle's *Physics* and in the treatise against Proclus Philoponus accepts the doctrine that the circular movement of the firesphere is supernatural (*huper phusin*). In the *Meteorology* commentary, the *contra Aristotelem*, and the *de Opificio Mundi*, on the other hand, Philoponus argues that this motion must be natural. Of these three latter works the *Meteorology* commentary seems to be nearest to the first group because it still *mentions* 'supernatural' motion.[24]

A.D.E. Cameron, 'The last days of the Academy in Athens,' in *Proceedings of the Cambridge Philological Society* 15, 1969, 21f.

[19] On the date see Hadot op.cit. 32; Cameron op.cit. 7.

[20] Cf *in Cael* 25,22-34

[21] See É. Evrard (1953).

[22] Evrard argued against Gudeman; see A. Gudeman and W. Kroll (1916).

[23] See Evrard (1953) 337f; 357. Cf also Gudeman (1916) 1770.

[24] See Evrard (1953) 305-12: 'In a matter of detail, the *Meteorology* commentary seems to be slightly closer to the first group than the two other works of its (i.e. the same) series: it still mentions the notion of supernatural motion, although it does so in order to criticise it', translated from p 321.

The weight of this argument is negligible. The problem of a relative chronology cannot be tackled by pointing out that a particular word or phrase, in this case *huper phusin*, occurs in one treatise but not in the other.[25] In this present context, the inadequacy of the argument is highlighted by the facts that we cannot be certain that the problem of supernatural motion was not discussed in the *contra Aristotelem* because of the fragmentation,[26] and secondly, the remark in the *Meteorology* commentary is one of criticism, as Evrard himself points out, and not one of approval.

2. In the *contra Proclum*, Evrard argues further,[27] Philoponus announces in several places his plan to write a refutation of Aristotle's doctrines of the eternity of the world and the existence of ether. In the same work, Philoponus does not yet seem to disagree with Aristotle on the question of the generation of heat by the sun through friction. In the *Meteorology* commentary, on the other hand, Philoponus severely criticises Aristotle's theory. He argues that the nature of the sun must be fiery and rejects Aristotle's doctrine of ether. According to Evrard, the two works appear to be steps towards the *contra Aristotelem*, in which treatise the theory of ether is thoroughly rejected in favour of a theory which assumes a fiery nature of the celestial bodies.[28]

Evrard takes this as evidence that the *Meteorology* commentary and the *contra Proclum* are closely related,[29] whereas what in fact it underlined is the affinity of the *Meteorology* commentary and the *contra Aristotelem*. For these treatises have, in so far as the problems of ether and the generation of heat by the sun are concerned, more in common than either of them with the *contra Proclum*. Evrard does not sufficiently justify his supposition that the *Meteorology* commentary must be understood as a step towards the *contra Aristotelem*.[30] And one could easily adduce a counterexample. On the question of the generation of heat by the sun the fragments of the *contra Aristotelem* suggest that Philoponus had not yet worked out an adequate theory based on the assumption that the sun consists of fire,[31] whereas the *Meteorology* commentary offers quite extensive explanations.[32]

The general point that the *Meteorology* commentary is on the whole less critical of Aristotle can of course not be taken as an indication for the relative date of that work. Such an argument overlooks the fact that the two treatises possess entirely different characters and do not fall into the same literary

[25] i.e. the phrase occurs at *in Meteor* 97,20 but not in any of Simplicius' quotations from the *contra Aristotelem*.

[26] Incidentally, certain remarks in Simplicius do suggest the contrary, see *in Cael* 35,12-20 and cf below.

[27] See Evrard (1953) 322-29.

[28] ibid. 'The two works constitute nothing but an intermediate step in a development of which the *contra Aristotelem* marks the final end', translated from p 334.

[29] ibid. 334.

[30] ibid. 334; 336f.

[31] Cf e.g. ap. Simplicium *in Cael* 82,14-18.

[32] Philoponus' remarks in the *Meteorology* commentary concerning the generation of heat by the rays of the sun accord with the cosmological assumptions of the *contra Aristotelem*, cf e.g. *in Meteor* 41,24-44,21; 47,26f; 49,22-34; 52,27-35; 53,22f. On the other hand, in his critical remarks on the *contra Aristotelem* Simplicius accuses Philoponus of ignorance, suggesting that the latter neglected to discuss the problem there, cf *in Cael* 82,26ff; 83,5.

category. Whereas the *contra Aristotelem* is a polemical treatise on Aristotelian cosmology, the *Meteorology* commentary is a lecture, an oral exegesis of an Aristotelian text. For this reason one should not be surprised to find that the commentary is less polemical and critical than the *contra Aristotelem*. It is significant, at any rate, that in the *Meteorology* commentary explicit criticism is often postponed or suspended.[33]

3. Finally Evrard adduces and discusses two cross-references.[34] After a brief rejection of Aristotle's etymology of the word *aithêr* in the *Meteorology*,[35] Philoponus says (*in Meteor* 16,30-32):

> The arguments which have been devised by <Aristotle> in the first book of the *de Caelo*, which show from circular motion that the heavens are a fifth substance of body, we will go through in detail (*dieleusometha*) in other <works?, lectures?> (*en heterois*).

Admittedly, this seems to be indeed an announcements of a rather unfavourable examination of the first chapters of the *de Caelo*, and for this reason Evrard took it as a forward reference to the *contra Aristotelem*.[36] However, this conclusion does not follow. Since Philoponus addresses an audience of students the remark may equally well be taken as an announcement of certain *lectures* on the *de Caelo* to be held in the near future. In that case the passage does not tell us anything about the relative dates of the *Meteorology* commentary and the *contra Aristotelem*. But even if one supposes that the above reference does have the *contra Aristotelem* in mind, one would have to concede that, although the verb *dieleusometha* is in the future tense, Philoponus may well refer to an already existing or almost completed work. For if one compares the passage with one of the typical announcements of the *contra Aristotelem* in the *contra Proclum*, the difference is striking.[37] There, the composition of the *contra Aristotelem* is no doubt envisaged as a distant future event, cf e.g. *contra Proclum* 7.6.258,22-26:

> But on the movement in a circle, which is the movement of the celestial bodies, it will be demonstrated more fully – *so God will* – in the objections to Aristotle on the eternity of the world that this motion is not eternal.

As compared to this, the remark in the *Meteorology* commentary, if it is understood as a reference to the *contra Aristotelem*, suggests that the polemic was, at any rate, already in the making.

After this supposed forward reference to the *contra Aristotelem* Evrard cites another passage which he takes to refer back to the *contra Proclum*,[38] *in Meteor* 24,38-25,2:

[33] Cf e.g. *in Meteor* 16,30-32; 24,38-25,2; 37,18-23; 91,18-20.
[34] See Evrard (1953) 339-45.
[35] Cf *Meteor* 1.3, 339b21-27.
[36] See Evrard (1953) 341.
[37] Cf with *contra Proclum* (ed Rabe) 155,19-24; 258,22-6; 396,23-25; 399,20-28, (461,1f); 483,18-21.
[38] See Evrard (1953) 341-5.

However, our opinions on all these things[39] have been stated elsewhere and lie ready for those willing to become acquainted with them (*gnônai prokeitai tois ethelousin*), in order that we may not repeat ourselves now.

Evrard rejects Hayduck's suggestion[40] that this is a reference to Philoponus' commentary on Aristotle's *de Generatione et Corruptione*, and he argues cogently that it must refer to one of the polemical treatises. Although Evrard admits that the *contra Aristotelem* is the more obvious candidate – for here the existence of the fifth element is explicitly denied – he decides that the treatise against Proclus must be meant, for he is already convinced that the *Meteorology* commentary predates the *contra Aristotelem*.[41]

Although Evrard has shown successfully that the *Meteorology* commentary is one of Philoponus' later treatises, his relative chronology *contra Proclum* – *Meteorology* commentary – *contra Aristotelem* remains dubious. The key to the understanding of the relation between the treatises lies in Philoponus' treatment of the problem of the movement of the firesphere. Historically, the problem originated in conflicting remarks made by Aristotle. When Aristotle wanted to demonstrate the existence of ether in the *de Caelo*, he denied that any of the four sublunary elements can move in a circle, either naturally or counternaturally.[42] But in the *Meteorology*, when he wanted to account for such phenomena as shooting stars and comets, which are, according to Aristotle, sublunary phenomena, he assumed without qualms that the firesphere and the upper air are moved in a circle by the agency of the celestial spheres. In Late Antiquity, Neoplatonists like Damascius, Simplicius and Olympiodorus tried to evade the concession that Aristotle contradicted himself. They upheld the idea that the firesphere moves in a circle and emphasised that its movement is neither natural nor counternatural, but belongs to it 'supernaturally'.[43] Philoponus himself accepted the doctrine of supernatural motion in his earlier works, until the *de Aeternitate Mundi contra Proclum*,[44] yet he arrived at the different solution in the *contra Aristotelem* that circular motion must belong to fire naturally.[45] Since the whole problem originated in passages of Aristotle's *Meteorology*, one must ask the question of how it is treated in the commentary on that work.

If one turns to the relevant passages, one is surprised to find that Philoponus' remarks are extremely economical. The first comment occurs in *in Meteor* 37,18-23 (on Aristotle *Meteorology* 1.3, 340b32-36):

[39] i.e. on the question of how many and which elements constitute the universe.

[40] See M. Hayduck, ed, *Ioannis Philoponi in Aristotelis meteorologicorum librum primum commentarium. CAG* XIV(1), Berlin 1901, 154 (index s.v. *Philoponos*).

[41] See Evrard (1953): 'The argumentation which I have just rehearsed does not exclusively belong to the *contra Proclum*; one finds it amplified in the *contra Aristotelem*. But the *Meteorology* commentary could not refer to that work because it (i.e. the MC) is earlier. Therefore, it is probably the *contra Proclum* to which it refers', translated from p 344.

[42] See *Cael* 1.2, 269a2-18.

[43] For Damascius see ap. Philoponum *in Meteor* 97,20f; Simplicius: *in Cael* 21,1-25; 35,12-20; 51,5-28; Olympiodorus: *in Meteor* 2,19-33; 7,21-30.

[44] See Evrard (1953) 305f; 309-14. Cf Philoponus *in Phys* 198,12-19; 198,32-199,12; 378,21-31; *contra Proclum* 240,28-241,10; 278,19-28.

[45] ap. Simplicium *in Cael* 34,5-11; 34,33-35,8; 35,12-20; 35,28-33.

One must know that the Platonists thought that the firesphere and the adjoining air are not carried along by the heavens, but that they possess this kind of movement naturally. For some of the totalities <of the elements>, they say, are unmoved, as earth and water, but the others move in a circle, as the totality of air and the firesphere. For none of the totalities move in a straight line. – But this is not the right moment (*kairos*) to discuss these things, for they have been fully decided (*diêgônistai*) by us elsewhere.

Curiously, Philoponus interrupts his exposition of relevant Platonistic ideas, which contradict Aristotle's, by the remark that the present lecture is unsuitable for going into further detail. The problem has been discussed sufficiently elsewhere. Accordingly, when circular notion in sublunary elements is alluded to in the next chapter of the *Meteorology*,[46] Philoponus does not pick up the problem at all. But in his comments on 1.7, 344a11-13, where Aristotle states once again that the firesphere and the upper air move in a circle, Philoponus says in *in Meteor* 91,18-20:

Now <Aristotle> wants the revolution of these two bodies <i.e. fire and air> to be forced (*biaion*). But how can something that is forced and counternatural be perpetual (*diênekes*)? – About these things, however, we have spoken (*eirêkamen*) sufficiently elsewhere.

The passages cited indicate that Philoponus discussed at length, and apparently proposed a solution to, the problem of the circular movement of fire and air in some earlier work. In particular the last quotation suggests that his discussion involved a criticism of Aristotle. The question arises: Where did Philoponus criticise Aristotle for holding that the movement of the firesphere is forced and contrary to nature? Two points have to be kept in mind.

First, the problem as it is presented here was not and could not have been raised in the treatise against Proclus because there Philoponus still accepted the doctrine of supernatural motion. And the point of that doctrine was precisely to *save* the Aristotelian position from contradiction. Secondly, the text of the *Meteorology* itself nowhere suggests or explicitly states that Aristotle regarded the circular movement of the spheres of fire and air as counternatural or forced. Aristotle merely says that these spheres are carried along (*sumperiagetai*) with the motion of the heavens.[47] Only an interpretation of such a phrase in the light of a particular passage in *de Caelo* 1.2 could lead to the conclusion that Aristotle regarded the motion as forced. The passage in question is *Cael* 1.2, 269b1f where Aristotle argues:

In consequence, it is necessary that circular motion, since (*epeidê*) it belongs to the (four sublunary elements) *counternaturally*, is the natural movement of some other element.

Although Aristotle's main view in the *de Caelo* is that the four elements cannot move in a circle at all, he appears to be supposing here that circular motion is

---

[46] See *Meteor* 1.4, 341b22-24; b35f.
[47] Cf e.g. *Meteor* 1.7, 344a12.

possible for them *qua* forced motion. Philoponus' criticism in the *Meteorology* commentary seems to be linked to this passage in Aristotle's *de Caelo*, and one may plausibly suggest that the treatise referred to in *in Meteor* 37,18-23 and 91,18-20 may in fact be the *contra Aristotelem*. For here it is clearly recognised that Aristotle allows sublunary elements to be moved in a circle counternaturaly,[48] and precisely the same criticism, i.e. that a counternatural movement cannot be perpetual, is made, ap. Simplicium *in de Caelo* 37,27-29:

> \<Philoponus\> frequently brings in the argument that if circular movement did not belong to the firesphere and the air by nature they would not last for a long time because the movement would be contrary to nature. For also Aristotle said that what is contrary to nature perishes very quickly.[49]

Here the same views are expressed as in the *Meteorology* commentary, and the discussion in the *contra Aristotelem* was no doubt elaborate enough to make further comments in the lecture superfluous. The evidence invites the conclusion that the *Meteorology* commentary may actually have been written a short while after important parts of the *contra Aristotelem* were composed.

IV

Apart from taking better account of the textual evidence, a further advantage of this revised chronology is that it is more economical: It is no longer necessary to postulate yet another lost reatise by Philoponus. Still discussing whether the movement of the firesphere is forced or natural, Philoponus states (*in Meteor* 97,12-16):

> In consequence, it is not the case that the firesphere falls behind the heavens on account of being carried along by force,[50] but it possesses this movement by nature, since if it were moved in this way by force and contrary to nature it must have been in a state contrary to nature all this time, never attaining a natural state, which is impossible. ...From this it is clear that the motion of comets is not supernatural either, as Damascius claims somewhere (*heterôthi*), which we have refuted.

Here Philoponus endorses the theory argued for in the *contra Aristotelem* that the movement of the firesphere is natural; then he attacks Damascius for arguing that the movements of the comets, i.e. of phenomena in the sphere of fire, ought to be regarded as supernatural. Damascius, like Simplicius, proposed and perhaps even originated the theory of supernatural motion. In the *Meteorology* commentary Philoponus does not discuss Damascius' views

---

[48] See ap. Simplicium *in Cael* 56,26-57,8.

[49] Cf also *in Cael* 34,7-10: 'For the firesphere and the air move in a circle, and they possess this movement by virtue of their own nature – just like the heavens. For the movement is either according to or contrary to nature, and it is better not to be at all than always to be in a state contrary to nature.'

[50] Cf Aristotle *Meteor* 1.7, 344b8-11.

on this to any extent, nor the doctrine of supernatural motion in general. His last remarks, therefore, suggest that he is referring to an earlier treatise of his. Since Evrard supposed that the *Meteorology* commentary was written before the *contra Aristotelem*, he chose to postulate the existence of a further, now lost treatise in which Philoponus refuted the idea of supernatural motion.[51] Evrard suggested that this treatise was composed between the *contra Proclum* and the *Meteorology* commentary.[52] There is, however, no independent evidence for such a treatise, and on the assumption that the absolute dates of the *contra Proclum* and the *contra Aristotelem* have been determined correctly it would be difficult indeed to find a place for it. On the other hand, if one reverses the traditional order of the *Meteorology* commentary and the *contra Aristotelem*, the above passage could well be interpreted as a reference to the latter work. It is not unreasonable to suppose that Philoponus' arguments *for* the naturalness of circular motion of the firesphere included an explicit refutation of the theory of supernatural motion. Although Simplicius gives no direct evidence of such a refutation in his discussion of the *contra Aristotelem*, a remark made by him indicates that it existed (*in de Caelo* 35,13-20):

> If the movement of the firesphere is indeed simple, it is better to say that it is supernatural, so that again one natural movement belongs to one elementary body. <Philoponus> debases (*paracharattei*) this by saying that fire has two natural movements, one in an upward direction belonging to the parts of fire which have become detached from the totality, the other, circular one belonging to the totality itself. ... And it is clear that in all these arguments he led himself astray into thinking that circular motion, which is the celestial motion, belongs to fire – not supernaturaly, but naturally.

It seems therefore that the *Meteorology* commentary refers back to the *contra Aristotelem* on several occasions, notably when Philoponus expresses his disagreement with Aristotle. There are not only good reasons for but also advantages in doubting the relative chronology proposed by Evrard. One can dispense with the awkward postulation of another lost treatise. We may conclude that the *contra Aristotelem* in fact predates the lecture on Aristotle's *Meteorology*. And a further conclusion may be drawn. If it is true that, in contrast to the *contra Proclum*, the *contra Aristotelem* comprised further books of a purely theological character, and if Philoponus continued to lecture on Aristotle even after he had worked out his full-scale attack on Aristotelian doctrine, then it seems to be too simplistic a view that the polemic against Aristotle marks Philoponus' final break with school-philosophy, triggering off his dedicated commitment to theology for the rest of his life. To be sure, the *contra Aristotelem* ought to be regarded too as a serious and philosophically

---

[51] Note that Evrard (1953) 316 alters the word order of the passage in his translation, taking *heterôthi* to refer to a treatise of Philoponus rather than Damascius: 'De là, il est évident que leur mouvement (sc. des comètes) n'est pas non plus surnaturel, comme le dit Damascius; nous avons réfuté cette thèse ailleurs.'

[52] See ibid. 321; 344f: 'All that one can say is that between the *contra Proclum* and the *Meteorology* commentary our author composed a work in which he showed that the circular movement of fire can neither be forced nor supernatural', translated from p 344f.

innovative contribution to cosmological theory of that time. This did not prevent but rather demanded that in the same treatise Philoponus should attempt to square his views with those of the religion he believed in. In his writings there is indeed no indication of a conversion, and the transition of emphasis from philosophy to theology seems to have been a gradual one. Most likely, Philoponus cherished his dual interest throughout his intellectual development, a conjecture which after all agrees with the picture of him as a theologian who supports his monophysite position with an array of argumentative rationality and bitterly rejects the naive intellectual framework of his Christian adversaries.[53]

[53] I would like to acknowledge gratefully my indebtedness to valuable comments made by G.E.R. Lloyd, David Sedley, and R.R.K. Sorabji on earlier versions of this paper.

CHAPTER TWELVE

Philoponus' Commentary on Aristotle's *Physics* in the Sixteenth Century

Charles Schmitt

As it is generally accepted, the term 'Renaissance' refers to an historical period in which there was a revival of interest in the literature, styles and forms of Classical Antiquity. Though the 'revival' is usually understood to refer specifically to ancient 'literary' texts, there can be no doubt that the specialised technical treatises of philosophy, natural science, mathematics and medicine played a role equally important, if not more important, in the cultural and intellectual life of the Renaissance. In addition to the rediscovery of the integral texts of Homer and the Greek dramatists, Cicero's *Letters to Atticus*, Quintilian, and Lucretius, the fifteenth century also saw the recovery of much of Galen, Theophrastus, Plato, Plotinus and Proclus, Pappus, Diogenes Laertius and Sextus Empiricus, as well as many additional classical authors of specialised literature. Indeed, the 'Renaissance' was a revival of the technical knowledge bequeathed by Antiquity as much as of works of recognised literary and rhetorical quality.

One aspect of the influence of ancient literature on the Renaissance which has received little attention until fairly recently is the role of the Greek commentators on Aristotle. In that vast corpus, most of which is conveniently assembled for us in the *Commentaria in Aristotelem Graeca*, is contained a wealth of interpretative and supplementary material, which is of great use not only for an understanding of the Aristotelian text itself, but also for understanding its historical context and the philosophical positions which were in competition with those of Aristotle in antiquity. A certain number of the Greek commentaries were known in the Middle Ages, both in the Islamic and in the Christian worlds, but such knowledge was very fragmentary. Only a small portion of the extant commentaries was available in Latin before the sixteenth century. Some of these attained a degree of importance and played a central role in the thirteenth- and fourteenth-century discussions of the soul, for example. These medieval versions are presently being edited in a critical fashion by a group of scholars at Louvain; this series should take its place

* In preparing this paper for publication I have benefited from the comments and suggestions of Jill Kraye, Charles Lohr and Richard Sorabji, to whom I am most grateful.

alongside the Greek texts produced in the last century by the Berlin Academy of Sciences. So far editions of commentaries by Themistius, Ammonius, Philoponus, Simplicius, Alexander and Eustratius have appeared.[1] But it remained for the sixteenth century to make accessible most of the material. For example, less than half of the works attributed to Alexander of Aphrodisias contained in the *CAG* and *Supplementum Aristotelicum* were available in the Middle Ages,[2] and, among the expositions of Philoponus, only the commentary on the *de Anima* was available.[3]

The need for a comprehensive publication of all of the Greek commentaries on Aristotle was already noted and made a programme for the future in Aldo Manuzio's prefatory letter to the first volume of his *editio princeps* of Aristotle in 1495.[4] Although Aldo himself did not live to achieve his aim, he did initiate it, and between that date and 1540 nearly the entire Greek *corpus* was made available to European scholars.[5] Parallel with the publication of the Greek texts – and generally delayed by only a few years – was the publication of Latin translations of the same texts, thus making the material accessible to a much wider readership than the rather restricted group who could cope effectively with the Greek text of the commentators. Most of the Greek editions themselves, as well as the majority of the translations, issued from Venetian presses, though Paris and Lyon served as secondary publication centres. By mid-century essentially everything could be read in Latin, and the impact of the new material can be traced in the Aristotelian literature of the period. In reading the many commentaries on Aristotle and other philosophical works of the sixteenth century one clearly discerns the rising tide of interest in these expositions along a spectrum of philosophical and scientific topics. Hitherto, the impact of these new sources of information has only imperfectly been charted, primarily with regard to discussions of the soul. Nardi's fundamental work on Simplicius,[6] the more recent studies on

[1] The series is entitled *Corpus Latinum commentariorum in Aristotelem Graecorum*, Louvain-Paris-Leiden 1957f. So far six volumes have appeared. There is also in progress the reprinting of the Renaissance Latin translations of the Greek commentators under the general title *Commentaria in Aristotelem Graeca, versiones Latinae*, ed. C.H. Lohr, Frankfurt 1978f. So far three volumes have appeared including the Venice 1554 edition of Dorotheus' translation of Philoponus' commentary on the *Physics*, reprinted 1984.

[2] F.E. Cranz, 'The prefaces to the Greek editions and Latin translations of Alexander of Aphrodisias, 1450-1575', *American Philosophical Society, Proceedings* 102, 1958, 510-46 and 'Alexander Aphrodisiensis', *Catalogus translationum et commentariorum*, ed P.O. Kristeller & F.E. Cranz, Washington D.C. 1960f, I, 77-135; II, 411-22.

[3] See M. Grabmann (1929) and G. Verbeke (1966).

[4] 'Habes nunc a me libros Aristotelis disciplinae. Habebis Deo favente et philosophicos tum morales tum physicos, et quoscunque ille divinus magister legendos posteritati reliquit, modo extent. Erunt deinde a me tibi et caeteris studiosis commentatores Aristotelis: Ammonius, Simplicius, Porphyrius, Alexander, Philoponus et Themistius paraphrastes.' I use the edition *Aldo Manuzio editore: dediche, prefazioni, note ai testi*, ed G. Orlandi, Milan 1975, 7.

[5] To the best of my knowledge there is no single study where the relevant bibliography and other information is assembled. For some information see my 'Alberto Pio and the Aristotelian studies of his time', *Società, politica e cultura a Carpi ai tempi di Alberto III Pio*, Padua 1981, 43-64, at 55-9 (reprinted in my *The Aristotelian Tradition and Renaissance Universities*, London 1984, §VI).

[6] B. Nardi, *Saggi sull'aristotelismo padovano dal secolo XIV al XVI*, Florence, 1958, 365-442 ('Il

Alexander by Cranz,[7] and on the general Neoplatonism of the commentaries by Mahoney[8] have served to draw attention to the rich vein of material there to be mined. The range of the impact – in logic, natural philosophy, metaphysics, and psychology – has scarcely been charted, nor has the interplay between Greek, Arabic, Hebrew, and medieval and Renaissance Latin interpretation of Aristotle been evaluated and analysed. During the second half of the sixteenth century, those who wanted to understand Aristotle – which for them meant philosophy *tout court* – frequently tried to relate the text of the Stagirite to the varying interpretations of Philoponus, Simplicius, Averroes (1126-98), Thomas Aquinas (*c*. 1225-74), John of Jandun (died 1328), Pomponazzi (1462-1525) and Soto (1494/5-1560), among many others.

Particularly little studied has been the impact of the newly available Greek commentators on the *Physics*. Here is meant primarily Simplicius and Philoponus, both of whom left behind extensive and detailed expositions of that work, neither of which was known directly to Latin writers of the Middle Ages, but which were to become available in the sixteenth century. As long ago as Wohlwill and Duhem[9] it has been known that some of the criticisms and alternative positions put forward in the commentaries on the *Physics* by the two sixth-century writers later attained importance in the history of the development of physical thought. Moreover, it was also realised by the same historians that the critiques of Aristotle put forward by Simplicius and Philoponus were very similar to some of the positions which became central in the formulation of the 'new science' of the seventeenth century. Thus far, however, there has been little systematic attempt to consider the reaction of the sixteenth century as a whole to the reorientation which became possible with the availability of Simplicius and Philoponus. The story is not simple and it cannot be covered comprehensively here, though I hope to be able to indicate some lines further research might take. What I shall do is to focus upon Philoponus, whose significance in the story is possibly less than that of Simplicius, but without a full story of the *fortuna* of the *Physics* of both authors a valid conclusion regarding their relative merits is not possible.

Before turning to a consideration of the impact of the Grammarian's partial commentary on the *Physics* (only the first four books are integrally extant),[10] I should like to deal briefly with two other points. First, I should like to sketch a portrait of Philoponus as a commentator, emphasising why what he had to say was of potential importance for the sixteenth century. Secondly, I shall say something general about the recovery and assimilation of his

---

Commento di Simplicio al *De anima* nelle controversie della fine del secolo XV e del secolo XVI'), originally published in 1951.

[7] See above, n 2.

[8] E.P. Mahoney, 'Neoplatonism, the Greek commentators, and Renaissance Aristotelianism', *Neoplatonism and Christian Thought*, ed D.J. O'Meara, Norfolk 1982, 169-77, 264-82. Mahoney is the author of several other papers, following on the initiative of Nardi, which discuss the role of the Greek commentators in Renaissance discussions of the nature of the soul.

[9] Esp E. Wohlwill (1906) and P. Duhem (1954) I, 380-4.

[10] Fragments of books 5-8 are found in MS Paris BN, gr. 1853 and are printed in *Ioannis Philoponi in Aristotelis physicorum libros* ed. H. Vitelli, Berlin, 1887-88 (=*CAG* XVI-XVII) 787-908.

philosophical works in the West down to the sixteenth century.

With regard to Philoponus' position as an Aristotelian commentator at least two things are of utmost importance. First, he was a Christian; secondly, he was a Platonist. The first, at least, was a two-edged sword, for it was known to the sixteenth century, through the *Suda* and other sources, that Philoponus was not a completely orthodox Christian and that he had been condemned by the Church.[11] None the less, it was known that he had taken a strong Christian position on the central issue of the creation of the world. In writing against both Aristotle himself and Proclus, he had argued, for the first time, from a rational and essentially philosophical basis that the world was created in time *ex nihilo*. Consequently, though his Christianity may have been tinged by a lack of orthodoxy, he still stood closer to the position of an ideal Christian philosopher than any other of Greek antiquity. For that reason, if for no other, he might be said to have had an appeal for the European philosopher of the sixteenth century.

A certain amount of evidence indicates this to have been the case. Pier Nicola Castellani, writing in 1519 in the introduction to his edition of the Pseudo-Aristotle *Theology* spoke of him in the same breath with Dionysius the Areopagite as one who, more than any other, deserved the epithet *Christianus philosophus*.[12] The rabid anti-Aristotelian, Gianfrancesco Pico (1469-1533) – of whom more later – perhaps as fundamentalist a Christian as one can find among Catholics of the sixteenth century, spoke of him as *germanissimus peripateticus*.[13] The sentiment perhaps comes out most clearly in the editions of Philoponus' *de Aeternitate Mundi contra Proclum*. The editor of the first printing of the Greek text in 1535, Vittore Trincavelli (1491-1563), praises the work for refuting the doctrine of the eternity of the world put forth by Proclus who was 'excited with great hatred and rage towards the Christians'.[14] He goes on to say that he rejoiced at having the good fortune to come across the work and to be able to present it to the public.[15] One of the translators of the work into

---

[11] *Suidae lexicon*, ed A. Adler, Leipzig 1928-38, II, 649 [I, §464]. This work was printed for the first time in 1499. A useful summary of information available on Philoponus before recent times is in I.A. Fabricius, *Bibliotheca Graeca*, ed G.C. Harles, Hamburg 1790-1809, X, 639-69.

[12] Castellani claimed that the *Theology* contained more sound theological tenets than any other pagan works, 'ut si illae paucae [sententiae] quae exorbitant, non inter extarent, minime Aristotelem aethnicumve alium, sed Christianum potius philosophum (ut Dionysium [Areopagitam] Philoponumque) merito existimaremus,' Aristotle, *Theologia sive mistica phylosophia secundum Aegyptios*, Rome 1519, fol Biir. I am indebted to Jill Kraye for this reference.

[13] Gianfrancesco Pico, *de Animae Immortalitate Digressio*, Bologna 1523, fol 7v.

[14] 'Itaque cum diebus superioribus huius florentissimae Reipublicae Bibliothecae indicem percurrerem, interlegendum oblatus est titulus libri in Proclum a Ioanne Grammatico de mundi aeternitate conscripti, in quo illius sententiam, cum maximo in Christianos odio ac furore, percitus composito ad hanc rem libro multis rationibus mundum ipsum nec principium habuisse, nec finem consecuturum ostendere contendisset, admirabili arte, neque non pari eloquentia vir ille divinus obtundit ac reiicit', *Ioannis Grammatici Philoponi Alexandrini contra Proclum de mundi aeternitate*, Venice 1535, fol Av.

[15] '[L]aetabar equidem summopere me in eius viri praeclara monumenta incidisse, qui preterque quod eruditissimis commentariis omnes fere liberaliores artes illustravit, ut non immerito teste Suida ob indefessam, quam bonis literis operam assidue, impendebat, Philoponi cognomentum adeptus sit, Christianam etiam religionem adversus impios acerrime tutatus est. In hoc enim libro Porphyrii et Procli de mundi aeternitate rationes adeo evidenter confutat,

Latin, Jean Mahot (fl.1557), in his dedication letter says that Philoponus is to be held in the highest esteem, for he defended the Christian religion with such great thoroughness against the lies of its maligners.[16] Though further work is necessary, it seems that the bulk of sixteenth-century Christian opinion was that Philoponus should be given favoured treatment because of his Christianity.

Though the majority of his surviving writings are commentaries on works of Aristotle, his basic standpoint can be described as 'Platonic'. He had a reputation in the Renaissance, albeit based in part upon the *Life of Aristotle*, now attributed to his teacher Ammonius rather than to Philoponus himself, for being one of the initiators of the attempt to bring about a fusion of Plato and Aristotle.[17] At any rate, he and his Latin contemporary Boethius made moves in the direction of bringing about a concord between the philosophies of Plato and Aristotle, which became a literary *genre* in its own right especially during the fifteenth and sixteenth centuries.[18]

Though some of Philoponus' most characteristic and influential physical doctrines were known to the West in a derivative fashion during the high Middle Ages,[19] full knowledge of the *Physics* commentary came at an appropriate historical moment. By the time we get to 1500 there was already a sizeable critical literature referring to some of the more questionable and internally problematic doctrines of Aristotelian physics. Largely forged in the Middle Ages, this critique took in such central Aristotelian teachings as place, void and the explanation of projectile motion.[20] Philoponus had taken exception to Aristotle on each of these points, and when his writings became available to the Latin philosophers of the sixteenth century, they provided a ready-made storehouse of weapons by which to question further the already

---

adeoque argute in authores proprios retorquet, ut nemo sit, qui (modo aequo legat animo) in illos non exclamet, quod in psalmis divinus cecinit vates, sagittae infantium factae sunt plagae eorum', ibid. This reflects Greek Psalms 63.8.

[16] 'Porro, quam sit amandus hic autor qui nostram religionem adversus sceleratorum linguam tanta diligentia defendit, pii omnes intelligunt, qui norunt quanta sedulitate iis sit occurrendum, qui captiosarum conclusionum fraude, veluti cuniculis quibusdam conantur ea subvertere quae nostrae religonis persuasit veritas, inter quos insignis fuit Proclus, qui in Christianos tetrum linguae suae virus eiaculatus est', *Ioannes Grammaticus Philoponus Alexandrinus in Procli Diadochi duodeviginti argumenta, de mundi aeternitate ... Ioanne Mahotio Argentenaeo interprete*, Lyon 1557, fol B2v-B3r.

[17] See, for example, Angelo Poliziano, *Opera quae quidem extitere hactenus omnia ...*, Basel 1553, 227-8, a passage called to my attention by J. Kraye, 'Cicero, Stoicism and textual criticism: Poliziano on katorthôma', *Rinascimento* 23, 1983, 79-110, at 84. On the work see I. Düring, *Aristotle in the Ancient Biographical Tradition*, Göteborg 1957, 120-39. For Renaissance discussions of the work see M. Wilmott, 'Francesco Patrizi da Cherso's Humanist Critique of Aristotle', Ph.D. thesis, Warburg Institute, University of London 1984.

[18] For a general orientation see F. Purnell, 'Jacopo Mazzoni and his comparison of Plato and Aristotle', Ph.D. thesis, Columbia University 1971, 31-92 ('The *Comparatio* Tradition').

[19] This is particularly true of the so-called *opinio Avempace*, on which see E.A. Moody, 'Galileo and Avempace: The dynamics of the leaning tower experiment', *Journal of the History of Ideas* 12, 1951, 163-93, 375-422 (reprinted in Moody's *Studies in Medieval Philosophy, Science and Logic*, Berkeley-Los Angeles-London 1975, 203-85); A. Maier, *An der Grenze von Scholastik und Naturwissenschaft*, Rome 1952, 219-54; M. Clagett (1959) *passim*; S. Pines, 'La dynamique d'Ibn Bajja', in *Mélanges Alexandre Koyré*, Paris 1964, I, 442-68; E. Grant (1981) *passim*.

[20] These themes are traced in Clagett (1959).

weak foundations of Aristotelian physics.

As already noted, knowledge of Philoponus in the West during the Middle Ages was incomplete. The *de Anima* commentary was partially translated by William of Moerbeke (died *c.* 1286); though it was relatively important in the circle of Thomas Aquinas, the translation is extant in only a handful of manuscripts, fewer, in fact, than the other medieval translations of Greek commentators.[21] The other commentaries of Philoponus were unknown, though there was some indirect testimony. The Latin translation of Averroes' *Long Commentary on the Physics* refers to a Joannes Grammaticus in accurate terms and this was picked up in the West,[22] for example, by Pseudo-Siger of Brabant in his questions on the *Physics*.[23] Philoponus, of course, was known to Arabic writers, furnishing the basis for the so-called *opinio Avempace* recited by Averroes and having such a brilliant *fortuna* in the West.[24] Somewhat surprisingly, perhaps, Philoponus gets one rather imprecise mention in Maimonides' *Guide for the Perplexed.*[25]

Only when we arrive at the sixteenth century is there a substantial change: during the first half of the century essentially all of the extant commentaries on Aristotle, as well as the *contra Proclum*, are seen into print. It is quite probable that Philoponus' works had a certain manuscript diffusion in fifteenth-century Italy, though the subject has not yet been pursued systematically. We do know, however, that during the second half of the fifteenth century a range of different works, usually including the commentary on the *Physics* were to be found in the Vatican library,[26] the library of Cardinal Bessarion (1402-72) which later went into the Biblioteca Marciana,[27] the library of S. Marco in Florence,[28] and the famous private library of Giovanni Pico (1463-94), which then went to Alberto Pio (1475-1531).[29] Moreover, at the turn of the sixteenth century Pietro Pomponazzi had the eminent Greek scholar Marcus Musurus (*c.* 1470-1517)

[21] See the introduction to the edition cited above in n 3.

[22] Aristotle-Averroes, *Opera*, Venice 1562-74 (repr. Frankfurt 1962) IV, 141r.

[23] [Pseudo-]Siger de Brabant, *Questions sur la physique d'Aristote*, ed P. Delhaye, Louvain 1941, 153, 169.

[24] See Aristotle-Averroes, *Opera* IV, 160r-162r and *passim*. For the history of the discussion see the references given above in n 19.

[25] Moses Maimonides, *The Guide of the Perplexed*, tr. S. Pines, Chicago 1963, 177 (I, 71).

[26] According to the evidence presented by R. Devreesse, *Le Fonds grec de la bibliothèque vaticane des origines à Paul V*, Vatican City 1965, various works of Philoponus are listed in the inventories of 1475, 1481, and 1484, though the *Physics* commentary is not among them.

[27] Nearly all of Philoponus' extant works, including the *Physics* (no. 565, p 221), are in the inventory of 1474 as listed in L. Labowsky, *Bessarion's Library and the Biblioteca Marciana. Six Early Inventories*, Rome 1979, 191-243.

[28] A variety of works, including the *Physics* (no. 1143, p 257), are listed in B.L. Ullman & P.A. Stadter, *The Public Library of Renaissance Florence*, Padua 1972.

[29] The relevant information is in P. Kibre, *The Library of Pico della Mirandola*, New York 1936, and G. Mercati, *Codici latini Pico-Grimani-Pio*, Vatican City 1938. Mercati's volume contains a number of useful details about the diffusion of Philoponus in the fifteenth and early sixteenth centuries. Other information, especially on the diffusion of Philoponus in the Veneto, is contained in E. Mioni, *Aristotelis codices graeci qui in Bibliothecis Venetis adservantur*, Padua 1958, and *Manoscritti e stampe venete dell'aristotelismo e averroismo (secoli X-XVI)*, catalogo di mostra, Venice 1958.

translate a key passage from Philoponus in which he was particularly interested.[30]

The first of the major philosophical works to appear in print was the commentary on the *Posterior Analytics* (1504).[31] This was followed by his expositions of *de Generatione Animalium* (1526, considered spurious by the modern editor of the Greek), *de Generatione et Corruptione* (1527), *de Anima* (1535), *Physica* (1535), *contra Proclum* (1535), *Prior Analytics* (1536), and *Meteorology* (1551). The *Metaphysics* exposition attributed to him was never printed, but appeared in a Latin translation (1583) by Francesco Patrizi (1529-97). An exposition of the *Categories*, now usually attributed to Philoponus, was also printed, but under the name of his teacher Ammonius.[32] In general, the Greek text was printed but once, usually serving as the only edition until the critical ones prepared by the Berlin Academy. The works were all translated into Latin, several more than once. The commentaries were obviously in demand, the Latin versions being reprinted frequently. The *Posterior Analytics*, for example, appeared in Latin twelve times between 1542 and 1569 and the *Physics* nine times between 1546 and 1581.

It is with the latter work that I shall be primarily concerned for the remainder of this paper, and I shall try to give some idea of its distribution and the various reactions to it in the sixteenth century, though at this stage research into the subject is still very incomplete. Moreover, I shall concentrate my attention primarily upon reactions to the first section of book 4 of the *Physics* which deal with the question of place (*topos*; *locus*) and void (*kenon*; *vacuum* or *inane*). While these are by no means the only interesting parts of Philoponus' expositions of Aristotle, they are useful weather-vanes for the winds of change which were blowing over sixteenth-century physics. In the new science which was to emerge in the seventeenth century two fundamental concepts were those of isotropic space and an assumed existence of void space in nature. Both these developments went against the Aristotelian view, and both bear striking similarities to the positions which Philoponus put forward in criticism of Aristotle. A full analytical study of the role of Philoponus in the development of early modern physical thought lies in the future, though a number of scholars have already made various suggestions and have considered various aspects of the question.[33]

[30] Pietro Pomponazzi, *Corsi inediti dell'insegnamento padovano*, ed A. Poppi, vol II, Padua 1970, 11.

[31] For bibliographical information see the Appendix at the end of this chapter, pp 227-9.

[32] The commentary on the *Categories* was first published at Venice in 1503 and is contained in *Philoponi (olim Ammonii) in Aristotelis Categorias commentarium*, ed A. Busse, Berlin 1898 (*CAG* XIII,1).

[33] See, inter alia, W. Boehm (1967) 337-87 (Wirkung des Johannes Philoponos auf die Nachwelt); P. Galluzzi (1979); E. Grant (1965) and (1981); M. Jammer, *Concepts of Space*, Cambridge, Mass. 1954; G. Lucchetta (1978a), (1983) 701-15, and 'Interpretazioni della dinamica pregalileiana in alcuni recenti studi', *Cultura e scuola* 77, Gennaio-marzo 1981, 133-41; C.B. Schmitt (1967) and 'A fresh look at mechanics in 16th-century Italy', *Studies in the History and Philosophy of Science* I, 1970, 161-75 (repr. in my *Studies in Renaissance Philosophy and Science*, London 1981, §XII): E. Wohlwill (1906) and *Galilei und sein Kampf für die Copernikanische Lehre*, Hamburg-Leipzig 1909, 87-107; M. Wolff (1978).

As already noted, the commentary on the *Physics* is only partially extant. The sixteenth-century edition and the Latin translations cover only the first four books of the work, though excerpts from the remaining four books are found in a Paris manuscript, and were published for the first time by Vitelli.[34] The commentary is quite lengthy, being at least four or five times as long as the Aristotelian text itself. Some comments are very long. Particularly interesting and important are the two long digressions on place and void which occur in book 4.[35] In the Marciana manuscript (gr. 230) alone, it seems, the Greek term *parekbasis* is added and this is printed in the *editio princeps*, but not in Vitelli's edition.[36] The Latin texts of the sixteenth century entitle them *digressiones*. These texts, which are more or less coherent treatises in their own right (covering 30 and 20 pages respectively in the Berlin edition), are essentially systematic criticisms of the Aristotelian positions put forward in the text being presented. As such they call to mind some of the 'expositions' of Aristotle printed in Giordano Bruno (1548-1600) *Opera Latine Conscripta*.[37] Among other things, these *digressiones* as well as other parts of the commentaries report a good deal of pre-Philoponus criticism of Aristotle, which could easily be used as ammunition by later anti-Aristotelian authors.

Before turning to how the sixteenth century dealt with Philoponus' critique it might be well first very briefly to recall the issues at stake here.

Place, as Aristotle defines it, is the 'limit of the surrounding body with respect to that which it surrounds' (212a5-6). Built into this definition, in conjunction with the doctrines of natural place and natural motion, is the concept that place is anisotropic, and that it belongs to a body surrounding anything which is 'in place'. Philoponus, on the other hand says that place is 'a certain interval (*diastêma*; *spatium* in the Latin translation), measurable in three directions, different from the bodies which occupy it, and incorporeal in its very nature. Place consists of the dimensions alone and is empty of every body. In fact, vacuum and place are essentially the same thing'.[38] This approaches our common-sense notion of space – or indeed that of Newton – and is radically different from Aristotle's. The properties Aristotle attributed to place (e.g. boundary, directionality and non-emptiness) are much more determinate than those of Philoponus, i.e. receptivity, incorporeality, three-dimensionality and emptiness. In other words, Philoponus' concept is a much more flexible one, which could be used in various physical systems, while Aristotle's is largely tied to a particular set of metaphysical and physical presuppositions. Perhaps most striking is Philoponus' insistence on identifying place and vacuum (*tauton gar ... to kenon kai ho topos*),[39] which was the direction that modern science took in the seventeenth century.

---

[34] See above, n 10.

[35] Vitelli ed, 557-85, 675-95, paraphrased in English by D.J. Furley in Chapter 6 of this volume.

[36] See ibid. 557, 675.

[37] For example, the work entitled *Acrotismus Camoeracensis* in *Jordani Bruni Nolani Opera Latine conscripta*, ed F. Fiorentino et al., Naples 1879-91, I, 53-190.

[38] Vitelli ed, 567, lines 30-3.

[39] ibid. lines 32-3.

After a long and complex discussion, Aristotle rejected the possibility of a void space existing in nature.[40] Among his reasons for doing so was that he believed that all motion must be through a medium. If that were not so, Aristotle argued, the motion of bodies through space would all be instantaneous, since the velocity of motion is restricted only by the interference it meets by the density of the medium through which it is moving. In a void the resistance would be nil, and hence the velocity of motion infinite. Philoponus took a different line on that issue (and on many others as well). He reasoned that the velocity of motion of any body through a medium is determined by the motive power of the moving body, which defines the maximum velocity. The density of the medium then *takes away* from the maximum velocity, so that when the density is nil (i.e. in a vacuum), the velocity will be a maximum, but non-infinite, figure.

On both issues therefore Philoponus came to radically different conclusions from Aristotle, conclusions which had fruitful implications for later developments. On both points, also, he had fixed upon some of the most evident weaknesses in the Aristotelian system of natural philosophy. Though his writings were not known in the West during the Middle Ages, it is interesting to note that similar sorts of criticisms had developed independently, the story of which has been reconstructed by Duhem, Maier, Clagett, Grant and others.[41] The central thrust of Philoponus' arguments and their full implications for Aristotelian philosophy were recognised for the first time among Latin writers by Gianfrancesco Pico della Mirandola (1469-1533).

Pico was the nephew of the much better known Giovanni Pico, who has been long recognised as a characteristic figure of the Italian Renaissance and a keen champion of man's dignity and of his ability to use his intellect to reach an ontological level equal to that of the angels. Gianfrancesco saw matters very differently, falling under the influence of Savonarola (1452-98) while still a young man. He expended his major energies upon showing that all 'human philosophy' was fruitless and misguided, and that man's only hope resided in the acceptance of Scripture with the concurrent rejection of all attempts at human learning. Since Aristotle still represented the norm of human knowledge, his major effort was directed towards showing that Aristotle's philosophy and science had no legitimate foundation. Heir to the fifteenth-century Italian humanists, who had turned up new classical sources in abundance, Pico was one of the first to make use of classical materials which could be directed towards refuting central tenets of Aristotelian doctrine. The two classical authors which he was the first to use in such a way were Sextus Empiricus and Philoponus. He also made full use, we might mention in passing, of the arguments against Aristotle found in the medieval Jewish thinker Hasdai Crescas (1340-1410), whose work was also later put to a similar use by Giordano Bruno. Here our purpose is to consider Philoponus, so let us focus upon Pico's adaptation of the arguments found in book 4 of the

[40] *Physics* 4.6-9.
[41] See the references given in nn 9 and 19.

commentary on the *Physics*. Taking his basically critical framework from Sextus Empiricus, Pico used Philoponus on an *ad hoc* basis to demolish particular Aristotelian doctrines. He did this not to build a new philosophy to replace the discredited Aristotelian one, but was satisfied merely to leave the Aristotelian one in shambles.

Pico used Philoponus principally in his major work *Examen Vanitatis Doctrinae Gentium et Veritatis Christianae Disciplinae* (1520), particularly on the key issues of place and void.[42] His summation of the nature of place is very similar to that of Philoponus: 'thus place is space, empty (*vacuum*) assuredly of any body, but still never existing as a vacuum alone of itself'.[43] The identification of *locus, spatium,* and *vacuum* is very close to what Philoponus had held, and indeed points the way both conceptually and terminologically towards the developments of the next century. Thus Pico wants to make a three-dimensional empty space the fundamental category rather than the Aristotelian determinate *locus*. As in Philoponus, the separate discussion of *vacuum* disappears after place and space are properly understood. For Aristotle place and vacuum are never identical, while for Philoponus, and Pico following him, they always are. Once the identification has been made, the discussion on vacuum can turn on other points, such as the one we mentioned above about motion in a vacuum. Against following Philoponus – here called 'that Greek author, illustrious in the Peripatetic family'[44] – Pico shows that motion in a void does not imply the contradiction claimed by Aristotle. In fact, he picks out the salient point that in a vacuum bodies of unequal weight will move with the same velocity characteristically, and with his usual anti-Aristotelian pugnacity, he says point blank 'the dogma of Aristotle cannot be true'.[45] Philoponus and Pico following him, of course, put their fingers on a major flaw in Aristotle's system, though neither of them drew the kind of implications which Galileo was later to draw from the situation. Pico, for his part, was not interested in formulating a 'new physics', but was perfectly happy to show the inconsistency and problematic nature of the old one. It is interesting to note that Pico pays no attention here to Avempace, whose opinion he surely knew, for he criticises him elsewhere in his work.[46] There are probably several reasons for this. First, throughout his writings there is a very strong dislike for Islamic thinkers, both on religious and on cultural (linguistic) grounds. Secondly, Philoponus' argument, once the *Physics* commentary was available, provided a more coherent and detailed starting point than did the second-hand account of Avempace which had been

---

[42] For a more detailed discussion see Schmitt (1967), esp V.

[43] 'Spatium itaque locus est, ex sese corpore quidem vacuum, sed nunquam tamen re ipsa vacuum, sicuti materia aliud est quam forma, nunquam tamen sine forma.' Gianfrancesco Pico, *Opera quae extant omnia* ... Basel 1601, 768 (*Examen Vanitatis* VI,4). The work was first published in 1520 and reprinted in 1573 and 1601. For information on this work, its context and influence, see C.B. Schmitt (1967).

[44] 'Diximus quid Graecus author et in Peripatetica nobilis familia Philoponus, adversus Aristotelem vacuum eliminantem disputaverit', ibid. 771 (*Examen Vanitatis* VI,6).

[45] 'Quod si verum est, verum esse non potest dogma Aristotelis, falsum certè erit, et sibi ipsi prorsus adversum', ibid. 770 (*Examen Vanitatis* VI,5).

[46] *Opera* 284 (*de Rerum Praenotione* II,4, first ed 1506).

provided by Averroes. Thirdly, being a good humanist, he was bound to prefer a Greek source to a medieval one. Finally, Philoponus was a Christian – possibly one cannot say a 'good' Christian – who had to wrestle with Aristotle's natural philosophy in the way in which Pico was doing, but nearly a thousand years earlier. His espousal of the Alexandrian was wholehearted, however, and he closed his discussion of Philoponus on the vacuum by saying: 'I have argued about the vacuum thus far from Philoponus whom I have used as interpreter and as it were paraphrast.'[47]

Pico fully saw the implications of the critiques of Aristotelian natural philosophy contained in Philoponus' commentary on the *Physics*, and he made full use of them in his attempt to undermine the foundation of the Aristotelian system. That having been accomplished before the text had appeared in print might lead us to expect a large-scale reliance on Philoponus' analysis of certain key issues once the work was edited and made easily available. Such, however, does not appear to have been the case. This is not to say that the *Physics* commentary went unread once it was printed, for it is mentioned, cited, and sometimes extensively paraphrased or quoted by many during the last two-thirds of the sixteenth century. It is fitted into the general interpretative framework, but seldom, if ever, did it attain the critical and polemical status which it held in Pico's *Examen Vanitatis*. As we might expect, many of the more conservative and committed Aristotelians found the extensive and biting criticisms of their master unacceptable. Sometimes the offending arguments went unmentioned in the commentaries, other times they were attacked and refutation was attempted. In general we can say that sixteenth-century reactions to Philoponus on the *Physics* fall into three broad categories. First, those who already had an axe to grind against Aristotle welcomed it with open arms, for it provided valuable arguments – all with an ancient pedigree – against some central Aristotelian doctrines. Second, the more committed Aristotelians, who found the out-and-out attacks on their master which it contained unacceptable, set about to refute its arguments. Thirdly, there was another broad group, not perhaps so committed as the other two, who gave it careful consideration in formulating their own positions.

Pico's lead in applying Philoponus' *digressiones* against Aristotle had little resonance. Those who took up the battle against peripatetic philosophy later in the century betray little evidence of deriving much of their critique of Aristotle's doctrine of place and vacuum from the sixth-century thinker. Three of the major critics were Cardano (1501-76), Patrizi, and Telesio (1509-88) and none relies very much on Philoponus. The first is very reliant on Strato's arguments in favour of a void derived from the *Spiritualia* of Hero of Alexandria, which also had become newly available and which was influential down to Gassendi at least.[48] In writing their highly innovative

---

[47] 'Et de vacuo hactenus ex Philopono vice interpretis et quasi paraphrastis functus disputavi', ibid. 771 (*Examen Vanitatis* VI,5).

[48] See W. Schmidt, 'Heron von Alexandria im 17. Jahrhundert', *Abhandlungen zur Geschichte der Mathematik* 8, 1898, 195-214; M. Boas, 'Hero's *Pneumatica*: a study of its transmission and

sections on space both Patrizi and Telesio were extremely critical of the Aristotelian position, but neither made discernible use of Philoponus, though Patrizi knew his writings well, since he translated the *Metaphysics* commentary attributed to him and made use of Philoponus in other contexts.[49] Two other critics made limited use of the *Physics* commentary, but scarcely in a decisive way. Though Peter Ramus (1515-72) mentions Philoponus fairly often in the relevant section of his posthumously published exposition of Aristotle's *Physics*, the central critical arguments play little role.[50] One of the most incisive and vehement critics of Aristotle in the sixteenth century was, of course, Giordano Bruno. Effectively combining Platonism and atomism, he makes many scathing remarks about Aristotle's spatial doctrines. Both in the Latin and in the Italian works he makes use of some of the arguments put forward earlier by Philoponus, but there is no direct reference to the sixth-century author in the Italian works. In his *de Immenso et Innumerabilibus* (1591) there is one revealing, but rather general, acknowledgment of indebtedness to Philoponus: 'Many of the peripatetics could not agree with Aristotle's rejection of the void; Philoponus was bolder in his attacks than the others ...'[51] Bruno then goes on to spell out the nature of the anti-Aristotelian polemic. In the final analysis, however, Bruno, like most other anti-Aristotelians of the sixteenth century, seems to have made restricted use of the rich vein of criticism found in book 4 of Philoponus' commentary.

A certain amount of detailed discussion of Philoponus' arguments is found among those who ended up rejecting them. For example, Jesuit textbooks writers such as Toletus (1532-96)[52] and Pereira (*c.* 1535-1610), whose expositions dominated the Jesuit schools for several generations, looked at the position of Philoponus seriously before discarding it. It is interesting to note, for example, that Pereira, following Simplicius, groups Philoponus with the Stoics, 'many Platonists' and Strato among others.[53] In fact he introduces

---

influence', *Isis* 40, 1949, 38-48; and A.G. Keller, 'Pneumatics, automata, and the vacuum in the work of Giambattista Aleotti', *British Journal for the History of Science* 3, 1967, 338-47.

[49] *Ioannis Philoponi ... in omnes XIIII Aristotelis libros eos qui vocantur metaphysici quas Franciscus Patricius de graecis, latinas fecerat, nunc primo typis excussae in lucem prodeunt*, Ferrara 1583. In Chapter 9 above Richard Sorabji considers the possible influence of Philoponus on Patrizi in putting space outside the categories.

[50] *P. Rami scholarum physicarum libri octo in totidem acroamaticos libros Aristotelis*, Frankfurt 1606. Philoponus is cited frequently throughout the book in a variety of different contexts, though he cannot be called one of the primary authorities. He is specifically opposed to Aristotle several times (e.g. pp 56, 116, where his views on the vacuum are summarised).

[51] 'Et ex peripateticis plurimi, Aristotelis sententiae contra vacuum acquiescere non potuere, quam prae caeteris audactius Philoponus est insectatus ...', Bruno, *Opera* I, 231.

[52] *Francisci Toleti ... commentaria una cum quaestionibus in octo libros Aristotelis de physica auscultatione ...* Venice 1580, fols 116v-120v (on place) and *passim*. The first edition seems to have been Cologne, 1574. It was reprinted at least five more times.

[53] 'Et quoniam ea opinio quae locum facit esse intervallum quoddam, non solum vulgo placet, sed etiam doctis viris, primo aspectu videtur probabilis; et ut refert Simplicius, habuit auctores et defensores multos ex philosophis, ut Stoicos, complures etiam Platonicos, Galenum, item Stratonem Lampsacenum, Syrianum, et Philoponum, qui in commentariis quos scripsit in 4. librum Physicorum longam digressionem facit hac de re', *Benedicti Pererii ... De communibus omnium*

much of the vocabulary from the Latin translation of Philoponus – e.g. *intervallum, spatium* – and prefers to consider his critical analysis of the Aristotelian position rather than the more traditional and better known one from Avempace.[54] Though he ultimately rejects the position, he gives it a fair, if highly condensed, hearing. By the time we get to the Coimbra commentary on the *Physics* (1592) little trace of Philoponus remains, though he is still mentioned in passing and there are several references to Pico's *Examen Vanitatis*.[55] The force of Philoponus is somewhat lost in a sea of references to other authorities. If Pereira favoured classical sources for his discussions, the opposite seems to have been true of the Coimbra commentators, at least on this issue. The discussion is framed in a way much more reminiscent of how it had been done in the fourteenth century – e.g. Avempace, but not Philoponus, is mentioned with regard to the famous problem of book 4, text 71,[56] rather than the way late sixteenth-century writers increasingly used classical sources either to replace or, at least, to supplement the medieval ones

The Coimbra Jesuits were not alone in seeming to continue on from the fourteenth-century formulation, which Koyré effectively drew attention to 35 years ago.[57] Girolamo Borro (1512-92), whose *de Motu Gravium et Levium* appeared in 1575 and who was one of Galileo's Pisan teachers, mentions Philoponus in appropriate contexts, but his whole analysis was based much more on Avempace and the traditional problem cluster, than that which presented itself with the introduction of new classical sources.[58] As I have argued elsewhere,[59] Borro is not easy to characterise – among other things he speaks of testing the gravity question by dropping balls of different weight some years before Galileo suggested it – but his failure to utilise Philoponus where appropriate should be kept in mind.

Borro's contemporary and rival, Francesco Buonamici (died 1603) has more to say on the issue. Also a teacher of Galileo, Buonamici was the author of the gigantic *de Motu* (1591) of more than a thousand pages, possibly the most comprehensive treatise on the traditional Aristotelian question of motion which took form in the sixteenth century.[60] In it he takes a rather

---

rerum naturalium principiis et affectionibus, libri XV, Venice 1591, 394b. Pereira's book first appeared in this form at Rome in 1576. Between that date and 1618 it was reprinted at least fifteen times.

[54] ibid. 394a-397a.

[55] *Commentarii Collegii Conimbricensis Societatis Iesu in octo libros physicorum Aristotelis Stagiritae ...,* Lyon 1594, II, 3-76. Both Pico and Philoponus are mentioned on p 23. The commentary was first published at Coimbra in 1592 and at least fifteen more times by 1625.

[56] ibid. II, 72-76, where the question 'Utrum corpus in vacuo moveri possit an non' is discussed with reference only to the medieval discussion.

[57] A. Koyré, 'Le vide et l'espace infini au XIVe siècle', *Archives d'histoire doctrinale et littéraire du moyen âge* 17, 1949, 45-91 (repr. in Koyré's *Etudes d'histoire de la pensée philosophique,* Paris 1961, 33-84).

[58] Girolamo Borro, *de Motu Gravium et Levium ...,* Florence 1575. Philoponus is not mentioned in the long discussion of whether there can be motion in the void (pp 124-32), though he is cited several times in passing (e.g. pp 104-41).

[59] C.B. Schmitt, *Studies in Renaissance Philosophy* §§IX, 267-71; XI.

[60] Francesco Buonamici, *De motu libri X quibus generalia naturalis philosophiae principia summo*

rigid Aristotelian stance, though he gives an airing to many diverse opinions and strictures on Aristotle. For example, there is a surprising interest in the mathematical analysis of natural philosophical questions, and he shows himself aware of – if not always sympathetic to – an immense range of different speculations on problems regarding motion (i.e. change).[61] He takes a somewhat unyielding position regarding the interpretation of Aristotle – one reminiscent of his Paduan contemporary, Cesare Cremonini (1550-1631), a colleague, but also both friend and professional adversary of Galileo.[62] Buonamici, somewhat surprisingly perhaps, reproves Philoponus for contaminating Aristotelian doctrine with that of Christianity. Speaking of Ammonius and Philoponus together he says: 'The Alexandrian philosophers wish to be seen as Christians when they deal with Aristotle, and they have fallen into being simultaneously pseudophilosophers and pseudochristians.'[63] Thus he seems to uphold a double-truth situation in which philosophy and religion are to be kept separate. He speaks harshly of the Grammarian on a number of different issues, among them the now familiar doctrine of identifying place with three-dimensional space.[64]

Other commentators on the *Physics* who take the work of Philoponus seriously include Lodovico Boccadiferro (1482-1545) Francesco Vimercato (*c.* 1512-*c.* 1571), and Federigo Pendasio (died 1603). The first two were working on their commentaries in the late 1540s and must have been rather early beneficiaries of the new Latin translations. Boccadiferro has extended presentations and paraphrases drawn from the Latin text of Philoponus.[65] If he ends by siding with Aristotle, he none the less allows Aristotle's critical commentator to have his say. Vimercato, in common with several others, joins the information contained in Philoponus with the information he had on Avempace drawn from Averroes.[66] Perhaps as much as any peripatetic of the

---

studio collecta continentur, Florence 1591. The fundamental study of this work is now M.O. Helbing, *Ricerche sul De motu di Francesco Buonamici* (thesis for 'Diploma di perfezionamento in filosofia', Pisa, Scuola Normale Superiore 1982).

[61] The richness of his analysis and the range of his sources are indicated in the study of Helbing.

[62] For a brief evaluation of the position of Cremonini with references to earlier literature see my *Cesare Cremonini, un aristotelico al tempo di Galilei,* Venice 1980 (repr. in my *The Aristotelian Tradition and Renaissance Universities,* London 1984, §XI). Among Cremonini's works is *Apologia dictorum Aristotelis de quinta caeli substantia adversus Xenarcum, Ioannem Grammaticum, et alios,* Venice 1616.

[63] '... namque Alexandrini philosophi, dum ita tractant Aristotelem, ut Christiani videri velint, in pseudo-philosophos simul et in pseudochristianos lapsi sunt.' Buonamici *de Motu* 810.

[64] ibid. 434. There are many passages in Buonamici's work discussing, usually in critical terms, various positions of Philoponus. I am indebted to Dr Helbing for supplying me with information on this subject. When his research has been published we should be in a much better position to evaluate Buonamici's general attitude towards Philoponus.

[65] *Explanatio libri I physicorum Aristotelis ex Ludovici Buccaferreae ... lectionibus excerpta,* Venice 1558, esp fols 45-49, but also frequently elsewhere. Boccadiferro's work was printed at least three more times by 1613.

[66] *Francisci Vicomercati ... in octo libros Aristotelis de naturali auscultatione commentarii ...,* Venice 1564

sixteenth century Vimercato joined a sensitivity to the new humanist method to an awareness of the tradition of philosophical incisiveness forged by the scholastics. In a characteristic passage which shows effectively how well he bridged the two cultures, he says:

> Among other arguments which Philoponus brought to bear [on the problem of the void] there is one which Avempace took as his own, and he [Avempace] drew his conclusions in opposition to Aristotle (as we gather from Averroes' commentaries). This argument, which was watered down in various ways by later peripatetics and was much debated in the schools, I now bring to bear on the question, while reserving for another place other arguments, including Philoponus' full critique of Aristotle as well as a more careful consideration of the point presently at issue.[67]

He then continues by expounding in a rather detailed way Philoponus' analysis of motion in a void and relating it to the then better known version of Avempace.

Pendasio has perhaps the most extensive discussion of Philoponus. In his long commentary on the *Physics*, published posthumously in 1604, more than sixty substantial pages are devoted to a discussion of the brief Aristotelian texts on *locus* and *vacuum*.[68] He shows a thorough knowledge and understanding of Philoponus' text, not hesitating to cite from the Greek text in crucial passages, as for example, '*energeian asômaton kinêtikon*, id est actum incorporeum motivum'.[69] He, too, joins Avempace's critique to that of Philoponus, saying that they are similar, but that Philoponus 'used different words'.[70] Also interesting is the attention given in the same context to the vacuist arguments derived from Hero of Alexandria.[71]

Vimercato worked at Paris, Buonamici at Pisa and Pendasio at Bologna. By the second half of the sixteenth century, Philoponus' critique of Aristotle on *locus* and *vacuum* was common equipment for all who wanted to understand Aristotle on those key issues. To round out the situation we might look briefly at Jacopo Zabarella, whose name appears as often as any in current discussions of Renaissance Aristotelianism.

Zabarella (1533-1589), one of the most intelligent of the secular philosophers who passed through Padua and a man fully committed to presenting a 'scientific' and usable Aristotle to his readers, knew and used several commentaries of Philoponus. The Christian commentator is certainly

---

171-95 *passim*. The first edition of this commentary was in 1550. It was printed at least five times by 1593.

[67] 'Atque inter caeteras quas attulit rationes, una est, quam Aven Pace ut propriam sumpsit, atque (ut ex Averrois commentariis colligimus) contra Aristotelem conclusit, quam unam, quia a posterioribus peripateticis aliter atque aliter diluitur, et in scholis plurimum agitatur, nunc afferam, caeteris totaque Philoponi ipsius contra Aristotelem disputatione, atque etiam huius ipsius, quae nunc afferetur, diligentiori exquistitione in alium locum reservatis', ibid. 195.

[68] *Federici Pendasii Mantuani ... physicae auditionis texturae libri octo ...*, Venice 1604, 418-542.

[69] ibid. 509.

[70] ibid. 513.

[71] This is dealt with in part 23 (of the discussion on book 4) entitled 'Rationes Heronis referentes ad confirmandum vacuum', ibid. 539-40.

not one of his dominant authorities. Indeed, he seems not to be mentioned at all in Zabarella's *de Motu Gravium et Levium* (1590),[72] which deals with some of the fundamental issues upon which Philoponus had much to say. Simplicius is considered attentively, as are Averroes, Duns Scotus (1266-1308), Nifo (1470-1538), Zimara (*c.* 1475-*c.* 1537) and Genua (1491-1563) among others. In the chapter entitled 'Nonnullorum opinio de motu elementi in vacuo et eius confutatio' Avempace's position is discussed and rejected, but Philoponus is not mentioned.[73] Throughout his works there are occasional references to the Alexandrian, but he is never given much of a voice: at one point when a specific problem of the *de Caelo* is in question he dismisses him in no uncertain terms, saying 'sed eius [Philoponus] sententia non est digna consideratione'.[74] This view, like that of his Pisan contemporary Buonamici, illustrates the general opinion of the peripatetic camp.

Zabarella's work appeared posthumously in 1590, that of Buonamici in 1591. About the same time a dialogue and other notes on motion were being put together by a young man still less than thirty. These jottings are a curious mixture of old and new. The traditional scholastic authorities, including Avempace and Averroes, as well as Scotus and Thomas, are there, but so too are Archimedes and Apollonius.[75] The work is, of course, the so-called *de Motu* of Galileo Galilei (1564-1642), whose earlier notes (the *Juvenilia*) were an expression of conventional late sixteenth-century scholasticism.[76] William Wallace has made a count of authorities in the *Juvenilia* which I think can be said to do no more than reflect the current state of university natural philosophy in Italy. None the less, Wallace's analysis reveals that there are a surprising number of references to Philoponus. The frequency of citation of Philoponus' name comes after that of Aristotle, Averroes, Thomas and the Thomists, and Simplicius, standing ahead of such authorities as Plato, Albert and Scotus.[77] This does not tell us a great deal, but is a reminder of the place to which Philoponus had risen during the half century after the first publication of his commentaries. In the *de Motu* notes, Galileo's attitude towards Philoponus seems somewhat confused and, perhaps, second-hand,

[72] I use the text contained in *Jacobi Zabarellae Patavini de rebus naturalibus libri XXX ...*, Frankfurt 1607, 307-74. After the first edition of 1590 the work was reprinted at least eight more times.

[73] ibid. 318-21.

[74] ibid. 282 (*de Natura Coeli*, cap VIII).

[75] This material is contained in *Le opere di Galileo Galilei*, ed A. Favaro, Florence 1890-1909, I, 243-419. A partial English version is available as Galileo Galilei, *On Motion and On Mechanics*, ed I.E. Drabkin and S. Drake, Madison 1960, 1-131.

[76] *Le opere* I, 15-177. There are also questions on logic by Galileo only partially published. In recent years there has been much discussion of the sources, meaning, and dating of these works. These intricate problems do not directly concern us here.

[77] Professor Wallace has treated this question in several places, and he seems to come to slightly different conclusions regarding the precise number of citations of various authorities. In any case, Philoponus comes fairly high on the list. See W.A. Wallace, *Prelude to Galileo. Essays on Medieval and Sixteenth-Century Sources of Galileo's Thought*, Dordrecht etc. 1981, esp 136, 196-7. In the latter, more detailed list, Philoponus stands above such important figures as Albert the Great, Duns Scotus, and William of Ockham.

but he does know the intellectual position the Alexandrian adopted and cites him along with Avempace and others with regard to the famous text 71 of book 4 of the *Physics*.[78] It is not clear to what extent Philoponus' views directly influenced Galileo's important innovations. Wohlwill's claims about the similarity of Philoponus' views on *impetus* to those of Galileo are difficult to deny. The point at issue is more whether Galileo learned directly from a reading of Philoponus or through more circuitous secondary routes. What is clear is that some of Galileo's positions bear a strong resemblance to those of the sixth-century Alexandrian. To determine the priority and importance of competing sources within Galileo's intellectual makeup is not an easy task, but there can be no doubt that Philoponus' critical attitude towards a number of central Aristotelian positions added to the evidence which late sixteenth-century thinkers had at their disposal, if and when they wanted to formulate a philosophy to replace that of Aristotle.

Undoubtedly much more can be found regarding the position of Philoponus in sixteenth-century thought. While he seems to have been almost wholly neglected in a number of authors where we would expect to find some interest,[79] we have seen that a concern with his physical thought (and other aspects of his philosophy) was present in the writings of a number of the central figures of sixteenth-century philosophy and science. In addition, the more philological side of the re-introduction of the Grammarian's works is likely also to be of some interest, both from the point of view of the accuracy of the translations which circulated[80] and from the point of view of the

---

[78] 'Tanta est veritatis vis, ut doctissimi etiam viri et Peripatetici huius sententiae Aristotelis falsitatem cognoverint, quamvis eorum nullus commode Aristotelis argumenta diluere potuerit. Nec certe ullus unquam argumentum, quod 4° Phys. t. 71 et 72 scribitur, evertere potuit: nunquam enim adhuc illius fallacia observata fuit; et quamvis Scotus, D. Thomas, Philoponus et alii nonnulli contrariam Aristoteli teneant sententiam, attamen veritatem fide potius quam vera demonstratione, aut quod Aristoteli responderint, sunt consecuti', *Le opere* I, 284. 'Philoponus, Avempace, Avicenna, D. Thomas, Scotus, et alii, qui tueri conantur in vacuo fieri motum in tempore, non bene discurrunt, ponentes in mobili duplicem resistentiam, alteram, nempe, accidentalem a medio provenientem, alteram intrinsecam a propria gravitate. Nam hae duae resistentiae una sunt, ut patet; idem enim medium gravius plus resistit, et facit mobile levius', ibid. I, 410.

[79] For example, I have found no reference to Philoponus on place and void in *Chrysostomi Javelli ... super octo libros Aristotelis de physico auditu quaestiones subtilissimae ...* Venice 1568, 88-116, where there is a good deal of discussion of the medieval criticisms of Aristotle, including that of Avempace; Julius Caesar Scaliger, *Exotericarum exercitationum liber quintus decimus, de subtilitate ad Hieronymum Cardanum*, Paris 1557; or Francisco Vallès, *Octo librorum Aristotelis de physica doctrina versio recens et commentaria ...*, Alcalà 1562. Perhaps more surprisingly, Philoponus is only very briefly mentioned (p 56) in Jacopo Mazzoni, *In universam Platonis et Aristotelis philosophiam praeludia, sive de comparatione Platonis et Aristotelis ...*, Venice 1597, though not in connection with place or void (pp 190f). Though Philoponus is mentioned once or twice in passing, there is not much attention paid to him in *Aristotelis ... naturalis auscultationis libri VIII*, ed Iulius Pacius, Hanau 1608, e.g. in the long discussion of text 71 of book 4 (pp 620f). Philoponus is singled out for special mention as an interpreter of the *Physics* in Sebastian Fox Morcillo, *De ratione studii philosophici libellus*, printed in P.J. Nuñez (Nunnesius), *De studio philosophico seu de recte conficiendo curriculo peripateticae philosophiae ...*, Leiden 1621, 167-203, at 191-2.

[80] Most of the works were translated more than once in the sixteenth century, the reason being that the first translation was found faulty in some ways. For example, J.B. Rasarius says in the preface to his new translation of the *Physics*: 'Ille [i.e. G. Dorotheus] enim, qui vertit, cum

introduction of new terminological distinctions into the Latin scientific and philosophical vocabulary.[81] Moreover, the precise extent to which some of his glosses on Aristotle became a part of the interpretative framework remains to be worked out. For example, I know of at least one case in which he was used, along with Simplicius and Averroes, to establish an interpretative structure for understanding Aristotle's natural philosophy.[82]

Of course the role of Philoponus in the development of sixteenth-century thought cannot be isolated from other tendencies. Besides Avempace, whose influence has been traced fairly carefully, the role of the reintroduction of other ancient thinkers – above all Hero and Simplicius – must be considered in conjunction with Philoponus, at least with regard to issues in physics. The importance of the reintroduction of Simplicius on the *Physics* must be investigated in tandem with Philoponus.[83] The two are often discussed in the same contexts in the sixteenth century. My impression, after looking at some of the Renaissance commentaries on the *Physics* and other relevant texts of natural philosophy is that, if anything, Simplicius was discussed even more extensively than was Philoponus. Whatever the true situation, it is clear that the impact of the *Physics* commentaries of both authors must be given very serious attention by anyone who wants to make sense of the development of sixteenth-century physical thought.

A fuller account of the *fortuna* of Philoponus in the sixteenth century and an analytical study of the place of his physical thought in the development of that subject during the Renaissance requires a much more concerted effort than I have been able to give here. I hope that I have at least been able to show that the subject is worthy of further investigation.

---

ineptissimus est, qui verba se annumerare lectori putavit, tum vero multo indoctior, quippe qui neque verba, nec sententiam videtur esse consecutus. Itaque praeter dicendi genus foedum et inquinatum, in quo ille regnat, perversa omnia sic me perturbabant, ut in proposito susceptoque consilio vix permanserim. Sed vicit publica utilitas privatam molestiam: veruntamen id praestare non potui, quod optabam: opus enim fuisset commentarios ipsos de integro in Latinam linguam convertere: quod sane in animum induxeram, sed aliorum studiorum tempus appetebam, quod me ad dicendum, non ad scribendum invitabat. Quare alius agere coactus sum', *Aristotelis physicarum libri quattuor cum Ioannis Grammatici cognomento Philoponi commentariis*, tr J.B. Rasarius, Venice 1558, fol *iiv. This is, of course, a commonplace sentiment among those producing a new translation to replace a previous one. See, however, the comments of Vitelli in XVI, pp xviii-xx.

[81] See Galluzzi (1979) 125-34.

[82] In the edition *Aristotelis ... physicorum libri VIII omniaque opera, quae ad naturalem philosophiam spectare videntur. Pars tertia. Summae et capitum divisiones explanationesque ex Simplicio, Joanne Grammatico, et Averroe ...*, Venice 1608. I have used this edition in the Biblioteca universitaria, Pisa, shelfmark: F.a. 10.97.

[83] After the *editio princeps* of Venice 1526, his commentary on the *Physics* was published in Latin translation about eight times between 1543 and 1587. The philological and historical side of the Latin translations of Simplicius' commentaries has been throughly studied by F. Bossier, 'Filologisch-historische navorsingen over de middeleeuwse en humanistische latijnse vertalingen van de commentaren van Simplicius', Ph.D. thesis, Leuven 1975, 3 vols.

Appendix

Check-list of sixteenth-century editions of Philoponus'
commentaries on Aristotle and of his contra Proclum

Those editions found in the printed catalogues of the British Library (BM), or of the Bibliothèque nationale in Paris (BN) or in H.M. Adams, *Catalogue of Books Printed on the Continent of Europe, 1501-1600 in Cambridge Libraries,* Cambridge 1967, are indicated appropriately. Locations are given for those editions not found in any of these sources.

Analytica Priora (CAG XIII, 2)

1535 Venice: Bart. Zanettus	Greek	Adams, BM, BN
1539 Venice: O. Scotus	tr. G. Dorotheus	BN
1546 Venice: O. Scotus	tr. G. Dorotheus	Adams, BN
1550 Venice: O. Scotus	tr. G. Dorotheus	BN
1554 Venice: O. Scotus	tr. G. Dorotheus	BN
1558 Venice: H. Scotus	tr. J.B. Rasarius	BN
1559 Venice: H. Scotus	tr. J.B. Rasarius	BM
1569 Venice: V. Valgrisius	tr. J.B. Rasarius	BN
1581 Venice: H. Scotus	tr J.B. Rasarius	BM

Analytica Posteriora (CAG XIII, 3)

1504 Venice: Aldus	Greek	Adams, BM, BN
1534 Venice: Heredes Aldi	Greek	Adams, BM, BN
1539 Venice: O. Scotus	tr. Phil. Theodosius	Vatican L
1542 Venice: H. Scotus	tr. Phil. Theodosius	Vatican L
1543 Paris: J. Roigny	tr. Phil. Theodosius	BM, BM
1544 Paris: J. Roigny	tr. Phil. Theodosius	Adams BN
1545 Venice: H. Scotus	tr. Phil. Theodosius	Firenze BN
1548 Venice: H. Scotus	tr. Phil. Theodosius	Firenze BN
1550 Venice: H. Scotus	tr. Phil. Theodosius	Padova B Antoniana
1553 Venice: J. Gryphius	tr. Phil. Theodosius	Adams, BN
1559 Venice: H. Scotus	tr. Phil. Theodosius	BN
1559 Venice: Off. Valgrisiana	tr. Martinus Rota	Adams
1560 Venice: Off. Valgrisiana	tr. Martinus Rota	Adams, BN
1569 Venice: H. Scotus	tr. Phil Theodosius	BM

Meteorologica 1 *(CAG* XIV, 1)

1551 Venice: Apus Aldi Filii	Greek with tr. J.B. Camotius	BM, BN
1567 Venice: H. Scotus	tr. J.B. Camotius	BM

de Generatione et Corruptione (CAG XIV, 2)

1527 Venice: Aldus	Greek	Adams, BM, BN
1540 Venice: H. Scotus	tr. H. Bagolinus	Adams, BM, BN
1543 Venice: H. Scotus	tr. H. Bagolinus	BM, BN
1549 Venice: H. Scotus	tr. H. Bagolinus	Adams
1558 Venice: H. Scotus	tr. H. Bagolinus	Firenze BN
1559 Venice: H. Scotus	tr. H. Bagolinus	BN
1564 Venice: V. Valgrisius	tr. Andr. Sylvius	BN
1568 Venice: H. Scotus	tr. H. Bagolinus	BN

de Generatione Animalium (CAG XIV, 3)

1526 Venice: J.A. de Sabio	Greek	BM, BN
1526 Venice: J.A. de Sabio	tr. N.P. Corcyraeus	BM, BN

de Anima (CAG) XV)

1535 Venice: B. Zanettus	Greek	Adams, BM, BN
1544 Lyon: A. & J. Huguetan	tr. G. Hervetus	BN
1544 Venice: H. Scotus	tr. Matt. a Bove	BM, BN
1547 Venice: H. Scotus	tr. Matt. a Bove	Adams
1551 Venice: H. Scotus	tr. Matt. a Bove	CB Schmitt
1554 Venice: H. Scotus	tr. Matt. a Bove	BM
1558 Lyon: Heredes I. Juntae	tr. G. Hervetus	Adams, BM, BN
1559 Venice: H. Scotus	tr. Matt. a Bove	Roma B Vallicelliana
1560 Venice: H. Scotus	tr. Matt. a Bove	BM, BN
1581 Venice: H. Scotus	tr. Matt. a Bove	Paris B Mazarine

Physica (CAG XVI-XVII)

1535 Venice: B. Zanettus	Greek	Adams, BM, BN
1539 Venice: O. Scotus	tr. Guil. Dorotheus	BN
1542 Venice: ??	tr. Guil. Dorotheus	Chantilly
1546 Venice: O. Scotus	tr. Guil Dorotheus	Adams, BN
1550 Venice: O. Scotus	tr. Guil Dorotheus	BN
1554 Venice: O. Scotus	tr. Guil Dorotheus	BN
1558 Venice: H. Scotus	tr. J.B. Rasarius	BM, BN
1559 Venice: H. Scotus	tr. J.B. Rasarius	BM
1569 Venice: V. Valgrisius	tr. J.B. Rasarius	BN
1581 Venice: H. Scotus	tr. J.B. Rasarius	BM

Metaphysica

1583 Ferrara: P. Mamarellus	tr. Fr. Patricius	Adams, BM, BN

de Aeternitate Mundi contra Proclum

1535 Venice: B. Casterzagensis	Greek	Adams, BM, BN
1551 Venice: H. Scotus	tr. G. Marcellus	Adams, BN
1557 Lyon: Nic. Edoardus	tr. J. Mahotius	Adams, BM, BN

Bibliography

Works of Philoponus

All works of Philoponus will be listed of which the whole or any part is known
to be extant, but only the most significant of those works which have been
wholly lost, or whose ascription to Philoponus is highly dubious. Where there
are several editions only the most recent will normally be cited. For
sixteenth-century Latin translations, see Chapter 12, pp 228-9.

I. Commentaries on Aristotle

The extant commentaries on Aristotle are published in the series *Commentaria
in Aristotelem Graeca* (*CAG*) ed H. Diels, Berlin 1882-1909, vols 13-17. The
bulk of these are soon to be translated into English, ed. Sorabji.

1. *in Categorias*, ed Busse, vol 13, part 1, 1898.
2. *in Analytica Priora*, ed Wallies, 13, 2, 1905. Wallies doubts the authenticity
   of the commentary on the second book.
3. *in Analytica Posteriora*, ed Wallies, 13, 3, 1909. Wallies' doubts about the
   authenticity of the commentary on the second book have been disputed.
4. *in Meteorologica 1*, ed Hayduck, 14, 1, 1901.
5. *in de Generatione et Corruptione*, ed Vitelli, 14, 2, 1897.
6. *in de Anima*, ed Hayduck, 15, 1897. The Prӧoemium is translated into
   English by J. Dudley in *Bulletin de la société internationale pour l'étude de la
   philosophie médiévale* 16-17, 1974-5, 62-85. Hayduck's claim that the
   commentary on Book 3 is by Stephanus and not by Philoponus is
   challenged in the present volume by Wolfgang Bernard. There are in
   addition other *de Anima* commentaries attributed to Philoponus.
   (i) Verbeke has edited William of Moerbeke's Latin translation of a
       commentary by Philoponus on *de Anima* 3, 4-8 (*Corpus Latinum
       Commentariorum in Aristotelem Graecorum* 3). This commentary concerns
       Aristotle's treatment of the intellect, and is distinct from the
       corresponding section in the Greek of the *CAG* series. Verbeke adds
       further Latin fragments, one representing a different Moerbeke
       translation of part of the commentary on book 3. There was an earlier
       edition of the Latin for 3,4-8 by M. de Corte, Liège 1934. Emendations
       to this were proposed by A. Mansion (1947), and Greek fragments of
       Philoponus' discussion of the intellect were assembled out of
       Sophonias' commentary by S. van Riet (1965).
   (ii) There is a *ms* of Gennadius Scholarius (Codex Laurentianus 19, plut.

86), in which at fol 269, he claims that Thomas Aquinas' *de Anima* commentary, which he (Gennadius) here translates into Greek, is substantially identical with a commentary by Philoponus. However, it is not close to any extant commentary by Philoponus – see Jugie (1930), Schissel von Fleschenberg (1932) and Verbeke (1966) lxxi-lxxxii.

7. *in Physica*, books 1 to 4, with fragments of the commentary on books 5 to 8, ed Vitelli, vols 16-17, 1887 and 1888.

8. There were probably further commentaries on Aristotle, now lost. I.A. Fabricius *Bibliotheca Graeca* 3, 218 and 10, 646 reports a commentary on the *Sophistici Elenchi*, and Philoponus may be referring to this, or to a commentary on the *Topics*, perhaps incorporating this, at *in An Post* 3,4. It has also been conjectured, on the basis of a reference at *in Meteor* 16,31, that there was a commentary on the *de Caelo*, in addition to the *contra Aristotelem*, although this is highly speculative.

9. The commentary *in Metaphysica* is apparently spurious. S. Ebbesen judges it to be later than Michael of Ephesus, i.e. after 1100 (1981, vol 3, pp 86-7). This commentary exists in a Latin translation by Patrizi and in two Greek *mss* which agree well enough with it, although they were not used by Patrizi.

II. Commentary on Porphyry on Aristotle

10. A commentary on Porphyry's Introduction (*Isagoge*) to Aristotle's *Categories*. There is a possible reference to such a commentary at *in Phys* 250,28. What purports to be a Latin translation, *Philoponi interpretatio in Quinque Voces*, Vat. Lat. *ms* 4558, 193r-230r, has not yet been investigated. A. Baumstark claimed to find Syriac fragments in Vatican Syriac *ms* 158. He supplied Syriac text and German translation in *Aristoteles bei den Syrern vom v-viii Jahrhundert* 1, Leipzig 1900, pp 167ff; 173ff.

III. Commentary on Plato

11. Philoponus refers to a commentary, now lost, on the *Phaedo*, at *in An Post* 215,5.

IV. Medical Writings

For medical writings in Arabic translation attributed to Philoponus, see
12. M. Steinschneider (1869) 163-5.
There are two Greeks *mss* purportedly by Philoponus, entitled:
13. *On Fevers*, Mosquensis Gr. 466, fol 157ff.
14. *On Pulses*, cod. vat. Gr. 280, fol 204ff.
But all medical attributions are the subject of controversy.

V. Mathematical writing (see below for geometrical theorems in the Summikta Theôrêmata)

15. *Commentary on Nicomachus' Introduction to Arithmetic*, ed Richard Hoche, Pars I, II, Leipzig 1864, Pars III, Berlin 1867.

VI. Astronomical writing

Philoponus has left the oldest extant Greek treatise on the astrolabe:
16. *du Usu Astrolabii eiusque Constructione Libellus*, ed H. Hase, *Rheinisches Museum für Philologie* 6, 1839, 127-71, repr. with French translation by A.P. Segonds, Paris 1981. English translation by H.W. Greene in R.T. Gunther, *Astrolabes of the World*, vol 1, Oxford 1932, 61-81.

VII. Grammatical writings

The following works are concerned with accents:
17. *de Vocabulis quae Diversum Significatum Exhibent Secundum Differentiam Accentus*, several divergent epitomes of Philoponus' work, ed Lloyd W. Daly (1983). Previous editions are worth listing, since they appeared under various titles: Aldus Manutius, Venice 1497; H. Stephanus, Geneva 1572; E. Schmidt, Wittenberg 1615; P. Egenolff, Breslau 1888.
18. *Tonika Parangelmata Ailiou Herodianou peri Schêmatôn*, ed W. Dindorf, Leipzig 1825, an epitome of Herodian's views on accents in his *Universal Prosody*. It may be an extract from a longer work, since this would explain other attributions to Philoponus.

VIII. Writings on the creation and destructibility of the universe

Philoponus claims at *in Phys* 55,26 to have examined in some earlier place 'theôrêmeta' concerning the creation of the universe. The reference has been variously taken as being to *Summikta Theôrêmata* (Evrard) or to *aet* (Verrycken). Infinity arguments in favour of creation are included in the *Physics* commentary, 428,14-430,10; 467,5-468,4, and in the later *Meteorology* commentary, 16,36ff. But beyond this, there is a series of works largely devoted to the problem. All but no. 22 are to be translated into English, ed. Sorabji.
19. *de Aeternitate Mundi contra Proclum* ed H. Rabe, Leipzig 1899. For the Arabic version, see G. Graf, *Geschichte der christlichen arabischen Literatur*, vol I, Vatican 1944, 417-18. The Greek lacks the opening which reported the first of the eighteen arguments by Proclus which are under attack. But of Proclus' arguments the first nine have been published in an Arabic version independent of Philoponus by A. Badawi, *Neoplatonici apud Arabes, Islamica* 16, Cairo 1954, and from here the missing first argument of Proclus has been translated into French by G.C. Anawati (1956) 21-5.
20. *contra Aristotelem*. The surviving fragments have been assembled in English translation by Christian Wildberg, and are to be published

shortly. Books I to III are on Aristotle's eternal fifth element, books IV to VI have to do with the eternity of the universe, and the missing books VII and VIII concerned the transformation of this world into one more divine. Most of the surviving fragments are preserved by Simplicius *in Cael* 25-201 and *in Phys* 1129-82, but there are a further four in Arabic, one in Greek and one, not previously utilised, in Syriac. There is a French translation of the fragments of book I by É. Evrard (1943).

21. There are excerpts from a distinct work on the subject in Simplicius *in Phys* 1326-1336. For its distinctness, see H.A. Davidson 1965, 358-9. The arguments cited turn on the finite power of the universe.

22. S. Pines has published an English translation of an Arabic summary of a work by Philoponus on the subject written after the *de Aeternitate Mundi contra Proclum* and the *contra Aristotelem*. It seeks to improve the arguments for creation, without attacking a particular opponent. Its relation to the preceding treatise is unclear, although Pines suggests it may incorporate it: (1972) 320-52. Arabic with French translation in Troupeau (1984).

23. *de Opificio Mundi*, ed W. Reichardt, Leipzig 1897, an avowedly theological treatment of creation in the book of Genesis.

IX. Further philosophical and mathematical writing

24. *Summikta Theôrêmata* is lost, but references at *in Phys* 55,26; 156,17; *in An Post* 179,11; 265,6, suggest that the 'theorems' may have covered the creation of the universe, three-dimensional matter, mathematical optics and geometry.

X. Theology: Monophysite writings

A number are printed in Syriac, with Latin translation, in A. Šanda, *Opuscula Monophysitica Ioannis Philoponi*, Beirut 1930. Some extracts are translated into German in Walter Böhm (1967).

25. *Arbiter* or *Diaetêtês*, no. I in Šanda, with two Greek fragments in John of Damascus, *de Haeresibus* 83, PG 94, cols 744-54, or ed Kotter 4, 1982, pp 50-5, partly preserved also in Nicephorus Callistus *Historia Ecclesiastica* book 18, ch 47, PG 147, cols 425-8, and F. Diekamp ed, *Doctrina Patrum de Incarnatione Verbi*, Münster 1907, 272-83. Extracts in German in Böhm. Fragment translated into English in Ebied, van Roey, Wickham (1981) 26.

26. *Epitome of the Arbiter*, II in Šanda, German extracts in Böhm.

27. *Apologies*, two defences of the Arbiter, *Dubiorum Quorundam in Diaetete Solutio Duplex*, III in Šanda, German extracts in Böhm.

28. *Against the Fourth Council* or *Four Tmêmata (Divisions) Against Chalcedon*, extracts in Syriac in the Chronicle of Michael the Syrian, translated into French by J.-B. Chabot, *Chronique de Michel le Syrien, Patriarche Jacobite d'Antioche* 1166-1199, Syriac vol 4, Paris 1910, 218-38, French vol 2, Paris 1901, 92-121.

29. *On the Whole and its Parts to Sergius*, IV in Šanda, translated into Latin by G. Furlani (1921-2), German extract in Böhm.
30. *On Difference, Number and Division*, V and VII in Šanda, Syriac text and Italian translation of VII in G. Furlani (1923).
31. *Letter to Justinian*, VI in Šanda; Latin translation in G. Furlani (1919-20b); German extracts in Böhm.

XI. *Tritheist writings*

32. *On the Trinity* or *On Theology*. Syriac fragments, with Latin translation, in A. van Roey (1980). Four of these fragments are preserved in the Chronicle of Michael the Syrian, in J.-B. Chabot (see 28), Syriac vol 4, 361-2, French vol 2, 331-2. Other fragments are translated into Italian by G. Furlani (1920) 679-736, in the first of six opuscula and (1923-4) 663-5; 667-9; 671, 674. Some fragments in English in Ebied, van Roey, Wickham (1981) 29; 30
33. *Against Themistius*. Syriac fragments, with Latin translation, in A. van Roey (1980). Some translated into Italian by G. Furlani (1923-4) 668-9; 671. Some fragments in English, Ebied, van Roey, Wickham (1981) 33; 51-2.
34. *Letter to a Partisan*. Syriac fragment, with Latin translation, by A. van Roey (1980). Italian translation by G. Furlani, 'Sei scritti' (1920).
35. Four fragments of uncertain origin, in A. van Roey (1980).

XII. *Writings on the resurrection*

36. *On the Resurrection*. Syriac fragments, with French translation, by A. van Roey (1984). Greek fragment in Timotheus of Constantinople, *de Receptione Haereticorum* PG 86, 61C (cf 44A), copied by Nicephorus Callistus *Ecclesiastica Historia* book 18, ch 47, PG 147, 424D-425A.
37. *Against the letter of Dositheus*. Syriac fragment, with French translation, no. 29, in A. van Roey (1984).

XIII. *Anti-Arian writing*

38. *Against Andrew the Arian*. Syriac fragments, with Latin translation, of writing against the Arians, possibly all from this same work, in A. van Roey (1979) esp 329-48.

XIV. *Other theological writings*

39. *de Paschate*, Greek text ed Walter, Jena 1899, authenticity disputed.
40. *Against Iamblichus' 'On Statues'*, lost, but mentioned by Photius, who at PG 103, 708B-D, adds a short description.

Secondary literature

Abel, A. 1963-4, 'La légende de Jean Philopon chez les Arabes', *Correspondance d'Orient* 10, 251-80.

Altaner, B. 1978, *Patrologie*, 8th edition Freiburg, s.v. 'Philoponos' (English translation of 5th edition by H.C. Graef, Freiburg and Edinburgh 1958).

Anastos, M.V. 1953, 'Aristotle and Cosmas Indicopleustes on the void', *Hellenika* 4. Prosphora eis S.P. Kyriakiden, Thessaloniki, reprinted in his *Studies in Byzantine Intellectual History*, London 1979.

Anawati, G.C. 1956, 'Un fragment perdu du *De Aeternitate Mundi* de Proclus', *Mélanges de philosophie grecque offerts a Mgr. Diès*, Paris 1956, 21-5.

Andersen, O. 1976, 'Aristotle on sense-perception in plants', *Symbolae Osloenses* 51, 81-6.

Anton, J.P. 1969, 'Ancient interpretations of Aristotle's doctrine of *homonyma*', *Journal of the History of Philosophy* 7, 1-18.

Bäck, A. 1986, 'A general solution to the fallacy of accident', *Journal of the History of Logic*, forthcoming.

Baltes, M. 1978, *Die Weltentstehung des platonischen Timaios nach den antiken Interpreten*, part 2, Leiden.

Bardy, G. 1924, 'Jean Philopon', in A. Vacant, E. Mangenot, E. Amann, eds, *Dictionnaire de Théologie Catholique* 8, Paris, 831-9.

Bernard, W. 1986, Chapter 8 in this volume.

Bernard, W. forthcoming, *Rezeptivität und Spontaneität der Wahrnehmung bei Aristoteles, Zu Aristoteles, De Anima B, Γ.*

Blomqvist, J. 1979-80, 'John Philoponus and Aristotelian cosmology (Swedish with English summary)', *Lychnos*, 1-19.

Blumenthal, H.J. 1976, 'Neoplatonic elements in the De Anima commentaries', *Phronesis* 21, 64-87.

Blumenthal, H.J. 1981, 'Some Platonist readings of Aristotle', *Proceedings of the Cambridge Philological Society* 27, 1-16.

Blumenthal, H.J. 1982, 'John Philoponus and Stephanus of Alexandria: two Neoplatonic commentators on Aristotle?' in D.J. O'Meara, ed, *Neoplatonism and Christian Thought*, Norfolk, Virginia, 54-66; 244-6.

Blumenthal, H.J. 1986, 'John Philoponus: Alexandrian Platonist?', *Hermes* 114, 314-35.

Blumenthal, H.J. 1986, 'Body and soul in Philoponus', *Monist*.

Blumenthal, H.J. in preparation, tentative title: *The Interpretation of Aristotle in Late Antiquity.*

Böhm, W. 1967, *Johannes Philoponos, ausgewählte Schriften*, Munich, Paderborn, Vienna.

Booth, E.G.T. 1983a, 'John Philoponos: Christian and Aristotelian conversion', *Studia Patristica* vol 17, part 1, 407-11.

Booth, E.G.T. 1983b, *Aristotelian Aporetic Ontology in Islamic and Christian Thinkers*, Cambridge 56-61.

Brinkmann, A. 1912, 'Scriptio continua und anderes', *Rheinisches Museum für Philologie* 67, 609-14.

Burkhard, K. 1912, 'Auszüge aus Philoponus als Randbemerkungen in einer Nemesiushandschrift', *Wiener Studien* 34, 135-8.

Chadwick, H. 1970, 'Philoponus', in N.G.L. Hammond and and H.H. Scullard, eds, *The Oxford Classical Dictionary*, Oxford, 824.

Chadwick, H. 1986, Chapter 2 in this volume.

Cheiko, L. 1913, 'Is John Philoponus the same person as Yaḥyā al-Naḥwī (a historical problem)?' (in Arabic), *Al-Mashriq* 16, 47-57.

Christensen de Groot, J. 1983, 'Philoponus on *De Anima* 2.5, *Physics* 3.3, and the propagation of light', *Phronesis* 28, 177-96.

Christensen de Groot, J. forthcoming, translation of Philoponus' discussion of light and colour, in *DA 2.7*, 324,25-342,16, to be published in *Ancient Philosophy*.

Chroust, A.-H. 1966, 'The psychology in Aristotle's lost dialogue *Eudemus* or *On the Soul*', *Acta Classica* 9,49-62.

Claggett, M. 1959, *The Science of Mechanics in the Middle Ages*, Madison Wisconsin, chs 7 & 8.

Cohen, M.R. and Drabkin, I.E. 1958, *A Sourcebook in Greek Science*, Cambridge Mass., translates sections of *in Phys*, 217-23.

Corte, M. de 1934, *Le Commentaire de Jean Philopon sur le troisième livre du traité de l'âme d'Aristote*, Liège.

Craig, W.L. 1979, *The Kalam Cosmological Argument*, London.

Cremonini, C. 1616, *Apologia dictorum Aristotelis de quinta caeli substantia, adversus Xenarchum, Ioannem Grammaticum et alios*, Venice.

Daly, L.W. 1983, *Iohannis Philoponi. De vocabulis quae diversum significatum exhibent secundum differentiam accentus*, American Philosophical Society Memoirs 151, Philadelphia.

Davids, T.W. 1882, 'Joannes (564) Philoponus', in W. Smith and H. Wace, eds, *A Dictionary of Christian Biography*, vol 3, London, 425-27.

Davidson, H.A. 1969, 'John Philoponus as a source of medieval Islamic and Jewish proofs of creation', *Journal of the American Oriental Society*, 89, 357-91.

Davidson, H.A. 1979, 'The principle that a finite body can contain only finite power', in S. Stein & R. Loewe, eds, *Studies in Jewish Religious and Intellectual History Presented to Alexander Altmann*, Alabama, 75-92.

Dijksterius, E.J. 1924, *Val en Worp*, Groningen.

Dindorf, W. 1825, *Joannes Philoponos*, Leipzig.

Drecker, J. 1928, 'Des Johannes Philoponos Schrift über das Astrolab', *Isis* 11, 15-44.

Ducci, E. 1964, 'Il *to eon* parmenideo nell' interpretazione di Filopono', *Rassegna di scienze filosofiche*, 253-300.

Duhem, P.L. 1954, *Le Système du monde* I and II, Paris 1913-17, repr. I Paris 1954, II Paris 1974, see Index s.v. 'Jean d'Alexandrie'.

Ebbesen, E. 1981, *Commentators and Commentaries on Aristotle's 'Sophistici Elenchi'*, Corpus Latinum Commentariorum in Aristotelem Graecorum 7, 3 vols, Leiden.

Ebied, R.Y., van Roey, A., Wickham, L.R. 1981, *Peter of Callinicum: anti-Tritheist Dossier*, Louvain.

Egenolff, P. 1902, 'Zu Lentz' Herodian II', *Philologus* 61, esp 77-101.

Endress, G. 1977, *The Works of Yaḥyā Ibn 'Adī*, Wiesbaden.

Evrard, É. 1943, 'Philopon contre Aristote, livre premier', Mémoire présenté à la Faculté de Philosophie et Lettres de l'Université de Liège (unpublished typescript).

Evrard, É. 1953, 'Les convictions religieuses de Jean Philopon et la date de son commentarie aux 'Météorologiques', *Bulletin de l'academie royale de Belgique, classe des lettres* 5, 299-357.

Evrard, É. 1957, *L'École d'Olympiodore et la composition du 'Commentaire à la Physique' de Jean Philopon*, dissertation, Liège.

Evrard, É. 1965, 'Jean Philopon, son 'commentaire sur Nicomaque' et ses rapports avec Ammonius', *Revue des études grecques* 78, 592-8.

Evrard, É. 1985, 'Jean Philopon, Simplicius et les ténèbres primitives', in C. Rutten and A. Motte, eds, *Aristotelica, mélanges offerts à Marcel de Corte*, Brussels.

Fabricius, J.A. and G.C. Harles, 1807, *Bibliotheca Graeca*, vol 10, Hamburg, 639-69 (chapter 34 = 'De Ioanne Philopono Grammatico').

Faggin, G. 1967, 'Giovanni Filopono', in C. Giacon et al., eds, *Enciclopedia filosofica* III, Florence, cols 189-90.

Furlani, G. 1919-20a, 'L'anatema di Giovanni d'Alessandria contro Giovanni Filopono', *Atti della Accademia delle scienze di Torino*, 55, 188-94.

Furlani, G. 1919-20b, 'Una lettera di Giovanni Filopono all' imperatore Giustiniano', *Atti del Reale Istituto Veneto di scienze, lettere ed arti* 79, 1247-65.

Furlani, G. 1920, 'Sei scritti antitriteistici in lingua siriaca', *Patrologia Orientalis* 14, 673-766, esp 679-736.

Furlani, G. 1921-2, 'Il trattato di Giovanni Filopono sul rapporto tra le parti e gli elementi ed il tutto e le parti', *Atti del Reale Istituto Veneto di scienze, lettere ed arti* 81, 83-105.

Furlani, G. 1922, 'Il contenuto dell' Arbitro di Giovanni Filopono', *Rivista trimestrale di studi filosofici e religiosi* 3, 385-405.

Furlani, G. 1923, 'Unità e dualità di natura secondo Giovanni il Filopono', *Bessarione* 27, 45-65.

Furlani, G. 1923-4, 'Un florilegio antitriteistico in lingua siriaca', *Atti del Reale Istituto Veneto di scienze, lettere ed arti* 83, 661-77.

Furlani, G. 1924, 'Sull' incendio della biblioteca di Alessandria', *Aegyptus* 5, 205-12.

Furlani, G. 1925, 'Giovanni il Filopono e l'incendio della biblioteca di Alessandria', *Bulletin de la société archéologique d'Alexandrie* 21, 58-77.

Furlani, G. 1926, 'Meine Arbeiten über die Philosophie bei den Syrern, *Archiv für Geschichte der Philosophie* 37 (ns 30), 3-25.

Furley, D.J. 1986, Chapter 6 in this volume.

Galluzzi, P. 1979, *Momento, Studi galileiani*, Rome.

Gottschalk, H.B. 1966, 'Lucretius on the "Water of the Sun" ', *Philologus* 110, 311-15.

Gottschalk, H.B. 1971, 'Soul as harmonia', *Phronesis* 16, 179-98.

Grabmann, M. 1929, 'Mittelalterliche lateinische Übersetzungen von Schriften der Aristoteles-Kommentatoren Johannes Philoponos, Alexander von Aphrodisias und Themistios', *Sitzungsberichte der Bayerischen Akademie der Wissenschaften, Philosophisch-historische Klasse* 7.

Grant, E. 1964, 'Motion in the void and the principle of inertia in the middle ages', *Isis* 55, 265-92 reprinted, along with his other articles, in his *Studies in Medieval Science and Natural Philosophy*, London 1981.

Grant, E. 1965, 'Aristotle, Philoponus, Avempace and Galileo's Pisan dynamics', *Centaurus* 11, 79-85.

Grant, E. 1978, 'The principle of the impenetrability of bodies in the history of concepts of separate space from the middle ages to the seventeenth century', *Isis* 69, 551-71.

Grant, E. 1981, *Much Ado About Nothing: theories of space and vacuum from the middle ages to the scientific revolution*, Cambridge.

Grumel, V. 1937, 'Jean Grammaticos et saint Théodore Studite', *Echos d'Orient* 36, 181-9.

Gudeman, A. and Kroll, W. 1916, 'Ioannes (No. 21, Ioannes Philoponus)', in G. Wissowa-W. Kroll, eds, *Realencyclopädie der klassischen Altertumswissenschaft* 9.2, cols 1764-1795.

Haas, A.E. 1906, 'Über die Originalität der physikalischen Lehren des Johannes Philoponus', *Bibliotheca Mathematica* 3(6), 337-42.

Haase, W. 1965, 'Ein vermeintliches Aristoteles-Fragment bei Johannes Philoponos', in H. Flashar and K. Gaiser, eds, *Synusia, Festgabe für W. Schadewaldt*, Pfullingen, 323-54.

Hermann, T. 1930, 'Johannes Philoponos als Monophysit', *Zeitschrift für neutestamentliche Wissenschaft* 29, 209-64.

Hiltbrunner, O. 1975, 'Iohannes Philoponos', *Der Kleine Pauly*, vol 2, Munich, cols 1430-31.

Hoffmann, P. 1986, Chapter 3 in this volume.

Hunger, H. 1978, *Die hochsprachliche profane Literatur der Byzantiner* vol 1 (= *Byzantinisches Handbuch*, part 5, vol 1), Munich, 25-41.

Joannou, P. 1961, 'Christliche Metaphysik in Byzanz, Johannes Philoponos (VI Jh.)', *Wissenschaftliche Zeitschrift der Martin-Luther-Universität, Halle-Wittenberg* 10, 1389-90.

Joannou, P. 1962, 'Le premier essai chrétien d'une philosophie systématique, Jean Philopon', *Studia Patristica* vol 5, 508.

Judson, L. 1986, Chapter 10 in this volume.

Jugie, M. 1930, 'George Scholarios et Saint Thomas d'Aquin', *Mélanges Mandonnet* I, Paris, 423-40.

Jugie, M. 1950, 'Filopono, Giovanni', *Enciclopedia Cattolica*, vol 5, Vatican City, cols 1349-50.

Kraemer, J.L. 1965, 'A lost passage from Philoponus' *contra Aristotelem* in Arabic translation', *Journal of the American Oriental Society* 85, 318-27.

Krafft, F. 1982, 'Zielgerichtetheit und Zielsetzung in Wissenschaft und Natur', *Berichte zur Wissenschaftsgeschichte* 5, 53-74.

Kremer, K. 1961-2, 'Die Anschauung der Ammonius (Hermeiou)-Schule über den Wirklichkeitscharakter des Intelligiblen. Über einen Beitrag der Spätantike zur platonisch-aristotelischen Metaphysik', *Philosophisches Jahrbuch (Görres-Gesellschaft)* 69, 46-63.

Kremer, K. 1961, *Der Metaphysikbegriff in den Aristoteles-Kommentaren der Ammonius-Schule (Beiträge zur Geschichte der Philosophie und Theologie des Mittelalters* 39, 1), Münster.

Kremer, K. 1965, 'Das "Warum" der Schöpfung: "quia bonus" vel/et "quia voluit"? Ein Beitrag zum Verhältnis von Neuplatonismus und Christentum an Hand des Prinzips "bonum est diffusivum sui" ', *Parusia: Festgabe für J. Hirschberger*, Frankfurt, 241-64.

Kuelb, P.H. 1843, 'Johannes der Grammatiker', in J.S. Ersch and J.G. Gruber, eds, *Allgemeine Encyklopädie der Wissenschaften und Künste*, 2nd section, ed A.G. Hoffmann, vol 22, Leipzig, 191-2.

Lebon, J. 1923, 'Fragments syriaques de Nestorius dans le *contra Grammaticum* de Sévère d'Antioche', *Le Muséon* 36, 47-65.

Lebon, J. 1952, *Contra Impium Grammaticum, Corpus Scriptorum Christianorum Orientalium* (1st ed 1929-38) 93, 101, 111.

Lee, T.-S. 1984, *Die griechische Tradition der aristotelischen Syllogistik in der Spätantike, Hypomnemata* 79.

Lucchetta, G.A. 1974-5, 'Ipotesi per l'applicazione dell' 'impetus' ai cieli in Giovanni Filopono', *Atti e Memorie dell' Accademia patavina di scienze, lettere ed arti* 87, 339-52.

Lucchetta, G.A. 1978a, *Una fisica senza matematica: Democrito, Aristotele, Filopono*, Trento.

Lucchetta, F.A. 1978b, 'Aristotelismo e cristianesimo in Giovanni Filopono', *Studia patavina-Rivista di scienze religiose* 25, 573-93.

Lucchetta, G.A. 1983, 'Dinamica dell' *impetus* e Aristotelismo Veneto', in *Aristotelismo Veneto e Scienza Moderna, Saggi e Testi* 18, Padua.

Ludwich, A. 1888, 'Commentatio de Philopono Grammatico', *Königsberger Universitäts Progr.* 11, 3-18.

MacCoull, L.S.B. 1982, 'A Trinitarian formula in Dioscorus of Aphrodito', *Bulletin de la Société d'Archéologie Copte* 24, 103-10.

MacCoull, L.S.B. 1983, '*monoeidês* in Dioscorus of Aphrodito: an addendum', *Bulletin de la Société d'Archéologie* 25, 61-4.

MacCoull, L.S.B. 1986, 'Dioscorus of Aphrodito and John Philoponus', *Studia Patristica* 18.1, Kalamazoo.

McGuire, J.E. 1985, 'Philoponus on *Physics* II 1: *phusis, hormê emphutos* and the motion of simple bodies', *Ancient Philosophy*.

McKenna, J.E. 1986, 'Christian theology and scientific culture', *Studia Biblica et Theologica*, published by Fuller's Seminary, 133-43.

MacLeod, C.K.M. 1964, 'Jean Philopon, Commentaire sur le de Intellectu, traduction latine de Guillaume de Moerbeke', Diss. Louvain, summary in *Revue philosophique de Louvain* 62, 1964, 727-8.

MacLeod, C.K.M. 1966, 'Jean Philopon, commentateur d'Aristote', and 'Le "De Intellectu" de Philopon et la pensée du XIIIe siècle', in G. Verbeke (1966) xi-xix and lxxi-lxxxvi.

Mahdi, M. 1967, 'Alfarabi against Philoponus', *Journal of Near Eastern Studies* 26, 233-60.

Mahdi, M. 1972, 'The Arabic text of Alfarabi's *Against John the Grammarian*', in S.A. Hanna, ed, *Medieval and Middle Eastern Studies in Honour of A.S. Atiya*, Leiden 268-84.

Maier, A. 1951, *Zwei Grundprobleme der scholastischen Naturphilosophie*, 2nd ed, Rome.

Mansi, J.D. 1759-1798, *Sacrorum Conciliorum Nova et Amplissima Collectio*, Florence, vol 9, cols 501-502C (repr. Paris, 1901-6).

Mansion, A. 1947, 'Le texte du *de Intellectu* de Philopon corrigé à l'aide de la collation de Mgr. Pelzer', in *Mélanges A. Pelzer*, Louvain, 325-46.

Martin, H. 1962, 'Jean Philopon et la controverse trithéite du VIe siècle', *Studia Patristica* 5, 519-25.

Mercati, G. 1914, 'Un codice non riconosciuto dello Ps.-Filopono sull' *Isagoge* di Porfirio', *Rheinisches Museum für Philologie* 69, 415-16.

Meyer, P. 1901, 'Johannes Philoponus' in A. Hauck, ed, *Realencyclopädie für prostestantische Theologie und Kirche* 9, Leipzig, 310-11.

Meyerhof, M. 1930, 'Von Alexandrien nach Bagdad', *Sitzungsberichte der preussischen Akademie der Wissenschaften, Philosophisch-historische Klasse*, 389-429.

Meyerhof, M. 1932, 'Joannes Grammatikos (Philoponos) von Alexandrien und die arabische Medizin', *Mitteilungen des deutschen Instituts für ägyptische Altertumskunde in Kairo* 2, 1-21.

Meyerhof, M. 1933, 'La fin de l'école d'Alexandrie d'après quelques auteurs arabes', *Archeion* 15, 1-15.

Minio-Paluello, L. 1957, 'A Latin commentary (?translated by Boethius) on the *Prior Analytics* and its Greek sources', *Journal of Hellenic Studies* 77, part 1, 93-102.

Moody, E.A. 1951, 'Galileo and Avempace: the dynamics of the leaning tower experiment', *Journal of the History of Ideas* 12, 163-93; 375-422.

Moraux, P. 1954, 'Notes sur la tradition indirecte du *De Caelo* d'Aristote', *Hermes* 82, 145-82.

Moraux, P. 1963, 'Quinta essentia', in G. Wissowa-W. Kroll, eds, *Realencyclopädie der klassischen Altertumswissenschaft* 24, cols 1171-1263, 1430-2.

Murru, F. 1981, 'Miscellanea linguistica I: Astiage e Filopono sul problema dei casi', *Vichiana* 10, 202-9.

Nauck, A. 1847, 'Philoponos', in J.S. Ersch, J.G. Gruber, eds, *Allgemeine Enzyklopädie der Wissenschaften und Künste* III 23, Leipzig, 465-73.

Neugebauer, O. 1975, *A History of Ancient Mathematical Astronomy*, Berlin, Heidelberg, New York, vol 2, 1041f.

O'Donnell, J.R. 1967, 'John Philoponus', *New Catholic Encyclopedia*, vol 7, San Francisco, 1066

Oehler, K. 1964, 'Aristotle in Byzantium', *Greek, Roman and Byzantine Studies* 5, 133-46.

Petschenig, M. 1881, 'Zu Johannes Philoponos' *Peri tôn diaphorôs tonoumenôn*', *Wiener Studien* 3, 294-7.

Pines, S. 1938a, 'Études sur Awḥad al-Zamān Abu' l-Barakāt al-Baghdādī', *Revue des Études Juives* n.s. 3 and 4, 3-64 and 1-33, reprinted in *The Collected Works of Shlomo Pines* I, Jerusalem and Leiden 1979.

Pines, S. 1938b, 'Les précurseurs musulmans de la théorie de l'impetus', *Archeion* 21, 298-306.

Pines, S. 1953, 'Un précurseur bagdadien de la théorie de l'impetus', *Isis* 44, 246-51.

Pines, S. 1961, 'Omne quod movetur necesse est ab alio moveri', *Isis* 52, 21-54.

Pines, S. 1964, 'La dynamique d' Ibn Baǧǧa' in *Mélanges Alexandre Koyré* I, Paris 442-68.

Pines, S. 1969, 'Saint Augustin et la théorie de l'impetus', *Archives d'histoire doctrinale et littéraire du moyen age* 44, 7-21.

Pines, S. 1972, 'An Arabic summary of a lost work of John Philoponus', *Israel Oriental Studies* 2, 320-52.

Praechter, K. 1909, 'Die griechischen Aristoteleskommentare', *Byzantinische Zeitschrift* 18, 516-38.

Rahman, F. 1975, 'The eternity of the world and the heavenly bodies in post-Avicennan philosophy', in G.F. Hourani, ed, *Essays on Islamic Philosophy and Science*, Albany N.Y., 222-37.

Reiner, H. 1954, 'Der Metaphysik-Kommentar des Joannes Philoponos', *Hermes* 82, 480-2.

Richard, M. 1950, '*Apo phônês*', *Byzantion* 20, 191-222 reprinted in his *Opera Minora*, vol 3.

Riet, S. van 1965, 'Fragments de l'originel grec du "De Intellectu" de Philopon', *Revue philosophique de Louvain* 63, 5-40.

Roey, A. van 1979, 'Fragments antiariens de Jean Philopon', *Orientalia Lovaniensia Periodica* 10, 237-50.

Roey, A. van 1980, 'Les fragments trithéites de Jean Philopon', *Orientalia Lovaniensia Periodica* 11, 135-63.

Roey, A. van 1981, see Ebied.

Roey, A. van 1984, 'Un traité cononite contre la doctrine de Jean Philopon sur la resurrection', in *Antidoron, hommage à Maurits Geerard*, Wetteren.

Rose, V. 1871, 'Ion's Reisebilder und Ioannes Alexandrinus der Artzt', *Hermes* 5 (1871), 205-15.

Rossi, P. 1978, 'Tracce della versione latina di un commento greco ai *Secondi Analitici* nel *Commentarius in Posteriorum Analyticorum* di Grossatesta', *Rivista di Filosofia Neoscolastica*, 433-9.

Saffrey, H.D. 1954, 'Le chrétien Jean Philopon et la survivance de l'école d'Alexandrie au VIe siècle', *Revue des études grecques* 67, 396-410.

Sambursky, S. 1958, 'Philoponus' interpretation of Aristotle's theory of light', *Osiris* 13, 114-26.

Sambursky, S. 1962, *The Physical World of Late Antiquity*, London.

Sambursky, S. 1967, 'Philoponus, John', in P. Edwards, ed, *The Encyclopaedia of Philosophy* 6, New York, 156f.

Sambursky, S. 1970, 'John Philoponus', in C.C. Gillispie, *Dictionary of Scientific Biography*, New York, 134-9.

Sambursky, S. 1972, 'Note on John Philoponus' rejection of the infinite', in S.M. Stern, A. Hourani, V. Brown, eds, *Islamic Philosophy and the Classical Tradition, Essays Presented to Richard Walzer*, Oxford, 351-3.

Sambursky, S. 1977, 'Place and space in late Neoplatonism', *Studies in the History and Philosophy of Science*, 8, 173-87.

Sambursky, S. 1982, *The Concept of Place in Late Neoplatonism*, Jerusalem.

Šanda, A. 1930, *Opuscula Monophysitica Ioannis Philoponi*, Beirut.

Scharfenberg, J.G. 1768, *Dissertatio de Joanne Philopono tritheismi defensore*, Leipzig.

Schissel, O. von Fleschenberg 1932, 'Kann die Expositio in libros de Anima des S. Thomas Aquinas ein Kommentar des Joannes Philoponos zu Aristoteles' *Peri psuchês* sein?', *Byzantinisch-Neugriechische Jahrbücher* 9, 104-110.

Schmidt, M. 1855, 'Textkritische Bemerkungen zu Johannes Philoponus peri kosmopoiias', *Zeitschrift für Wissenschaftliche Theologie*, 289-97.

Schmitt, C.B. 1967, *Gianfrancesco Pico della Mirandola (1469-1533) and his Critique of Aristotle*, The Hague, ch 5.

Schmitt, C.B. 1986, Chapter 12 in this volume.

Schrenk, L.P. 1987, 'Philoponos', *Encyclopedia of Early Christianity*, New York.

Sedley, D. 1986, Chapter 7 in this volume.

Segonds, A.P. 1981, *Jean Philopon, traité de l'astrolabe*, Paris.

Sheldon-Williams, I.P. 1967, 'The Greek Christian Platonist tradition from the Cappadocians to Maximus and Eriugena', in A.H. Armstrong, ed, *The Cambridge History of Later Greek and Early Medieval Philosophy*, Cambridge, 420-533.

Sodano, A.R. 1962, 'I frammenti dei commentari di Porfirio al *Timeo* di Platone nel *de Aeternitate Mundi* di Giovanni Filopono', *Rendiconti dell' Accademia di Archeologia, 'Lettere e Belle Arti di Napoli*, 97-125.

Sorabji, R.R.K. 1982, 'Infinity and the creation: a turning point in the history of philosophy', inaugural lecture, King's College, London, reprinted as Chapter 9 in this volume.

Sorabji, R.R.K. 1983, *Time, Creation and the Continuum*, London.

Sorabji, R.R.K., 1986, Chapter 1 in this volume.

Sorabji, R.R.K. forthcoming, 'Johannes Philoponus', in G. Krause and G. Müller, eds, *Theologische Realenzyklopädie*, Berlin.

Sorabji, R.R.K. *Matter, Space and Motion*, forthcoming.

Sparty, A. 1975, 'The *de Intellectu* of John Philoponus and the work of St Thomas Aquinas', *Rocziniki filozoficzne, Lublin* 23, 81-99.

Steinschneider, M. 1869, 'Johannes Philoponus bei den Arabern', in 'Al-Farabi-Alpharabius-des arabischen Philosophen Leben und Schriften', *Mémoires de l'academie impériale des sciences de St. Petersbourg*, série 7, XIII, 4, 152-76; 220-4; 250-2.

Stoeckl, A. 1889, 'Johannes Philoponus', *Wetzer und Welte's Kirchen-lexicon*, vol 6, Freiburg, cols 1748-54.

Tannery, P. 1896, 'Sur la période finale de la philosophie grecque', *Revue philosophique* 42, 226-87, repr. in his *Mémoires scientifiques*, ed. J.-L. Heiberg, Toulouse-Paris, vol 7, 1925, 211-42.

Tannery, P. 1888, 'Notes critiques sur le traité d'astrolabe de Philopon', *Revue de philologie* 12, 60-73, repr. in *Mémoires scientifiques*, vol 4, 1920, 241-60.

Tannery, P. 1925, 'Philoponus', *Mémoires scientifiques*, vol 7, 318-20.

Taran, L. 1969, 'Asclepius of Tralles' commentary to Nicomachus' *Introduction to Arithmetic'*, *Transactions of the American Philosophical Society* n.s. 59, part 4.

Taran, L. 1984, '*Amicus Plato sed magis amica veritas*, from Plato and Aristotle to Cervantes', in *Antike und Abendland, Beiträge zum Verständnis der Griechen und Römer und ihres Nachlebens* 30, Berlin, New York, 93-124.

Tatakis, B.N. 1949, *La Philosophie byzantine*, Paris, 39-50.

Temkin, O. 1962, 'Byzantine medicine: tradition and empiricism', *Dumbarton Oaks Papers* 16, 105 n 58.

Tiftixoglu, V. 1969, 'Philoponos', *DTV-Lexicon der Antike*, part 1, vol 3, Munich, 314-15.

Todd, R.B. 1980, 'Some concepts in physical theory in John Philoponus' Aristotelian commentaries', *Archiv für Begriffsgeschichte* 24, 151-70.

Todd, R.B. 1984, 'Philosophy and medicine in John Philoponus' commentary on Aristotle's *de Anima*', Dumbarton Oaks Papers 38, 103-10.

Totok, W. 1973, 'Johannes Philoponos', in his *Handbuch der Geschichte der Philosophie* II, Frankfurt 162, 173f.

Trabucco, F. 1958, 'Il problema del "De philosophia" di Aristocle di Messene e la sua dottrina', *Acme* 11, 97-150.

Trechsel, F. 1835, 'Johannes Philoponos: Eine dogmenhistorische Erörterung', *Theologische Studien und Kritiken* 8, 95-118.

Troupeau, G. 1984, 'Un epitome arabe du *de Contingentia Mundi* de Jean Philopon', in Memorial A.J. Festugière = *Cahiers d'orientalisme* 10, Geneva, 77-88.

Tsouyopoulos, N. 1969, 'Die Entstehung physikalischer Terminologie aus der neoplatonischen Metaphysik', *Archiv für Begriffsgeschichte* 13, 7-33.

Ullmann, N. 1970, 'Die Medizin im Islam', *Handbuch der Orientalistik* I vi, 89-91.

Vancourt, R. 1944, *Les Derniers Commentateurs d'Aristote*, Lille.

Verbeke, G. 1951, 'Guillaume de Moerbeke, traducteur de Jean Philopon', *Revue philosophique de Louvain* 49, 222-35.

Verbeke, G. 1966, *Jean Philopon, Commentaire sur le de Anima d'Aristote, Corpus Latinum Commentariorum in Aristotelem Graecorum* 3.

Verbeke, G. 1982, 'Some later Neoplatonic views on divine creation and the eternity of the world', in D.J. O'Meara, ed, *Neoplatonism and Christian Thought*, Norfolk, Virginia, 45-53 and 241-4.

Verbeke, G. 1986, 'Levels of human thinking in Philoponus', in *After Chalcedon: Studies in Theology and Church History offered to Professor Albert van Roey for his Seventieth Birthday*, eds. C. Laga, J.A. Munitiz and L. Van Rompay, Louvain, 451-70.

Verrycken, K. 1985, 'God en wereld in de wijsbegeerte van Ioannes Philoponus', dissertation, Catholic University of Louvain (a shortened English version is to be published by the Belgian Royal Academy and a chapter will be contributed to a collection of articles, ed. Sorabji, on the ancient commentators on Aristotle).

Wallace, W.A. 1981, 'Galileo and scholastic theories of impetus', in A. Maierù and A. Paravicini Bagliani, eds, *Studi sul XIV secolo in memoria di Anneliese Maier*, Rome.

Wallies, M. 1891, *Die griechischen Ausleger der aristotelischen Topik (Wissenschaftliche Beilage zum Programm des Sophien-Gymnasiums Berlin, Ostern 1891)*, Berlin.

Wallies, M. 1916, Review of Gudeman and Kroll (1916), *Berliner Philologische Wochenschrift* 36, cols 596-90.

Walzer, R. 1962, 'New studies on al-Kindi', in his *Greek into Arabic*, Oxford, repr. from *Oriens* 10, 1957, 203ff.

Westerink, L.G. 1962, *Anonymous Prolegomena to Platonic Philosophy*, Amsterdam, Introduction (new Budé edition in preparation).

Westerink, L.G. 1964, 'Deux commentaires sur Nicomaque: Asclépius et Jean Philopon', *Revue des études grecques* 77, 526-35.

Wickham, L.R. 1981, see Ebied.

Wieland, W. 1960, 'Die Ewigkeit der Welt (der Streit zwischen Joannes Philoponus und Simplicius)', in D. Heinrich, W. Schulz, K.H. Volkmann-Schluck, eds, *Die Gegenwart der Griechen im neueren Denken, Festschrift H.-G. Gadamer*, Tübingen 291-316.

Wieland, W. 1967, 'Zur Raumtheorie des Johannes Philoponus', in E. Fries, ed, *Festschrift J. Klein*, Göttingen 114-35.

Wildberg, C. 1984, 'John Philoponus' Criticism of Aristotle's Theory of Ether', Ph.D. Diss, Cambridge.

Wildberg, C. 1987, *Philoponus. Against Aristotle on the Eternity of the World*. Fragments assembled and translated into English, London and Ithaca N.Y.

Wildberg, C. 1986, Chapter 11 in this volume.

Wohlwill, E. 1905, 'Ein Vorgänger Galileis im 6 Jahrhundert', *Verhandlungen der Gesellschaft deutscher Naturforscher und Ärzte* 77, part 2, reprinted in *Physikalische Zeitschrift* 7, 1906, 23-32.

Wolff, M. 1971, *Fallgesetz und Massebegriff*, Berlin.

Wolff, M. 1978, *Geschichte der Impetustheorie*, Frankfurt.

Wolff, M. 1986, Chapter 4 in this volume.

Wolfson, H.A. 1929, *Crescas' Critique of Aristotle*, Cambridge, Mass., 410-23; 452-5.

Wolska-Conus, W. 1968-73, Introduction to Cosmas Indicopleustes, *Topographie Chrétienne*, Sources Chrétiennes, 3 vols, 141, 159, 197, Paris.

Wolska, W. 1962, *La Topographie Chrétienne de Cosmas Indicopleustès (Bibliothèque Byzantine, Études*, 3), Paris.

Zahlfleisch, J. 1897, 'Die Polemik des Simplicius gegen Alexander und andere in dem Commentar des ersteren zu der aristotelischen Schrift *de Caelo*', *Archiv für Geschichte der Philosophie* 10 (n.s. 3) 191-227.

Zahlfleisch, J. 1902, 'Einige Corollarien des Simplicius in seinem Commentar zu Aristoteles *Physik*', *Archiv für Geschichte der Philosophie* 15 (n.s. 8) 186-213.

Zimmermann, F. 1986, Chapter 5 in this volume.

Index locorum to the writings of Philoponus

Numbers in parentheses after titles refer to the Bibliography; numbers in bold type refer to the pages of this book.

General index